Structured Writing

Rhetoric and Process

Mark Baker

Structured Writing

Rhetoric and Process

Credits

Cover Design:	Douglas Potter, https://sites.google.com/site/douglascpotter/
Cover Image:	Copyright © 2005, Procsilas Moscas, CC BY-2.0
Figure 1.1:	Copyright © 2012, Tjo3ya, CC BY-SA 3.0

Disclaimer

Trademarks

XML Press
Laguna Hills, California 92637
http://xmlpress.net

First Edition
978-1-937434-56-4 (print)
978-1-937434-57-1 (ebook)

Table of Contents

Preface

All writing is structured. Writing without grammatical structure would be incomprehensible. All writing software is structured as well. Software that did not produce reliable, consistent data structures would be unreliable and unworkable.

What then does the structured writing community mean by structured writing? As a generality, it means approaches to writing that add a little more structure, over and above the basic requirements of grammar, to exercise some control over the rhetoric or processing of the content. And it also means the use of software that uses more specific data structures, either to support the rhetorical structures of a structured writing method or to support specific writing processes, such as publishing, single sourcing, or content reuse.

So when the industry talks about structure writing methods and tools, it is actually talking about *differently structured* methods and tools. But you won't find one agreed definition of structured writing across the industry. Differently structured writing methodologies are often tied to differently structured software tools. This often leads the marketers of these methods and tools to suggest that their particular methodology or tool is the very essence and definition of structured writing. Some definitions see it as a writing method with no software component at all. Others identify it with a single software tool or standard.

This often leads people who see no virtue in these particular methods or tools, or who have had an unsatisfactory experience with them, to conclude that structured writing is bunk, or at very least that it is not for them, for their content, or for their company.

The truth is, it's all structured writing. Even the mainstream word processors and desktop publishing tools you use every day are structured writing tools. But not all structured writing methods and tools are equally effective for individual organizations. There are many ways to structure both the rhetoric and the process of creating content, and one of these may be vastly more productive for you and your organization than what you are doing now.

The purpose of this book is to present structured writing as a whole, to take a step back from the specific structures of individual tools and systems and show that all structured writing approaches share a few basic principles and operations. Understanding those principles and operations will help you choose the structured writing approach that is optimal for your content and your organization.

Whether you are considering a move to a differently structured writing system or trying to figure out why the one you have already implemented is not working as well as you hoped, this book will help you figure out how structured writing works, what is possible, and what is not possible, and it will help you figure out which techniques, structures, processes, and tools are going to work best for you.

In the age of the web, organizations produce and deliver ever more content on ever shorter deadlines. To keep up and maintain quality, they need tools and techniques that support rapid and reliable delivery of consistent, high-quality rhetoric. Many organizations are turning to structured writing solutions (that is *differently* structured solutions) to keep up and meet demand. However, without a clear and comprehensive understanding of what is possible, they often choose solutions that are sub-optimal or even worse than what they were doing before.

I have been in the structured writing industry for nearly 25 years. In that time I have worked with and designed structured writing systems that improved both process and rhetoric. I have also built tools and systems myself (some of which I will talk about in this book), because I have often felt that existing approaches did not do enough to balance the demands of process and rhetoric or to fully exploit the capacity of structured writing to promote both.

One of the things I have learned over my career is just how much our minds tend to follow the ruts laid down by our familiar tools and processes. I can't count the number of requirements documents I have read over the years that insisted that any new system must work exactly like the old system in almost every particular. I don't believe this is so much a reluctance to change as simply a difficulty imagining how things can be done differently. Making sure that a new system does everything you need often means specifying that it does everything the way you are doing it now.

Every tool encapsulates a methodology – a set of choices about how problems should be partitioned and addressed. This can lead people using a particular tool to view that tool's methodology as if it were the methodology of the craft itself, rather than simply one set of choices about how problems should be partitioned and addressed. Thus, individual tools create ruts in the mind, constraining our ideas about how things could be done.

By separating the principles and practices of structured writing from the implementations of particular tools, this book seeks to break out of the ruts created by particular tools, so you can see how to accomplish your objectives using these basic principles and practices before you select or build a tool set to implement those principles and practices in your organization.

It has been my particular privilege and opportunity to work with some of those rare people whose minds seem to be immune to such ruts and who were able to jog me out of my own ruts and help me see how different approaches could be both simpler and more effective. To name a dozen would be to neglect a score, but one name in particular deserves mention: Sam Wilmott, principal architect of the OmniMark programming language and all round markup language savant. Sam taught me to see the relationship of text and algorithms in a fundamentally new light, and everything I have done in my career leading up to writing this book has been working out the implications of what I learned from Sam. The SAM markup language I use for most of the examples in this book is an homage to Sam, and not just in its name.

Introduction

There are only six reasons for an organization to create content:

- To meet regulatory requirements
- To ensure the correct performance of internal processes
- To generate leads and support the sales process
- To lobby governments and affect social and political trends
- To improve customer retention through post-sales support
- To deliver content as a product in its own right

Each of these goes straight to the bottom line of revenue and profitability, and each depends directly on the quality of content. To create a content process with any goal in mind other than to maximize content quality is foolish, shortsighted, and almost certainly guaranteed to have negative effects on revenue and profitability even if it reduces costs in a particular division.

Rhetoric is the way in which content works to meet its goals. Each of the content goals outlined above requires a specific approach to rhetoric to make sure it does the job it is supposed to do. At the same time, like any business function, the creation of content needs to be done efficiently, and it needs to meet its deadlines. To achieve your business goals, you need great rhetoric and efficient processes.

Rhetoric

Aristotle defined rhetoric as "the faculty of observing in any given case the available means of persuasion." To put it another way, rhetoric is figuring out what to say and how to say it to persuade, inform, entertain, or enable the reader to act.

The word rhetoric is sometimes used dismissively, to describe content that is hollow or empty of meaning: "mere rhetoric." And content can use rhetorical tricks yet entirely lack substance. However, if there can be mere rhetoric there can also be substantive rhetoric, and if you want to communicate your substance effectively, you need sound rhetoric. Rhetoric without substance lacks value. Substance without rhetoric hides value. Substance plus rhetoric unlocks value.

Rhetoric takes many forms. In some cases it is a matter of choosing the right metaphor to communicate a particular point, or the right word to trigger an emotional reaction. But there are more mundane aspects to rhetoric, such as making sure you give the right information, use the right terms, and organize and label content so that people find what they need. These aspects are just as essential to persuading, informing, entertaining, or enabling the reader to act. They are just as much part of rhetoric as *le mot juste* or the striking metaphor.

These more mundane aspects of rhetoric are often quite structured, which means that they are more or less repeatable. If something is structured, it conforms to some shape or set of rules that you can observe, define, and reproduce independent of the particular instance. This means that you can create multiple instances of those things that all conform to the same structure.

A house is made of a particular set of bricks, but the bricks are put together according to an architectural plan. You can create a new house next door using the same architectural plan but different bricks. A recipe is made of particular words, but those words are organized according to a rhetorical plan. You can create a new recipe on the next page using the same rhetorical plan but different words, describing a different dish. Structured writing, in the purely rhetorical sense of the word, means capturing, defining, and implementing repeatable rhetorical structures.

Process

This ability to repeat structure is at the heart of process in any field. A process finds and defines the most efficient way to do something, which is pointless unless you plan to do the same thing more than once. Writing down how to do something you never plan to do again would serve no purpose.

Process is all about repetition. A process that aims to produce rhetoric is fundamentally concerned with the repeatability of rhetoric. Thus process and rhetoric are intimately combined in the production of content. Structured writing holds the key to repeatability in rhetoric and process.

Too often, however, structured writing systems and implementations focus on process issues alone, leaving rhetoric to take care of itself. There are three problems with this approach:

- Creating effective rhetoric is a complex task requiring the whole of the writer's attention. Yet many tools designed to support process goals such as publishing and reuse introduce complex tasks and concepts that pull attention away from the rhetoric that ought to be a writer's chief concern. This book shows how you can remove most of these distractions from writers.
- Some of the structures and processes enforced to meet process goals can be harmful to rhetoric. Although they may make the process more efficient, they can prevent writers from pro-

ducing the best rhetoric, thereby reducing the quality and effectiveness of content. Fortunately, as this book shows, there are structured writing techniques that can achieve most of the same process goals without harming rhetoric and often improving it. In many cases they also improve the process as well.

■ There are large deficits in rhetoric in many existing content sets because producing first-class, thorough, consistent rhetoric is a difficult task in its own right. (The principal cause of this is the curse of knowledge, a concept I explore in Chapter 9.) This book shows how you can use structured writing techniques to address the challenge of creating consistently good rhetoric.

Process and rhetoric, therefore, should never be treated independently. The whole purpose of the content process is to produce good rhetoric. It is, if you like, a rhetorical process. Process and rhetoric are intimately connected; you cannot hope to successfully treat one while ignoring the other.

This cuts both ways, of course. The things you do to support rhetoric must also be compatible with or contribute to efficient process. An approach to rhetoric that brings your production schedule to a halt does no one any good. Indeed, timeliness has become a key attribute of good rhetoric. Readers expect information to be up to date and available whenever they need it. If your process cannot produce good rhetoric in a timely manner, it cannot produce good rhetoric at all.

Complexity

The chief impediment to a process producing good rhetoric is complexity. Creating content is a complex business. There are rhetorical issues:

■ You have to research your subject matter and your audience.

■ You have to figure out what to say and how to say it to persuade, inform, entertain, or enable your reader to act.

And there are process issues:

■ You have to record your content, manage it, and publish it, potentially to many different audiences and many different media.

■ If your organization creates a lot of content, you have to coordinate with everyone else creating content and make sure it all works together and serves the organization's goals.

- As your subject matter and business needs change, you need to figure out how your content needs to change and implement those changes cleanly and efficiently.
- You have to create good consistent letter shapes with your pen.

Okay, we haven't had to deal with that last one for a while now. But something has to create letter shapes, line them up nicely, and fit them within the margins. Before we had machines to do it for us, that task was part of the writer's life and an essential skill if you wanted to create a readable document. And if you wanted more than one copy, you had to write out each copy by hand or hire copyists to do it for you.

Fortunately, Gutenberg invented a machine that took that burden off the writer's hands. The printing press allowed people who were not trained in calligraphy to become writers. It also put a lot of copyists out of business. (Yes, the loss of white-collar jobs to automation is not a new thing.) But it also made books much cheaper, revolutionizing civilization and creating many more white-collar jobs.

Forming letter shapes and making copies is part of the content creation process. As long as writers had to perform these tasks, they needed a particular set of skills and a great deal of time in order to create a new work. Following the introduction of the printing press many more new works were created. These basic process improvements had a profound impact on rhetoric, freeing time and energy to create new rhetoric.

Partitioning complexity

Process improvement of this kind can be described as a partitioning of a task or process. The printing press partitioned the task of physical book making from the task of language composition, thus freeing writers to spend more time and energy writing.

Of course, the writer's handwriting still had to be good enough for the printer to read. But then came the typewriter, the computer keyboard, and the touch screen with predictive typing, and now many children are not even taught cursive writing in school. Lament the withering of handwriting skills if you like, but it represents a new partitioning of the writing process, one in which manipulating a pen and shaping letters has been entirely partitioned from the writer.

Of course, this introduces a new skill for writers to master: typing. When a task is partitioned, there is always a new skill to learn. Partitioning a task creates the need to transmit information from one partition to another. Typewriters (or, today, the computer and monitor) take over the task of forming letters on paper or screen, but they need to know which letters the writer wants. The writer tells them by tapping keys on the keyboard. The keys are the interface between the partitions of the writing system. Using new interfaces is part of adapting to the newly partitioned system.

Structured writing enables you to partition the content production process by capturing information that must be passed from one part of your process to another. For instance, you can partition the creation of a list from the formatting of a list by using list markup (such as HTML `<ol>` and `<li>` tags) and a separate formatting language (such as CSS) to specify how lists are formatted. The markup language serves as an interface between the writing partition and the formatting partition, ensuring that the required information – this is a list, this is a list item – gets passed from one partition to another.

By partitioning the process correctly, you can allocate tasks that require complex skills and knowledge among different workers, allowing each to develop and practice their individual expertise while ensuring that all the pieces come together in a reliable and efficient manner.

Every content organization partitions and distributes content creation. Newspapers do not make reporters drive delivery trucks. Publishing houses do not let authors design book covers (much as they might want to). Wise organizations partition the job of editing and proofreading content from the job of writing it. When they need to deliver a large body of content to the web, organizations employ information architects and content strategists. All of these are examples of partitioning the complexity of the content process to improve quality and efficiency. Partitioning a modern content process is complex, often involving multiple partitions with multiple complex interfaces.

There are two potential benefits to partitioning the content process. First, it can make the writer's job easier, meaning that more people can write, and writers can focus more attention on subject matter and rhetoric and less on the tools. Second, you can improve efficiency by assigning partitioned tasks to people or processes better equipped to handle them. A typewriter or a computer makes much more consistent letter shapes than a human hand. The printing press makes identical copies much faster than a human copyist.

You can't reduce the fundamental complexity of content creation. All that stuff has to get done, and it is all difficult in one way or another. Different tasks require different skills and knowledge. Rhetoric alone – figuring out what to say and how best to say it – is a complex task that deserves the writer's whole attention. You cannot eliminate the complexity of everything that goes on around the writing task, but you can make sure that it is partitioned and directed to the people or processes that are best equipped to deal with it.

The complex relationship of process and rhetoric

Process and rhetoric are intimately connected. In the previous section, I divided the complexity of content creation into rhetorical complexity and process complexity. But if you look again, you will see that rhetorical issues pervade the process issues.

- You have to record the content, manage it, and publish it, potentially to many different audiences and many different media. But when you manage content and distribute it to different audiences, what are you doing but ensuring that you are delivering the right rhetoric? *Content management* is the management of rhetoric.
- If your organization creates a lot of content, you have to coordinate with everyone else creating content and make sure it all works together and serves the organization's goals. And what makes content work together? It must conform to the same rhetorical rules and goals. *Content strategy* is the strategy of rhetoric.
- As your subject matter and business needs change, you need to figure out how your content needs to change and implement those changes cleanly and efficiently. And when you determine which changes are needed and how to make them, what are you managing and changing but rhetoric? The *content process* is the process of creating, managing, and delivering rhetoric.

What happens if you don't manage the inherent complexity of content creation properly? Complexity cannot be destroyed. It has to go somewhere. If your content lacks needed information, the burden of finding that information falls on your readers. If your readers can't understand the terms used, read the font chosen, learn about an unfamiliar concept, get information when they need it, or tell current information from outdated, that is a failure of rhetoric.

Another name for the failure of rhetoric is poor content quality. Poor content quality is the direct result of the complexity of content creation not being handled properly within the content producing organization. The only product of the content process is rhetoric. All content problems are rhetoric problems and all rhetoric problems are the result of process problems. To consistently create good rhetoric, you must ensure that every piece of the inherent complexity of content creation is directed to a competent and adequately resourced person or process.

The relationship between complexity and quality has never been better understood or more important. Some of the defining technology of our age is mind-bogglingly complex, but the most successful products – those with the highest quality and value – partition complexity successfully, creating simple interfaces not only for consumers but also for those who design and build those products. Without a sophisticated partitioning of complexity throughout the supply chain and the design and manufacturing process, many modern products could not be brought to market.

But in many organizations, the current tools and processes fail to assign each part of the complexity of content creation to the person or process with the best combination of knowledge, skills and resources to handle it. The result is poor content quality, inefficient processes, and reduced revenue and profitability. The cure is to adopt a partitioning that better handles the complexity the organization is experiencing. Structured writing is a tool for implementing that cure. Note that structured writing is not the cure itself. It is a tool for implementing the cure. The cure is the correct distribution of complexity in your organization.

Complexity is complex

Managing complexity in your content process is complex. There is no magic elixir that will make the complexity of your content processes disappear, and adopting this year's gotta-have-it content technology will not magically put all your complexity into the right buckets.

Structured writing is itself a complex technique and some of the most popular structured writing methods of the day are highly complex. That is not necessarily bad. The aim is not to eliminate complexity altogether – that is impossible – but to partition it so that each part of that complexity is handled by the person or process with the knowledge, skills, and resources to handle it. Complex tools are fine if they help you achieve a better partitioning of complexity. But you can't ignore the complexity that these tools bring, nor can you ignore the impact of the additional complexity the wrong tools or techniques can dump on writers and others in your content system.

The aim of this book is to help you achieve a better partitioning of the complexity of your content system in order to improve your rhetoric and make your process more efficient and reliable. You cannot accomplish this without understanding the difficulties that come with specific structured writing techniques and tools.

Whenever I talk about techniques for supporting process, I will warn about any pitfalls those techniques hold for rhetoric. Whenever I talk about techniques for supporting rhetoric, I will warn about any pitfalls those techniques hold for process. Part of this involves examining methods

I don't recommend or recommend only in selective cases. Because some of these techniques are widely used, I think it is important to point out their pitfalls, if only to answer the question of why you should choose the alternate techniques that I recommend.

This may appear to give this book a split personality: part survey, part advocacy for a set of techniques that I think should be more widely used. But the truth is, almost all of the techniques described in this book are appropriate in some situations. Some are overused, and I will critique their use. Some are underused, and I will advocate their use. But my aim is to move toward balance, not to tilt the scale so much the other way that people are again using techniques and tools that do not fit their needs, circumstances, and resources.

The optimal partitioning of complexity for your organization is unique to your organization (and may well change over time). The aim of this book is to show you how you can use structured writing techniques to develop the partitioning that is optimal for your organization, not to recommend one form of partitioning over another.

How this book is organized

This book is organized into seven parts, as follows:

- **Structured Writing Domains:** Describes the four main ways you can apply formal structure to content, what I call the four domains of structured writing: the media, document, subject, and management domains. This lays the foundation for discussing how structure is used to partition complexity through the remaining sections of the book.

- **Process, Rhetoric, and Structure:** Describes how rhetoric can be represented in structured writing and how this affects the partitioning of the content system. It then looks at how to represent rhetorical structure at the level of both the individual document and the information set as a whole and how writers can work with rhetorical structures in structured writing.

- **Algorithms:** Describes the main content processing algorithms used in structured writing, how they affect the partitioning of the content system, and how they work in each of the structured writing domains.

- **Structures:** Describes the structures used in structured writing in more detail and looks at how to handle structural issues at every scale from full documents to individual phrases.

- **Languages:** Describes the various markup languages (and alternative technologies) used to capture and express content structure, including an overview of some of the major public markup systems, such as Markdown, DocBook, and DITA.

- **Management:** Describes how structured writing fits into the larger picture of managing the overall content process and shows how structured writing can enable some important management functions that are difficult or impossible to implement efficiently with less structured content.

- **Design:** Describes an approach for designing the right content partitioning for your content process and selecting the right tools and technologies to implement it.

The relationship between complexity and quality has never been better understood or more important. Some of the defining technology of our age is mind-bogglingly complex, but the most successful products – those with the highest quality and value – partition complexity successfully, creating simple interfaces not only for consumers but also for those who design and build those products. Without a sophisticated partitioning of complexity throughout the supply chain and the design and manufacturing process, many modern products could not be brought to market.

But in many organizations, the current tools and processes fail to assign each part of the complexity of content creation to the person or process with the best combination of knowledge, skills and resources to handle it. The result is poor content quality, inefficient processes, and reduced revenue and profitability. The cure is to adopt a partitioning that better handles the complexity the organization is experiencing. Structured writing is a tool for implementing that cure. Note that structured writing is not the cure itself. It is a tool for implementing the cure. The cure is the correct distribution of complexity in your organization.

Complexity is complex

Managing complexity in your content process is complex. There is no magic elixir that will make the complexity of your content processes disappear, and adopting this year's gotta-have-it content technology will not magically put all your complexity into the right buckets.

Structured writing is itself a complex technique and some of the most popular structured writing methods of the day are highly complex. That is not necessarily bad. The aim is not to eliminate complexity altogether – that is impossible – but to partition it so that each part of that complexity is handled by the person or process with the knowledge, skills, and resources to handle it. Complex tools are fine if they help you achieve a better partitioning of complexity. But you can't ignore the complexity that these tools bring, nor can you ignore the impact of the additional complexity the wrong tools or techniques can dump on writers and others in your content system.

The aim of this book is to help you achieve a better partitioning of the complexity of your content system in order to improve your rhetoric and make your process more efficient and reliable. You cannot accomplish this without understanding the difficulties that come with specific structured writing techniques and tools.

Whenever I talk about techniques for supporting process, I will warn about any pitfalls those techniques hold for rhetoric. Whenever I talk about techniques for supporting rhetoric, I will warn about any pitfalls those techniques hold for process. Part of this involves examining methods

I don't recommend or recommend only in selective cases. Because some of these techniques are widely used, I think it is important to point out their pitfalls, if only to answer the question of why you should choose the alternate techniques that I recommend.

This may appear to give this book a split personality: part survey, part advocacy for a set of techniques that I think should be more widely used. But the truth is, almost all of the techniques described in this book are appropriate in some situations. Some are overused, and I will critique their use. Some are underused, and I will advocate their use. But my aim is to move toward balance, not to tilt the scale so much the other way that people are again using techniques and tools that do not fit their needs, circumstances, and resources.

The optimal partitioning of complexity for your organization is unique to your organization (and may well change over time). The aim of this book is to show you how you can use structured writing techniques to develop the partitioning that is optimal for your organization, not to recommend one form of partitioning over another.

How this book is organized

This book is organized into seven parts, as follows:

- **Structured Writing Domains:** Describes the four main ways you can apply formal structure to content, what I call the four domains of structured writing: the media, document, subject, and management domains. This lays the foundation for discussing how structure is used to partition complexity through the remaining sections of the book.
- **Process, Rhetoric, and Structure:** Describes how rhetoric can be represented in structured writing and how this affects the partitioning of the content system. It then looks at how to represent rhetorical structure at the level of both the individual document and the information set as a whole and how writers can work with rhetorical structures in structured writing.
- **Algorithms:** Describes the main content processing algorithms used in structured writing, how they affect the partitioning of the content system, and how they work in each of the structured writing domains.
- **Structures:** Describes the structures used in structured writing in more detail and looks at how to handle structural issues at every scale from full documents to individual phrases.
- **Languages:** Describes the various markup languages (and alternative technologies) used to capture and express content structure, including an overview of some of the major public markup systems, such as Markdown, DocBook, and DITA.

Structured Writing Domains

Structured writing is a method for constraining and recording the structure of content to improve its rhetoric and manage the content creation process. Structure can be applied to content in several different domains. In this section, I introduce four domains of content structure, each with a fundamentally different approach to the relationship between process and rhetoric. Those domains are: media, document, subject, and management.

None of these domains is new. People have been using these approaches to markup for decades. Dividing these techniques into domains simply gives us a consistent way to talk about them and explore the virtues of each. Through the rest of the book I will show how choosing a markup language in each of these domains profoundly affects how you can manage process and rhetoric in your content system.

> Note: The term *semantic* is commonly used to characterize markup languages. Semantics is the study of meaning, so semantic markup is markup that tells us what the content means. Unfortunately, people can, and do, understand the phrase "semantic markup" in different ways, leading to confusion about what is and is not semantic markup. Therefore, I do not use the term semantic to describe markup languages, though I do use it in other contexts.

CHAPTER 1
How Ideas Become Content

Structured writing is an approach to content creation that can help you improve both your rhetoric and your process. To understand how structured writing can help, let's begin by looking at how content gets from ideas in your head to dots on a page or screen.

From ideas to dots

The process of creating and delivering content consists of translating ideas (stuff someone thinks or knows) into a form that can be read (dots of ink on a page or pixels on a screen).

Structured writing applies a structured methodology to that process. It is a long road from ideas to dots, and structured writing techniques can be applied at many points along that road. Almost all writing done today uses structured writing techniques to one extent or another. The principles of structured writing apply across the spectrum, from the tools and techniques used in most offices today to the most sophisticated structured writing systems.

All writing has structure in the linguistic sense of the word. Every comprehensible sentence has a grammatical structure. You may even have learned to diagram that structure in school.[1]

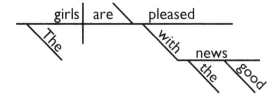

Figure 1.1 – Diagrammed sentence

---

[1] Copyright © 2012, Tjo3ya, CC BY-SA 3.0, [https://commons.wikimedia.org/w/index.php?curid=18312612]

Just as all writing has at least a basic grammatical structure, all writing on a computer involves creating basic data structures. Thus, the only case in which no structured writing techniques are involved is when a writer writes down ideas with pen and paper and gives that paper directly to the reader. In that case, the entire writing process, from an idea to words on paper, takes place in the writer's head and hands.

Writers rarely record their ideas directly in the final physical form these days. For instance, the writer may write in a word processor, edit the text on screen, and press **Print** to send the content to a printer or **Send** to have the content rendered on someone else's screen. The point, along the journey from ideas to dots, where content is recorded has moved back just a little bit.

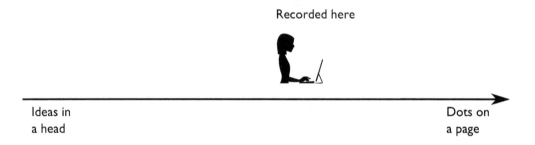

Word processing, desktop publishing, and various approaches to structured writing all establish a point between ideas and dots where the content is recorded and then provide algorithms to complete the journey from that point to dots on a page.

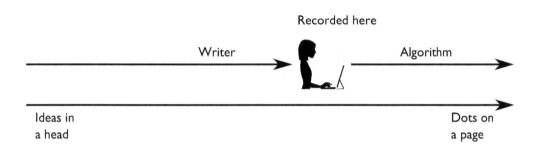

You can describe this process of earlier recording in terms of three domains, each domain reflecting a stage in the progress from ideas to dots. The domains are the media domain (which is concerned with lines and dots on paper or screen), the document domain (which is concerned with the expression and organization of documents), and the subject domain (which is concerned with the subject matter that we write about).

The three domains

Let's suppose you want to write a recipe for chicken noodle soup. You start out with the idea of a soup made with chicken and noodles. This is an idea about the subject matter and not yet any form of content.

You then decide to name the dish "Chicken Noodle Soup." You figure out which ingredients to use and how to make the dish. This is all information about making chicken noodle soup but it is not yet part of a document. You have not yet made any rhetorical decisions about how to communicate this information effectively. It is information in the *subject* domain.

Then, you decide how to present this information to help other people make Chicken Noodle Soup. You decide to have a title, a picture, an introduction, a list of ingredients, and a set of preparation steps. At this point, you are no longer gathering information, you are focused on presenting the information you have gathered, applying a specific rhetoric to your information. These are decisions in the *document* domain. Documents are how you organize and present information. (As you will see, however, some document types, such as recipes, are specific to a particular subject, and there is considerable overlap between document and subject domain considerations in their design, which may or may not be reflected in how they are recorded.)

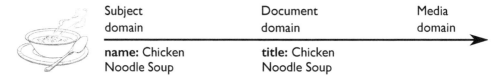

Then, you consider how the document will look on screen or on paper. What font will you use for the heading and the body text? How large will the heading and the body be? Will the quantity of the ingredients be flush right? Will there be leading dots? Will the presentation steps be numbered or just presented sequentially? How big will the picture be? Will the text wrap around it? These are decision in the *media* domain.

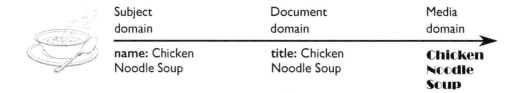

All content passes through the three domains. Content always begins with thinking about subjects in the real world. You decide to express ideas about those subjects in words. You collect those ideas and determine an order and structure to express them. Finally, you decide how they will be formatted in a particular medium. The question is, where in this process do you start recording the content?

Do you format your content as you write? Then you are working in the media domain.

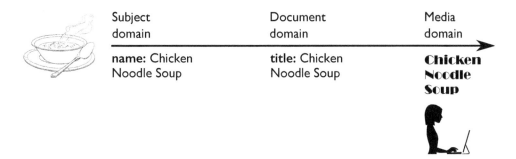

Do you record presentation units such as lists, headings, and steps without associating formatting? Then you are working in the document domain.

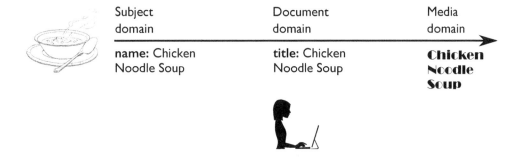

Do you record the raw information as data, for instance, identifying ingredients as ingredients and quantities as quantities, rather than as list items? Then you are working in the subject domain.

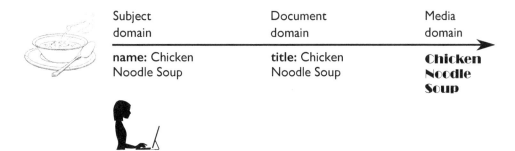

If you recorded your content in the media domain, it is ready to publish.

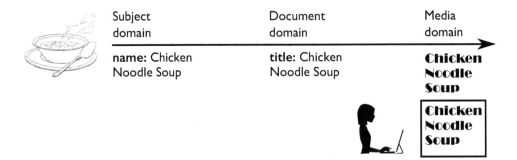

If you recorded it in the document domain, then it needs to be formatted for each target medium before it can be published. This is done by an algorithm that translates the document domain structures and into the appropriate media domain structures for the target media.

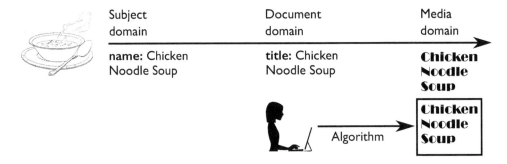

If you recorded it in the subject domain, then it needs to be organized into a document before it can be published. This is done by an algorithm that translates the subject domain structures into the appropriate document domain structures for the intended document.

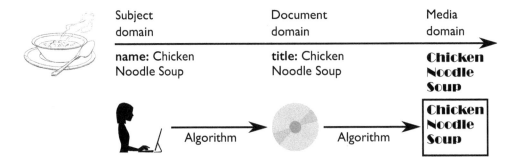

This does not mean that a structured writing system is always wholly in one domain. In fact, that is more the exception than the rule. Most systems are hybrids with structures from more than one domain, or structures that themselves have a foot in more than one domain. However, the differences between how things work in each of the domains is significant, and identifying which domain you are working in at any given time will help you understand what is possible and determine if moving to a different domain might make your process or rhetoric better.

At the beginning of this section I described the three domains in terms of the decisions that must be made in each domain. All of these decisions, from all three domains, have to be made for every document that you produce. When you record content in the media domain, you make decisions in all three domains as you write. When you record content in the document domain, you make decisions in the subject and document domains and defer media-domain decisions to algorithms. When you record content in the subject domain, you make decisions in the subject domain and defer document- and media-domain decisions to algorithms.

Recording content earlier in the process reduces the number of decisions you have to make while writing (reducing the complexity of the writing task), while preserving the ability to make different decisions later. This shift can have profound effects on the efficiency of your process and the usefulness of your content.

That is the promise of structured writing. However, it is not always as clear and simple as that makes it sound, since structured writing introduces decision-making requirements of its own. The next three chapters look at what it is like to write in each domain.

CHAPTER 2
Writing in the Media Domain

In the media domain, structures relate to the medium in which the content is displayed. Such content is often considered unstructured, but all content has structure, and you can find all the patterns and techniques of structured writing in the media domain. This makes it a good place to begin our study of the fundamentals of structured writing.

At its most basic, a hand guiding a pen over paper or a chisel into stone is working in the media domain through direct physical interaction with the medium. The writer guiding that pen both makes and executes decisions about the shape of letters, along with all the other decisions that fall on a writer.

The closest you get to pen and paper in the computer world is to use a paint program to directly place dots on the screen. You select the pen tool and use your mouse or a stylus to write. This records the text as a matrix of dots.

There is little structure here. There's just a pattern of dots, which represents text characters only in the sense that humans can recognize the pattern as text. The computer records the effect of your decisions about letter shapes, but it has no idea that they are letters.

This is an inefficient way to write. You can work faster if you use the paint program's text tool.

Dog

The text tool is our first step into structure. It partitions the complexity of forming letter shapes from the task of writing letters and directs that complexity to an algorithm in the paint program, which does a better and more efficient job of it than the writer does, as you can see from the much neater letter shapes in the sample above. However, those letters are still recorded as a set of dots, not as characters, so you can't go back and edit your text as text, only as dots. The paint program forms letter shapes, but it records dots.

This is a common problem in writing tools. For example, some wiki systems allow you to define content templates but don't record the structure of the template in the resulting page. This reduces your ability to maintain the rhetorical structure expressed by the template over time.

Recording characters as dots means you can't edit the text or change its formatting. To fix this you need to move away from dots and start using a tool that records characters as characters. You could go to a text editor, but a text editor does not keep any formatting information,[1] thus dropping the formatting completely. For most publishing purposes, plain text is inadequate. We need to maintain the ability to format the document.

One type of program that records text with formatting attached is a vector graphics program. A vector graphics program creates graphics as a collection of objects. For example, to represent a circle, a vector graphics program records an abstraction: a mathematical formula that encodes the essential properties of a circle – the center, diameter, line weight, etc. – rather than a set of dots. The computer then lets you manipulate that abstraction as an object, only rendering it as actual dots when the graphic is displayed on screen or paper.

---

[1] Unless you add markup to your text, but that would be getting ahead of ourselves.

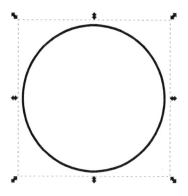 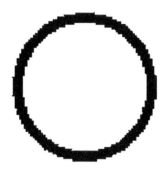

Figure 2.1 – Objects vs. dots

In Figure 2.1 you see a circle as an object displayed in a vector graphics program (Inkscape) on the left and a circle as a set of dots in a graphics program (Microsoft Paint) on the right.

In a typical vector graphics program, a shape is rendered into dots on screen instantly as you draw or edit the shape. Nonetheless, the computer is storing data describing the shape, not a circular pattern of dots, as it would in a paint program. In structured writing we call this separating content from formatting. The mathematical abstraction of a circle is the content; the dots that represent it on screen are the formatting or, rather, the result of applying formatting to the object. All the principles of structured writing are present in this basic piece of computer graphics.

Partitioning font information

Just as a vector graphics program represents a circle as a circle object, it represents text as a text object. A text object is a rectangular area that contains characters. It has numerous media-domain properties, such as margins, background and foreground colors, the text string, and the font face, size, and weight used to display that text (see Figure 2.2).

Figure 2.2 – A vector graphics text object

A vector graphics program displays text in a chosen font. If you change the value of the text object's font attribute, it immediately redraws the text in the new font, meaning you can change the font and font attributes as much as you like without editing the text itself.

The vector graphics program needs to know the shape of each character to render a text object in the media domain. However, that information is not stored as part of the text object. The representation of the text in the paint program includes the shape of the letters but in the vector graphics program it does not. That information has been moved out into a font file.

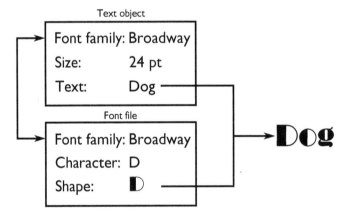

Figure 2.3 – Merging text and font information

The shapes of the letters (technically, *glyphs*) that make up the font are stored in font files. Font files consist of a set of shape objects that describe each glyph together with metadata such as the name of the font and the name of each glyph. To display the text, the graphics program uses information in the text object to identify the correct font file, locate the right glyph in that font file, and then draw the appropriate glyph on the current medium (see Figure 2.3).

Every modern operating system partitions the rendering of fonts into a separate font system that any application can use. By transferring font rendering complexity to the font system, the OS designers make it easier for developers to create applications that work with formatted text. Rather than programming font handling themselves, programmers call operating system APIs (Application Programming Interfaces) to do it for them. It also makes things easier for font designers, since it partitions off the problems of installing fonts and making them available to applications.

Font handling is a particularly powerful example of structure simplifying process, because it means that professionals working in the content industry – font designers, tool developers, and writers – no longer need to communicate or coordinate with each other to contribute to the content delivery process.

This is a recurring pattern. We partition information that is constant for a particular application into a separate file – the shape of the capital letter D is consistent, no matter how many times it occurs in a text – so we can keep that information separate from the text. This simplifies the format of the information and keeps the downstream presentation more consistent.

Designing a content structure, regardless of the domain you choose to work in, essentially consists of identifying the places in the content where we can partition out these invariant properties into separate structures.

CHAPTER 3
Writing in the Document Domain

A vector graphics program does a great job of partitioning formatting complexity from text. But what happens when you want to create a document that spans multiple pages or if you edit the text and need to change how it flows from one page to the next? A vector graphics program will force you to handle these pagination issues by hand. Pagination is complex, requiring many decisions about how and where to place page breaks. It is tedious and time consuming to execute those decisions by hand. An algorithm can make those decisions, but to do that, you need to move your content into the document domain.

The document domain is concerned with the presentation of information as a document. In the document domain you use structures such as title, list, and table, without specifying how they will be formatted. You make document-domain decisions and defer media-domain decisions to someone else, usually a designer who creates an algorithm to execute the decision.

> Note: The words *formatting* and *presentation* are commonly used as synonyms, but here I make a distinction between, for example, the decision to present a piece of information as a list (presentation), and the decision to format list items in 12-point Palatino with 1-em indent and a square bullet (formatting). In this sense, presentation and formatting are distinct and can be partitioned from each other. However, it's easy to lose this distinction because most of today's tools combine these two operations in a WYSIWYG display.

Word processors and desktop publishing programs partition the pagination process from the writing process by introducing some document-domain constraints. A document is made up of a series of pages that have margins and contain text flows. Text flows are made up of blocks (paragraphs, headings) inside of which text can flow, even from one page to the next. Common features, such as tables, are supported as objects that can exist in text flows. New pages are created automatically as text expands. In other words, the program creates a bunch of text containers and decides how to fit text into those containers. This leaves writers free to focus on writing and gives pagination decisions to the program.

Sections, paragraphs, headings, and tables, are all document-domain objects. Rather than working on a blank slate, as they do in a graphics program, writers now work within the constraints of these document-domain objects. These constraints remove or limit decisions about positioning of elements, which makes creating documents faster and more consistent.

But while these constraints are powerful, word processing and desktop publishing impose few other constraints on formatting. While they offer a basic set of document-domain objects – sections, paragraphs, tables, and so forth – they use a WYSIWYG display, which unifies, rather than partitions, writing and formatting. The writer worries less about pagination but is still thinking mostly in terms of styles and formatting – the concerns of the media domain. If you want the writer to think less about formatting you need to factor out the media-domain concerns. For that you need to move more fully into the document domain where you can take the formatting decisions out of the writer's hands entirely.

Consider a list. You may want the spacing above the first item of a list to be different from the spacing between other items of the list. This is a media-domain constraint – it is about formatting, not the structure of the document. However, this constraint is hard to enforce in the media domain.

Most media-domain writing applications create lists by applying styles to ordinary paragraphs. To format a list with an extra space above the first item, you need to create two different styles: a `first-item-of-list` style and and `following-item-of-list` style. The result might look like Figure 3.1.

```
p{first-item-of-list}: Carrots
p{following-item-of-list}: Celery
p{following-item-of-list}: Onions
```

Figure 3.1 – List structure in the media domain

This requires the writer to apply the `first-item-of-list` style to the first item of a list and the `following-item-of-list` style to the following items. The word processor does not enforce this constraint. The writer has to remember it, which creates two problems:

1. It makes writing a bit harder (and all the bits add up). The writer needs to decide which style to use for every list item, even though the design decision has already been made. And new writers must learn, and remember to apply, this rule.
2. If the writer gets it wrong, the problem can easily go unnoticed.

Structured writing works by factoring out invariants. Most constraints are invariants – that is, they are rules that apply to all instances of the same content structure. The constraint that all lists must have extra space before the first item is an obvious example. The best way to enforce an invariant constraint is to partition it out altogether.

To do this, you create a list structure – not a styled paragraph, but a structure that is specifically a list. A document-domain list structure looks like Figure 3.2:

```
list:
    list-item: Carrots
    list-item: Celery
    list-item: Onions
```

Figure 3.2 – List structure in the document domain

A list structure partitions the idea of a list from the physical formatting of a list by creating a container that did not exist before – the `list` structure. By creating a `list` structure, the writer tells the formatting algorithm that this is a list. Since a list is a feature of a document, it is a document-domain structure. The `list` container has no media-domain analog.

Taking the formatting decisions out of the writer's hand obviously reduces complexity for writers. But moving those decisions away from writers also ensures that they are executed more consistently. All of the media-domain constraints – all the design elements – that writers previously had to remember and execute are now executed by an algorithm. This means that you can enforce media-domain constraints that were difficult to enforce in the media domain itself.

This illustrates a consistent pattern in structured writing: you often move to the next domain to enforce – or factor out – constraints in the previous domain. This is an important design principle that you will see several times as we look at the structured writing algorithms. Wherever possible, look for a way to factor out a constraint rather than enforce it. Enforcing a constraint still requires the writer to think and act. Factoring it out removes it from the writer's concerns entirely.

Here is where specifically rhetorical structures enter the picture. In creating a list structure, writers record some of their rhetorical intent – though not all of it by any means – namely, the intent to express information as a list. Just as recording text as a text object rather than as a set of dots captures its nature as text, recording a list in a list structure rather than as lines preceded by bullet characters captures its nature as a list.

Once you have a `list` structure, you can create rules – in a separate file – about how lists are formatted. You may be familiar with this from HTML and CSS. Figure 3.3 shows a list structure in HTML (actually, this is a slightly more specific structure, but I'll get to that).

```
<ol>
    <li>Carrots</li>
    <li>Celery</li>
    <li>Onions</li>
</ol>
```

Figure 3.3 – List structure using HTML

Now that you have a distinct list object, you can factor out the invariant list formatting rule into a separate file. For HTML, you can do this with a CSS style sheet (see Figure 3.4):

```
li:first-child
{
    padding-top: 5pt;
}
```

Figure 3.4 – CSS for a list structure

This factors out the spacing-above-lists decision and assigns it to an algorithm, partitioning the complexity of list formatting by asking the writer to express a simple idea in familiar terms – this is a list – instead of executing a complex set of formatting instructions.

But wait! That's fine if all lists are formatted exactly the same way, but they're not. At the very least, some lists are bulleted and some are numbered. And then there are nested lists, which are formatted differently from their parents, and specialized lists, such as lists of ingredients, definitions, or function parameters. How do you make sure each type of list gets formatted appropriately?

A word processor or desktop publishing application addresses this problem using styles. You can create a style for each type of list formatting you want. A document-domain environment must make similar provision for different types of lists. However, the document domain deals with presentation (this is a list), not formatting (this paragraph starts with a bullet and a tab and is in 12pt New Century Schoolbook). So the question becomes, how many different kinds of list presentations do you need to support?

One obvious presentation difference between lists is that some express an ordered sequence of items while others present a collection of items with no necessary order (a rhetorical distinction). These two options are generally formatted as numbered and bulleted lists, respectively, but in the document domain, you express the presentation difference – the rhetorical difference – not the formatting difference.

The common way to handle this is to create two different list structures: an ordered list and an unordered list. Different markup languages call them by different names – `ol` and `ul` in HTML, `orderedlist` and `itemizedlist` in DocBook, for example – but they are conceptually the same thing. (Thus the HTML example in Figure 3.3 is a little more specific than just being a list structure. It is an ordered-list structure.)

The choice of the terms "unordered" and "ordered" is important, because those terms focus on the rhetorical properties of a list – whether the order of list items matters – rather than on its media-domain properties – bullets or numbers.

Note that both the decision to create a list and the decision to use numbers, letters, or Roman numerals, have rhetorical aspects to them. It is not that rhetoric is absent from the media domain, but that the rhetorical intent is not recorded in the media domain, whereas some part of the rhetorical intent is recorded in the document domain, which allows us to factor rhetorically significant decisions about list numbering styles into the algorithm that processes document domain content into the media domain.

Context and structure

Does the need for separate ordered and unordered list objects imply that you need a separate document-domain list structure for every possible way you might format a list? No. In fact, that would be working in the media domain by proxy. When you work in the document domain, you think in terms of document structures, not formatting, which means that each document-domain object must make sense in the document domain, not the media domain. Otherwise, the partitioning falls apart.

For example, consider nested lists. Items in a nested list are formatted differently from the list that contains them. At a minimum, they are indented more, and they usually have different number or bullet styles. In the media domain, you would need a different style for each level.

However, you don't need a separate nested-list structure in the document domain. Instead, you express nesting by actually nesting one list inside another (expressing the rhetorical intent that they should be nested). For instance, in Figure 3.5 one ordered list is nested inside another.

```
<ol>
    <li>
        <p>Dogs</p>
        <ol>
            <li>Spot</li>
            <li>Rover</li>
            <li>Fang</li>
            <li>Fluffy</li>
        </ol>
    </li>
    <li>
        <p>Cats</p>
        <ol>
            <li>Mittens</li>
            <li>Tobermory</li>
        </ol>
    </li>
</ol>
```

Figure 3.5 – Nested list structure in HTML

In the document domain the inner and outer list are identical ol/li structures. In the media domain, one might be formatted with Arabic numerals and the other with letters (see Figure 3.6).

1. Dogs
   a. Spot
   b. Rover
   c. Fang
   d. Fluffy
2. Cats
   a. Mittens
   b. Tobermory

Figure 3.6 – Nested list rendered

Both the inner and outer lists are ordered lists in the document domain, but in the media domain they are formatted differently based on context.

In this case, the algorithm that formats the page distinguishes the inner and outer lists by looking at their parents. For instance, Figure 3.7 shows a CSS rule for a list item nested one deep.

```
ol>li>ol>li
{
    list-style-type: lower-alpha;
}
```

Figure 3.7 – CSS for nested list items

The ability to distinguish structures by context enables you to reduce the number of structures you need to define, particularly in the document and subject domains. It also allows you to name structures more logically and intuitively, since you can name them for what they are – what rhetorical role they play – not how they are to be formatted or for where they reside in the hierarchy of the document.

It also points out another important difference between the way media-domain and document-domain structured writing is usually implemented. The media domain almost always uses a flat structure with paragraphs, tables, etc., following each other in sequence. For instance, Microsoft Word constructs a nested list as a flat sequence of paragraphs with different styles. Inner and outer lists are expressed purely by the indent applied to the paragraphs. (Word tries to maintain auto-numbering across nested list structures, but it does not always get it right.)

Document-domain structures are usually implemented hierarchically. List items are *inside* lists. Nested lists are *inside* list items. Sections are *inside* chapters. Subsections are sections *inside* other sections. Where the media domain typically only has before and after relationships (except in tables), the document domain adds inside and outside relationships to the mix. This use of nested, rather than flat, structures helps to create context, which helps to reduce the number of different structures you need. Just as we saw with the basic list structure, nested structures help partition logical document structures.

Some document-domain languages are more hierarchical than others. HTML is relatively flat; it has six different heading styles: H1 through H6. DocBook, a widely used document-domain structured writing language, is much more hierarchical and has only one element for the same purpose: title. However, DocBook's title element can occur inside 84 different elements and, therefore, can potentially be formatted in 84 different ways based on context. In fact, it can be formatted in more ways than that, since some of the elements that contain titles can also be nested in other elements, creating a hierarchical structure that provides even more contexts.

There is a balance to be struck here. Hierarchical structures are harder to create and can be harder to understand. They may require you to find just the right place in a hierarchy to insert a new piece of content, which is more difficult than simply starting a new paragraph in a word processor. This introduces complexity, and you need to make sure you don't introduce more complexity than you take away.

Constraining document structure

In addition to factoring out media-domain constraints, another important reason for working in the document domain is to constrain how documents are structured (the rhetoric they use). For example, suppose you want to make sure that all graphics have a figure number, a title, and a caption. In the media domain, you can make styles available for figure numbers, titles, and captions, but you can't enforce a rule that says all graphics must have these structures. In the document domain, you can express these constraints. You can make it illegal to place an image structure anywhere in the document structure except within a figure that has a title and caption. A structure to implement this constraint might look like Figure 3.8.

```
figure:
    title: Cute kitty
    caption: This is a cute kitten.
    image: images/cute.jpg
```

Figure 3.8 – Document-domain structure for a figure

If the only way to include an image is to use the image structure, and you only allow the image structure inside the figure structure, and you require the figure structure to contain the title and the caption structures, then the writer can't add a graphic without a figure, title, and caption. A document without these structures would be rejected by the conformance checks (see Chapter 29).

In the case of list structures, the document-domain allows writers to capture some of their rhetorical intent. This takes things a step further. This markup does not merely allow writers to express their rhetorical intent, it constrains their rhetorical choices.

The list object simply says, if you want to create a list, don't bother about formatting, just say it is a list and leave the formatting to the formatting algorithm. It does not force the writer to create a list in any particular location, it merely provides a higher-level tool for creating lists without having to worry about how they are formatted. The rhetorical choice to create a list or not remains entirely with the writer.

But the figure object introduces a requirement, a constraint that says, if you want to insert an image, you have to supply a title and a caption. Writers have no control over these rhetorical choices. They can decide to insert an image or not, but they cannot decide to insert an image without a title and a caption. Thus, you can impose a consistent rhetoric for images across all of your content.

Let's look at another example. A bibliography is a document structure for listing works cited in a document. It generally consists of the heading "Bibliography" followed by a set of paragraphs listing the cited works. In the media domain, you can easily create such a structure. It is just a sequence of paragraphs with some bold and italic formatting for author names, book titles, etc.

In your media-domain style sheet, you may have some character styles for things like `author-name` or `book-title`. You may even have a paragraph style for bibliography entries, but you typically don't have much more specific support.

But these few styles can't cover all the rules a publisher or organization places on the rhetoric of bibliographies. Writers are generally required to follow rules that detail how to handle titles, author names, and so forth. These constraints are not enforced by media-domain styles. Writers have to learn and follow these constraints themselves. If you want to remove these decisions from writers (and thus achieve rhetorical consistency among all writers), you need a document-domain structure for a bibliography. DocBook has one (see Figure 3.9).

```
<biblioentry id="bib.xsltrec">
  <abbrev id="bib.xsltrec.abbrev">REC-XSLT</abbrev>
  <editor>
     <firstname>James</firstname>
     <surname>Clark</surname>
  </editor>
  <title>
    <ulink url="http://www.w3.org/TR/xslt">XSL Transformations
    (XSLT) Version 1.0</ulink>
  </title>
  <publishername>W3C Recommendation</publishername>
  <pubdate>16 November 1999</pubdate>
</biblioentry>
```
Figure 3.9 – DocBook bibliography entry

Figure 3.9 is in XML, which can be hard to read, so here (Figure 3.10) is the same structure using the markup notation I have used in earlier examples.

```
biblioentry:(#bib.xslttrec)
    abbrev:(#bib.xsltrec.abbrev) REC-XSLT
    editor:
        firstname: James
        surname: Clark
    title: XSL Transformations (XSLT) Version 1.0
    publishername: W3C Recommendation
    pubdate: 16 November 1999
```

Figure 3.10 – Bibliography entry using simpler markup

This structure does not just constrain how bibliography entries are presented and formatted, it also factors out many of those constraints by breaking down the components of a bibliography entry into separate labeled fields. Given a `biblioentry` structure like this, you could create an algorithm to present and format a bibliography entry almost any way you wanted to. This structure not only partitions the formatting of the bibliography from the presentation of the bibliography, it also partitions the presentation from the underlying bibliographical data. This means you could write an algorithm to extract bibliography information from a document by looking for `biblioentry` structures and extracting the desired information from them. For instance, if you want to build a list of authors cited in the document, you could do so by searching the `biblioentry` records and extracting the name of the authors from the structures that record them in the bibliography structure.

Bibliographies are an interesting piece of structure because while the bibliography is a standard document structure, and therefore part of the document domain, it is also specific to a particular subject: books and periodicals. When we break down the bibliography structure into a set of data fields, we are actually factoring out the rhetoric of a bibliography entry – the decisions about the order and construction of each entry – and capturing subject data that could be used to construct different styles of bibliography. This means that this bibliography entry is also in the subject domain. It is time, therefore, to look at the subject domain in more detail.

CHAPTER 4

Writing in the Subject Domain

If the media domain is concerned with what the content looks like and the document domain is concerned with how the content is organized into a document, the subject domain is concerned with the subject matter of the content. Subject-domain structures tell you what the content is about.

A recipe is a useful example for illustrating the subject domain. Figure 4.1 is a recipe written in reStructuredText, a lightweight general purpose document-domain markup language:

```
Hard-Boiled Eggs
================
A hard-boiled egg is simple and nutritious.
Prep time, 15 minutes. Serves 6.

Ingredients
-----------
======  ========
Item    Quantity
======  ========
eggs    12
water   2qt
======  ========

Preparation
-----------
1. Place eggs in pan and cover with water.
2. Bring water to a boil.
3. Remove from heat and cover for 12 minutes.
4. Place eggs in cold water to stop cooking.
5. Peel and serve.
```
Figure 4.1 – Recipe for hard-boiled eggs marked up in reStructuredText

In reStructuredText, text underlined with equals signs is a major heading and text underlined with dashes is a minor heading. You create a table by using equals signs to mark the beginning and end of the table and the boundary between the table head and the table body. You place each row on a new line and mark columns by putting spaces in the rows of equals signs. You create ordered lists by putting numbers in front of lines of text. The equivalent HTML document to Figure 4.1 would look like Figure 4.2.

```
<html>
    <h1>Hard-Boiled Eggs</h1>

    <p>A hard-boiled egg is simple and nutritious.
    Prep time, 15 minutes. Serves 6.</p>

    <h2>Ingredients</h2>
    <table>
        <thead>
            <tr>
                <th>Item</th>
                <th>Quantity</th>
            </tr>
        </thead>
        <tbody>
            <tr>
                <td>eggs</td>
                <td>12</td>
            </tr>
            <tr>
                <td>water</td>
                <td>2qt</td>
            </tr>
        </tbody>
    </table>
    <h2>Preparation</h2>
    <ol>
        <li>Place eggs in pan and cover with water.</li>
        <li>Bring water to a boil.</li>
        <li>Remove from heat and cover for 12 minutes.</li>
        <li>Place eggs in cold water to stop cooking.</li>
        <li>Peel and serve.</li>
    </ol>
</html>
```

Figure 4.2 – Recipe for hard-boiled eggs marked up in HTML

The document shown in Figure 4.2 follows the normal rhetorical pattern of a recipe. That is, it has all the pieces of information a recipe normally has, in the order they normally occur in a recipe: introduction, list of ingredients, and preparation steps. However, it does not record the fact that it follows this rhetorical pattern. There is nothing in the markup to say that this is not a novel, a car manual, or a knitting pattern. Nor does the markup constrain a writer to follow the normal rhetorical pattern of a recipe.

Moving this document to the subject domain allows you to impose these rhetorical constraints and requires writers to record that they have followed them. Neither reStructuredText nor HTML gives you a way to impose or record rhetorical constraints, so you need a different markup language. Figure 4.3 shows what such a language might look like.

```
recipe: Hard-Boiled Egg
    introduction:
        A hard-boiled egg is simple and nutritious.
        Prep time, 15 minutes. Serves 6.
    ingredients:
        * 12 eggs
        * 2qt water
    preparation:
        1. Place eggs in pan and cover with water.
        2. Bring water to a boil.
        3. Remove from heat and cover for 12 minutes.
        4. Place eggs in cold water to stop cooking.
        5. Peel and serve.
```

Figure 4.3 – Recipe for hard-boiled eggs marked up in a subject-domain language

This structure breaks the document into a collection of named structures – introduction, ingredients, and preparation – which are contained in an overall structure called `recipe`. This is the basic rhetorical structure of a recipe. This markup clearly identifies this as a recipe (not a novel, a car manual, or a knitting pattern), and the writer is explicitly guided to follow this pattern. Also, the writer must present the ingredients as an unordered list and the preparation as a numbered list. (Chapter 29 looks at how to express and enforce such constraints.)

In other words, in this example, someone other than the writer has made certain basic rhetorical decisions about how to write a recipe – let's call that person an information architect, though this is not always the responsibility of people with that title. By making this rhetorical structure explicit, the information architect communicates this requirement to writers, which simplifies their task because they no longer need to make this decision for each recipe they write. It also ensures that all recipes follow this pattern.

The subject domain overlaps with the document domain in that the structure for discussing a particular subject is also a document type. That is, a recipe is a specific type of document. In fact, the word recipe is specifically a type of document. The word for the dish you make by following the recipe is "dish." A generic document-domain language like reStructuredText lets you write a recipe or any other kind of document you like. A specific recipe markup language constrains the writer to follow a particular rhetorical pattern when writing about how to prepare a dish.

Thus the use of the subject domain allows you to place specific rhetorical constraints on specific document types.

(I should note here that there are rhetorical structures that are not based on individual subjects but on other factors, such as methods of exposition that can be used with multiple subjects. I will look at these in Chapter 7.)

This does not mean we are entirely in the subject domain, any more than moving from a vector graphics program to a word processor meant we were entirely in the document domain. This recipe markup language enforces a couple of basic rhetorical decisions about recipes, but leaves plenty of others to the writer. Few markup languages are entirely in one domain. Later, I will move this recipe markup further into the subject domain, capturing more rhetorical decisions in the markup language design.

But the subject domain is not just about enforcing rhetorical decision on writers. It is better to factor out a decision than to enforce it, and even this very simple subject-domain markup factors out a couple of rhetorical decisions. This means that the writer neither makes nor implements those decisions. An information architect makes the decisions and then designs an algorithm to implement them.

Suppose you decide that the ingredients and preparation sections of each recipe should be titled "Ingredients" and "Preparation." If your writers write recipes in the document domain using a language such as Markdown, they have to remember to execute this decision each time. But in the subject-domain markup in Figure 4.3, the titles are factored out. Instead there are markup structures called `ingredients` and `preparation` that don't contain titles. A publishing algorithm adds the titles later.

This partitioning of tasks requires writers to record enough information for the publishing algorithm to do its job. The presence of the `ingredients` and `presentation` sections in the recipe structure provides the information the algorithm needs to insert the appropriate titles.

By factoring out titles, you also factor out the constraint on what those titles must be. You don't need to remember the title text, and a whimsical writer can no longer decide to re-title these sections "Stuff you need," "Stuff you do," or some other variant. And you can change these titles across all recipes, including those that are already written, by changing the algorithm.

This illustrates why it is generally preferable, where possible, to factor out a decision rather than enforce it. An enforced decision may be made correctly and consistently, thanks to the constraint, but it has still be made by the writer and it is recorded in the content as a *fait accomplis*. It can only be changed by changing the content. A decision that has been factored out can be made and changed later without affecting the content, opening up a number of possibilities that we will explore in later chapters.

That said, enforcing constraints is a powerful tool. If you publish a lot of recipes, you probably have many more rhetorical decisions you would like writers to make consistently. For instance, you might want every recipe to state its preparation time and the number of people it serves. With subject-domain markup, you can enforce and record that constraint by moving that information from the introduction section to separate fields (see Figure 4.4).

```
recipe: Hard Boiled Egg
    introduction:
        A hard boiled egg is simple and nutritious.
    ingredients:
        * 12 eggs
        * 2qt water
    preparation:
        1. Place eggs in pan and cover with water.
        2. Bring water to a boil.
        3. Remove from heat and cover for 12 minutes.
        4. Place eggs in cold water to stop cooking.
        5. Peel and serve.
    prep-time: 15 minutes
    serves: 1
```

Figure 4.4 – Recipe with preparation time and servings in separate fields

Now, writers no longer have to remember that this information is required and where it should appear. The information is partitioned off into separate fields. The system prompts the writers for those fields and will raise errors if they forget. And the decision about where to include this information in output formats has been transferred to an algorithm.

This does not mean that the `prep-time` and `serves` information must be displayed as separate fields in the output. You could display this information in separate fields so that readers can find it more easily, but you could also have the publishing algorithm construct sentences such as "Prep time, XX minutes. Serves YY." using the `prep-time` and `serves` field values for XX and YY.

Once again, factoring out a decision allows you to make other rhetorical decisions down the road. But in this case, the possibilities gained are much broader. For instance, you can easily create a cookbook of recipes that take 30 minutes or less, because you can query your recipes and select just those where the `prep-time` field has a value of 30 minutes or less.

Something interesting has happened here. Enforcing a rhetorical constraint – you must specify prep time and number of servings – moves you from markup that specifies presentation to markup that merely records data. In other words, `prep-time` and `serves` are data-oriented subject-domain structures that do not specify presentation at all. By enforcing one rhetorical decision (always include prep time and servings), you factor out another rhetorical decision (how and where to express prep time and servings). Partitioning this required information from the presentation of the recipe shifts responsibility for conformance from the writer to the constraints imposed by the structures, making the design more testable and repeatable.

Factoring out a constraint rather than enforcing it can be a difficult idea to adjust to. Our natural first instinct when trying to achieve a particular presentation is to specify it in detail. But this can be difficult, especially for rhetorical constraints such as requiring a recipe to contain certain pieces of information. Nor is it flexible if you want to vary the presentation. When you look at these kinds of problems, the first question should be, "Is there any way to factor out this constraint?" Impose a constraint only if you cannot factor it out.

When you *enforce* a constraint, the information architect has made the decision, but the writer must execute that decision in the content. When you *factor out* a constraint, the information architect has still made the decision, but an algorithm executes the decision, rather than the writer. Not only does this make the writer's job simpler, but it helps ensure conformance to the constraint and allows you to change the decision without changing the stored content.

However, there is another pattern here that recurs in future chapters: in the act of factoring out one constraint, you often impose another. When we factored out list formatting (Chapter 3), we introduced a constraint on how to create lists in the document domain. When we factored the need to remember to include prep time in the introduction, we created an additional required field in the recipe structure. This makes designing a structured writing system something of an iterative process. So let's perform another iteration of this refactoring into the subject domain.

The reStructuredText version in Figure 4.1 presents ingredients in a table, and the recipe structure in Figure 4.3 uses a simple list. The block that contains the list is labeled "ingredients," but the list is just an ordinary unordered list. The ingredient list in a recipe has rhetorical constraints, but the markup doesn't impose or record those constraints.

The rhetoric of an ingredient listing includes three pieces of information: the name of the ingredient, the quantity, and the unit of measure used to express the quantity. These can be presented as a list or a table. To factor out the presentation choice, you can create an ingredient structure that calls out each piece of information separately (see Figure 4.5).

```
ingredients:
    ingredient:
        name: eggs
        quantity: 12
        unit: each
    ingredient:
        name: water
        quantity: 2
        unit: qt
```

Figure 4.5 – Subject-domain markup for ingredients

Figure 4.6 shows a shortcut that makes this markup less verbose. (This is a markup syntax named SAM that I will talk about later).

```
ingredients:: ingredient, quantity, unit
    eggs, 12, each
    water, 2, qt
```

Figure 4.6 – Subject-domain shortcut for ingredient markup

This markup turns the ingredients into a set of records with named fields: ingredient, quantity, and unit. This ensures that writers capture all three pieces of information, and because the markup is independent of any one form of presentation, you are now free to use an algorithm to present them as a table, as a list, or in some other form.

But in the subject domain you get another set of benefits. The explicit markup of the subject matter lets you access this content as data and run algorithms to produce new and different forms of content (something I will explore in more depth in Chapter 14). For instance, it would allow you to convert measurements from imperial to metric units so you could publish your recipes in countries that use the metric system.

Using subjects to establish context

In Chapter 3, I noted that you can use context to identify the role that structures play in a document, which allows you to get away with fewer structures. For instance, you can use a single `title` tag for all titles because you can differentiate different types of titles from the context in which they occur. The same is true with subject-domain structures. They can provide context that allows you to treat basic text structures differently.

For example, consider the markup for the preparation steps in Figure 4.7.

```
recipe: Hard Boiled Egg
    introduction:
        A hard boiled egg is simple and nutritious.
    ingredients:: ingredient, quantity, unit
        eggs, 12, each
        water, 2, qt
    preparation:
        1. Place eggs in pan and cover with water.
        2. Bring water to a boil.
        3. Remove from heat and cover for 12 minutes.
        4. Place eggs in cold water to stop cooking.
        5. Peel and serve.
    prep-time: 15 minutes
    serves: 6
```

Figure 4.7 – Subject-domain markup for recipes

To factor out the decision about whether the list of ingredients is presented as a list or a table, you need additional structure. In Figure 4.7, the preparation steps are currently marked up as a numbered list. Suppose you want to present the steps as steps, rather than just as a generic numbered list (for instance, by labeling them as **Step 1.**, etc, rather than just **1.**). Do you need to create an additional `step` structure to do this? Not necessarily. In this case, you can distinguish an ordinary ordered list from a set of preparation steps based on context (the list is part of a `preparation` structure). You can then write a rule in the publishing algorithm that creates special formatting for ordered lists that are the children of `preparation` structures that are children of `recipe` structures. This uses the same method as used in Chapter 3 to format a nested list differently from its parent list based on context. This is another example of how structure creates context that you can use to simplify processing.

CHAPTER 5
The Management Domain: an Intrusion

So far I have talked about three domains: the media domain, the document domain, and the subject domain. But there is a fourth domain that intrudes into this picture: the management domain.

Why do I call the management domain an intrusion? Because while the subject, document, and media domains are about recording the content itself, the management domain is not about the content itself; it's about the process of managing content.

Suppose you run a company that publishes magazines, and you want to create a common store of recipes for use in all the magazines. However, different magazines have different requirements. *Wine Weenie* magazine needs to have a wine match with every recipe. *The Teetotaler's Trumpet*, naturally, wants a non-alcoholic suggestion.

Figure 5.1 shows how you could handle this in the document domain:

```
<section publication="Wine Weenie">
    <title>Wine match</title>
    <p>Pinot Noir</p>
</section>
<section publication="The Teetotaler's Trumpet">
    <title>Suggested beverage</title>
    <p>Lemonade</p>
</section>
```

Figure 5.1 – Conditional text markup (XML)

This is an example of conditional text. The `publication` attribute on the `section` element says, display this text only in this publication. Conditional text structures are management metadata, which means they are in the management domain. They do not specify the formatting, organization, or subject matter of the document. In this case, they specify which publication the content should appear in, which is a content management decision.

Content management decisions are always external to the writer's core task. They are not decisions about subject matter or rhetoric. They are decisions imposed on writers from elsewhere in the process. As such, the management domain is an intrusion not only into the content but also into the writer's time and attention. Some management-domain decisions should be made by writers, but putting the burden on writers increases the complexity of the writing process and the knowledge and skills required to write successfully. Wherever you can, look for alternatives to the management domain.

Any decision imposes two burdens: making the decision, which requires knowledge, and executing the decision, which requires skills. A conditional text system removes the execution of the decision (including or excluding the text at build time), but leaves the decision making (where and what to include) with the writer. But it also introduces new knowledge and skill requirements, since the writer has to know how to apply conditions and apply them correctly, which is not always a simple matter.

This is a common pattern with management-domain structures; the knowledge requirements they introduce are often greater then the ones they factor out. When we factored out list formatting in the document domain, it was a simple reduction in requirements. Writers still had to decide whether to create a list and record the fact that they were creating a list, but we removed the requirement to format the list and added nothing new to learn. But in this case, writers must learn something new: conditional markup and its logic and behavior. Let's look at another example that exhibits the same pattern.

Including boilerplate content

Suppose you must include a standard warning statement with any content that describes a dangerous procedure. Structured writing partitions complexity by factoring out invariants, and the invariant here is that this warning statement must appear whenever you describe a dangerous procedure.

Just as you extracted formatting information into a separate file to move content from the media domain to the document domain, you now extract the invariant warning from the document and place it in a separate file. Any place you want this warning to occur, you insert an instruction to include the contents of the file at that location. (see Figure 5.2).

```
procedure: Blow stuff up
    <<<(files/shared/admonitions/danger.sam)
    step: Plant dynamite.
    step: Insert detonator.
    step: Run away.
    step: Press the big red button.
```

Figure 5.2 – Include a warning in a dangerous procedure (SAM markup)

In the SAM markup in Figure 5.2, `<<<` is a command that includes the content of the file located at `files/shared/admonitions/danger.sam` in the source file replacing the `<<<` command.

The equivalent in XML would look something like Figure 5.3.

```
<procedure>
    <title>Blow stuff up</title>
    <xi:include href="files/shared/admonitions/danger.xml"/>
    <step>Plant dynamite.</step>
    <step>Insert detonator.</step>
    <step>Run away.</step>
    <step>Press the big red button.</step>
</procedure>
```

Figure 5.3 – Include a warning in a dangerous procedure (XML markup)

Why is this operation part of the management domain, rather than the document domain? Because it deals with a system operation: locating a file in the system and loading its contents. If you were purely in the document domain, the writer would perform this operation by finding the file with the warning in it, opening it, and copying the contents into the document. The `include` instruction is just that: an instruction. It is not a declaration about the subject matter or structure of a document, such as you would find in subject-domain or document-domain markup. It is an instruction to a machine to perform an operation. The management domain consists of either instructions or the declaration of data required to perform specific management instructions. It requires the writer to understand these instructions and other features of the content management system, such as the location of files.

Different structured writing systems have different instruction sets for handling the situation described above. In DITA, for instance, this use case is handled using something called a `conref` or a `conkeyref`. DocBook uses a generic XML facility called XInclude, which is shown in Figure 5.3. Thus writers need to learn the specific management features of the languages and tools they are using.

One downside of the management-domain include instruction is that writers must have system knowledge to use it correctly. However, the use of conditional expressions or include instructions is not just a problem for writers, it also complicates change management and content management by distributing management-domain structures such as conditions and file paths throughout your content. Management-domain structures encode management decisions in the content itself. Changing those decisions means changing management structures all over the content set.

Is there a way to partition management complexity from authoring and keep management decisions out of the content? In many cases this can be accomplished using the subject domain.

An alternative approach in the subject domain

Figure 5.4 shows how you might approach the two-magazines problem in the subject domain.

```
<wine-match>Pinot Noir</wine-match>
<beverage-match>Lemonade</beverage-match>
```

Figure 5.4 – Subject-domain alternative for conditional content

This markup says nothing about which documents should contain either of these pieces of information. Nor does it contain the subheadings that would introduce either of them in the appropriate publication. All these decisions are now left to algorithms. This allows you to do far more with this content without having to rewrite the source files in any way. Writers don't have to understand how the conditional text systems works. They just record two pieces of information.

I noted in Chapter 4 that the subject domain can factor out rhetorical decisions by simply recording facts about the subject. This is another example of this. By recording two different beverage matches, you can make different rhetorical decisions for two different audiences. The management-domain markup also attempted to capture two different rhetorical alternatives, but rather than factor them out, it embedded both alternatives into a single file. This approach not only makes the writer responsible for making the decision, it also means that changing the decision requires the writer to reopen the file and every file in which a similar decision is made. Thus the subject-domain approach not only makes the writer's life easier, it makes the change management process much simpler. (I will look more at change management in Chapter 40.)

Every time you move a decision out of the content and into an algorithm, you accomplish two things: you make authoring easier, and you preserve the ability to make different decisions without changing content. Factoring out the management domain is a big win on both fronts.

Let's factor out the decision involved in including a warning in dangerous procedures. As described in Chapter 4, factoring out invariant content is a feature of the subject domain. In this case, the invariant content comes from the rule: A dangerous procedure must have a standard warning.

The management-domain approach allows writers to insert a standard warning that is stored in one place.[1] Notice that the management-domain markup does not encapsulate the invariant rhetorical rule that dangerous procedures must have a standard warning. It simply provides a generic mechanism for inserting content from an external file. The writer must still remember and execute the rule about dangerous procedures.

The subject-domain approach, on the other hand, is all about the rhetorical rule itself. Specifically, it expresses the aspect of the subject domain that triggers the rule: whether a procedure is dangerous or not. Figure 5.5 shows how you might mark up this information.

```
procedure: Blow stuff up
    is-it-dangerous: yes
    step: Plant dynamite.
    step: Insert detonator.
    step: Run away.
    step: Press the big red button.
```

Figure 5.5 – Subject-domain markup for a warning notice

This markup identifies the procedure as dangerous, a fact about the subject matter that your rhetorical standards require you to mention. Rather than making the writer remember that a warning is required, locate the warning file, and include a reference to the file, you delegate those decisions to the publication algorithm. The algorithm, not the writer, must remember to include the material in `files/shared/admonitions/danger` when the value of the `is-it-dangerous` field of a `procedure` structure is set to `yes`. Algorithms are much better at this sort of task than humans are.

Of course, writers still must set `is-it-dangerous` to `yes` or `no` when they write a procedure (another example of creating a new constraint when you factor out an old one). But you can make it much easier for them to remember this requirement by making `is-it-dangerous` a mandatory field in the procedure structure and raising an error if the field is not included. This transfers the complexity of remembering the requirement and fulfilling it from the writer to the conformance and publication algorithms.

With this approach, writers cannot complete a procedure without supplying the required information. This removes a decision, simplifying the writer's job. In addition, writers don't need to know how the content management system works, what warning text is required, or where the

---

[1] This approach is a form of content reuse, an algorithm discussed in Chapter 13.

text is located. They simply record a fact about the subject – factoring out the complex rhetoric and all of its permutations.

One the other hand, this approach only factors out the reuse of one particular piece of content – the warning for dangerous procedures. If you have multiple invariant rhetorical rules about different kinds of subject matter, you need a separate subject-domain structure for each of them, whereas a single management-domain include instruction would let the writer handle them all.

On the other other hand, if you have many invariant rules, and you expect writers to remember all of them, you are asking your writers to carry an awful lot of rhetorical constraints in their heads, and you are going to limit your pool of writers to a few highly trained individuals. And those individuals are still likely to miss some instances, leading to omissions – rhetorical failures – that can have dangerous consequences for the reader.

Hybrid approaches

It is not always an either/or decision to use pure management-domain or pure subject-domain approaches. Management-domain structures tend to be used in generic document-domain languages, since such languages are not designed to be specific to any particular subject matter. Nonetheless, such languages often have roots in particular fields and sometimes include subject-domain structures from those fields. Both DocBook and DITA, for instance, originated in the field of software documentation, and both include structures, such as code blocks, that are related to the subject of software.

Some languages mix subject-domain structures into their management structures. One example is the product attribute, which is part of DITA's conditional-text processing system.

DITA lets you add the product attribute to a wide variety of elements. Your build system can then include or exclude elements in a particular output based on the value of that attribute. Figure 5.6 shows an example of the DITA product attribute.

```
<p>
The car seats <ph product="CX-5">5</ph><ph product="CX-9">7</ph>
</p>
```
Figure 5.6 – DITA product attribute example

DITA can afford to use this bit of subject-domain markup because product variations are a common reason for using conditional text processing in technical communication, the area for which DITA was created.[2]

I call this a hybrid approach because the DITA product attribute does not exist merely to declare that a piece of text applies to a particular product. It is a conditional-processing attribute. That is, it is an instruction, even though it is phrased as a subject-domain declaration.

To appreciate the difference, consider another approach to documenting multiple versions of a product. Rather than generating a separate document for each product variant, you could create a single document that covers all product variants and highlights the differences. Pure subject-domain markup would support either approach by simply recording the data for each variant:

```
seats:
    CX-5: 5
    CX-9: 7
```

This information could be presented as data similar to its source format or it could be used to algorithmically construct a sentence like this:

> The CX-5 seats 5 and the CX-9 seats 7.

The product attribute does not provide this flexibility:

```
<p>The car seats
<ph product="CX-5">5</ph><ph product="CX-9">7</ph>
</p>
```

This markup is designed to produce only a CX-5 or a CX-9 specific document. It is not designed to produce a document that covers both cars at once, because it does not specify that the values 5 and 7 are numbers of seats. That information is in the text, but not in a form that a publishing algorithm could reliably locate and act on.

Also, no writer would expect this management-domain markup to be used to create a single document covering both cars. No writer would interpret the markup to mean this. Rather than

---

[2] DITA can add other subject-domain attributes for conditional processing through a process called *specialization*. Chapter 34 has additional information about DITA specialization.

declaring facts about each car, this markup was designed to produce a document about one car or the other, not both. It is conditional-text markup and, therefore, an instruction.

While the introduction of subject-domain names into management-domain structures provides an appropriate bit of semantic sugar for writers, this hybrid approach really remains firmly in the management domain.

The following chapters contain many more examples of the management domain, including cases where there is and is not a subject-domain alternative.

Process, Rhetoric, and Structure

The purpose of the content process is to produce good rhetoric. Part I described how structured writing can enforce, or factor out, certain rhetorical rules in your content, and it hinted at how you can use structured writing to improve process. Now, it is time to take a deeper dive into process and rhetoric and look at how structure supports them.

Rhetoric is not just about how individual documents do their jobs. Documents are elements of a larger information set that serves a multitude of complex information needs. Organizing that larger information set is the domain of information architecture. Information architecture is a part of rhetoric, and the rhetoric of the information architecture has a direct impact on the rhetoric of individual documents. Thus, this part of the book looks at information architecture and the relationship between the rhetoric of documents and the rhetoric of information sets.

Ultimately, rhetoric is the domain of the writer. Rhetoric is what writers do. Structured writing does not usurp the writer's role in rhetoric. Rather, writers use structured writing as a tool to create rhetoric with greater precision and consistency. Thus, this part concludes with a look at how a structured approach to rhetoric fits into the writing process.

CHAPTER 6

Partitioning Complexity

Understanding how to successfully partition the complexity of a content system starts with asking what complexity is. The complexity of a task in the content process (or any other process) can be measured by the number of decisions people have to make and execute. Every decision requires knowledge (to make the correct choice) and skill (to execute that choice). The more decisions a task involves, the more knowledge and skill it requires, and the more knowledge and skill a task requires, the more complex it is. You can reduce the complexity of a task by partitioning it into separate, less complex tasks that each require less knowledge or fewer skills than the original task.

It is helpful to divide the decisions involved in a task into two kinds:

- **Core:** decisions that are fundamental to the task. These decisions are central to the task and cannot be removed by changing your processes.
- **Extended:** decisions that are introduced into the task by your processes or tools. Theoretically, at least, these decisions can be removed from the task by changing your processes or tools.

The more extended decisions you remove from a task, the simpler you make that task. The purpose of partitioning the content system is to remove these decisions from key tasks, making them less complex and, therefore, improving productivity and quality.

Of course, those decisions don't just vanish into thin air. They still must be made and acted on somewhere in the system. Complexity cannot be destroyed, but it can be moved around to ensure that every decision is made by the person or process with the information, skill, and resources to make and execute it best.

Simplifying the tasks of individual people and processes is not the only reason to change where decisions get made. Making decisions earlier or later in the content process can also have profound effects on the efficiency of your process and the kinds of rhetoric it can produce. For instance, partitioning decisions about document formatting away from the writer means that those decisions are not recorded in the content itself. If you later need to change how your content is formatted, you can change the formatting algorithm, without touching the content. The possibilities in this area go well beyond formatting, as we will see in later chapters.

All modern content systems are partitioned, but not all are partitioned in a way that fully handles the complexity of the content development and delivery process. As a result, many organizations produce content of poor or variable quality, and lose money. Why are so many content systems poorly partitioned? Some answers can be found by looking at the evolution of content development.

Desktop publishing

One of the biggest changes in the partitioning of the content system was ushered in by the desktop publishing (DTP) revolution of the 1980s. Prior to the introduction of desktop publishing, creating a formatted printed document generally meant handing off a manuscript from a writer to a typesetter for re-keying, followed by mechanical pasteup by a page-layout artist, preparation of proofs by a printer, correction of proofs, and printing.

This was in many ways a well-partitioned process. Professionals in each trade had a piece of the decision making and the tools to implement those decisions. However, the overhead of passing information from one trade to the next was cumbersome, resulting in a process that was time-consuming and expensive with many potential points of failure.

Desktop publishing eliminated much of that overhead by putting all of these functions – writing, design, layout, and proofing – in the hands of a single operator: the writer. This greatly sped up the publishing process and generally reduced costs by eliminating most of the trades traditionally involved in the process.

However, this radical change in partitioning introduced three problems:

First, it put a huge amount of complexity on the writer. In a desktop publishing environment, writers must make many more decisions, which require knowledge and skills in addition to those required to research and write content. The demand for writers who could handle this complexity changed hiring practices. Desktop publishing skill became a major hiring criteria for writers and remains so today. The focus in hiring moved from writing skills and knowledge of subject matter to the ability to manage the publishing process.

Second, because writers did layout and design while writing, their attention was divided between these tasks. Attention is a finite resource; when writers pay attention to layout and design rather than rhetoric, the quality of the writing suffers. And since the layout and design are only getting partial attention, their quality suffers as well.

Third, although desktop publishing removed the need for a lot of vertical communication between writers, designers, and typesetters, it made no provision for horizontal communication between writers, leaving every writer on an island. Every book was a separate project. The division of the writer's attention and the lack of horizontal coordination left huge amounts of complexity unhandled in large content systems. Duplication, omissions, and inconsistencies became difficult to detect and fix, and providing effective navigation between books was virtually impossible. All of this rhetorical complexity was dumped on the reader in the form of poor quality content. Desktop publishing did not create this problem, but it did nothing to fix it.

Style sheets

One of the most visible early signs of unhandled complexity creating quality problems came in the form of the dreaded "desktop publishing look" – an explosion of poorly designed and poorly laid out documents, characterized by bizarre font combinations, poor use of whitespace, poorly designed and placed graphics, and a seemingly random profusion of lines, shades, colors, and other decorative elements. (Insofar as these design defects hinder the reader's reception of the content, or detract from the image of the organization, they are also rhetorical problems.)

To help contain this, desktop publishing systems introduced style sheets, which partially partition content from formatting. Style sheets allow a writer to format text using predefined, named styles. Each style encapsulates a set of decisions about content formatting that have been made by a style sheet designer. Style sheets allow you to partition decisions about what the text looks like from decisions about what the text says.

Style sheets do not completely partition formatting decisions from writers, however. Writers still must pass information to the partition that makes formatting decisions, which means writers must decide which styles to apply (knowledge) and know how to apply them (skill). Although this is less complex and requires less design skill than formatting by hand, experience shows that many writers do not use styles correctly or even at all.

Although style sheets remove some formatting decisions, the WYSIWYG (What You See is What You Get) display used by desktop publishing systems has another flaw. A key aspect of any interface between partitions is the feedback it gives users to let them know if they have done their work correctly. A WYSIWYG display tells you only one thing: does it look right? It does not tell you whether you have used styles correctly, or at all. And what looks right to you may not look right to the next person.

Another problem is that style sheets partition the design of individual elements of a document, such as a title or a list, but not the overall design of the document. Document design remains on the writer's side of the partition. The style sheet restricts the formatting pallet, but it does not tell writers how to use that pallet to achieve an effective overall document design. Since writers are not document designers by trade, they may not always do this well, and multiple writers working on a common project will almost certainly do it inconsistently.

Many organizations use style guides to tell writers how to use the style pallet to design documents. (For instance, which styles to use when creating a bibliography.) But the style guide only provides instructions, it does not partition the task away from the writer. The style guide merely adds another knowledge requirement, adding complexity.

Styles, therefore, provide only limited relief from the complexity dumped on the writer, and they are only moderately successful in promoting consistent document appearance.

Content management systems

Meanwhile, the web brings a new set of challenges. A modern website is not a library of independent volumes; it's a complex hypertext consisting of many smaller pieces of content related in much more complex ways than paper documents ever were. Search engines and social networks have profoundly changed how readers seek and use content. Meeting those needs requires a level of coordination between writers – and between the pieces of content they create – that desktop publishing cannot provide.

To help manage these challenges, most organizations have adopted content management systems (CMS). This has led to new forms of partitioning and new roles, such as webmaster, information architect, and, more recently, content strategist.

Some of this new partitioning is useful, but sometimes content management systems can make things worse. For instance, some content management systems give control of the page header and sidebar to an information architect but let writers design the document part of the page pretty much as they did with a desktop publishing system. Unfortunately, this approach just wraps a frame around all of the shortcomings of desktop publishing, highlighting those shortcomings by closely juxtaposing the work of different writers.

Increasingly, therefore, content management systems have begun to incorporate more structured techniques and rely on a technology that actually predates desktop publishing: markup languages.

Markup languages

A markup language is a system for indicating the structure of a text via marks in the text itself. The structured writing examples we have looked at so far have all been written in markup languages. Markup languages were used for publishing on mainframe computers long before desktop computers had the processing power required for desktop publishing. A markup language – HTML – gave birth to the web by enabling writers to include in a document the information needed to format and display that document in a web browser.

The HTML document includes both information to be displayed to the reader and information to be used by the browser to display the document – information that is not shown to the reader. This combination of two different streams of information in a single stream of characters is at the heart of how a markup language works and the role it plays in a content process. It allows you to partition tasks within the content process by passing the information those tasks need along with the content intended for the reader.

HTML is probably the single most widely used markup language in the world today, but there are thousands of others, and people create new markup languages all the time. Each markup language represents a specific interface between partitions in a content system.

Markup languages are by far the most general type of content interface, and they can be used to implement nearly any kind of partitioning you might need. In fact, markup languages are usually found behind the scenes in WYSIWYG and forms-based interfaces. However, markup languages can also stand alone as interfaces in their own right. Markdown, for instance, has become a popular markup language for writing simple web content.

There can be (and usually are), multiple partitions in a content system. The interfaces between those partitions are frequently markup languages, including interfaces that writers never use themselves. For instance, an organization may use Markdown to author some of their content but then convert Markdown to HTML for publishing and use a CSS style sheet to specify the formatting for that content.

Today the term "structured writing" is often taken as a synonym for the use of markup languages. This is not really fair, since any use of computers for writing and publishing necessarily involves applying structure to text. The reason we associate markup languages so strongly with structured writing, however, is that markup languages allow organizations to create their own structures and, thus, change the way their content systems are partitioned. Any serious attempt to better manage the complexity of the content system by changing how the system is partitioned is likely to involve the use of markup languages, usually several markup languages in different roles.

Most of the discussion in this book focuses on the use of markup languages, because markup languages provide the greatest range of possibilities for effectively partitioning the content system. You may be able to hide some of the markup techniques discussed here behind WYSIWYG or forms-based interfaces. However, in this book I use markup to illustrate them all, because markup provides the clearest view of the structure and interfaces that define partitions.

Algorithms

To take content created using a markup language and turn it into a formatted document requires an algorithm. Algorithms are also used to manage the content process. An algorithm is simply a regular, codified, and repeatable method for doing a task.

Most complex tasks have repeatable elements in them. A design question that has been settled once can be implemented over and over again without having to redo the design work. If one piece of content is formatted a certain way, chances are many similar pieces can be formatted the same way. Algorithms are great at doing the same task the same way over and over. Therefore, partitioning such tasks away from writers and directing them to algorithms is a great way to reduce the complexity of the writer's task.

And while algorithms are repetitive by nature, they can also incorporate considerable knowledge, make sophisticated decisions, and execute complicated processes, which makes them capable of meeting both the knowledge and skill requirements to make useful decisions. Thus, you can use algorithms to execute many of the extended decisions in a content process, leaving writers and designers free to focus on the core decisions.

A computer program is an encoding of an algorithm that a computer can execute. A program describes an algorithm to a computer, but you do not need to be a programmer to design an algorithm. The real trick is to discern which parts of a process can be defined as algorithms. That is why an understanding of the main structured writing algorithms is vital to effectively partitioning your content system.

Algorithms are fundamental to structured writing. Algorithms and structures work together to support rhetoric and process, and you can't design one without the other. The heart of this book describes the principal structured writing algorithms and the structures that support them.

Partitioning and roles

Partitioning a content system often means changing the roles that people play in the organization. The introduction of desktop publishing, for instance, merged the role of writer and typesetter and, to a certain extent, the roles of document designer and prepress operator. Introducing more sophisticated (or simply more appropriate) structured writing techniques will likely mean introducing new roles or modifying existing ones.

Writing an algorithm redistributes complexity away from the person who used to do the task, but it directs that complexity to the person who writes and maintains the algorithm. For example, using a CSS style sheet to format lists redistributes the task of designing list formatting from writers, whose job is to know what they are talking about and how to say it, to a publication designer, whose job it is to know how to attractively format lists and how to code good CSS.

Using CSS partitions the complexity of formatting a page, distributing that complexity away from the writer. Now, the writer does not have to know about formatting or design, and the designer does not have to know about writing or the subject matter. This allows writers to focus on writing and designers to focus on design. The result is better writing and better, more consistent, design supported by two distinct roles with different skill sets.

As I discuss the major structured writing algorithms in Part III, you will see how these algorithms partition tasks between writers and professionals such as information architects, content engineers, and content strategists.

Quality and rhetoric

Many structured writing systems focus only on the separation of content from formatting (Chapter 10) and on content management functions (Chapter 36). Separating content from formatting can help improve content quality by freeing more of the writer's time and attention to focus on writing. However, you can do much more to improve the quality of content by placing explicit constraints on the rhetorical structures that writers create.

Organizations are becoming more aware of the impact that rhetorical quality has on their business. The web has made it easier for users to access your information, but it has also made it easier for them to detect content that is inconsistent in tone and style, outdated, redundant, incomplete, or just plain useless. The web did not create these flaws, but it made them painfully apparent. It has also given organizations a much more direct way to see and to measure the impact of their content and of its flaws. Organizations can no longer ignore major rhetorical faults. This has led to the emergence of the discipline of content strategy as organizations feel the impact of poor content quality and see content quality as a strategic value.

In the age of physical typesetting, there were no automated tools to help with the rhetorical quality of content, and the desktop publishing revolution was mostly concerned with digitizing the previously mechanical design and layout process. The limitations of the available tools forced organizations to partition their processes in ways that did not always support, encourage, or even permit, the highest quality rhetoric. The result was poor-quality content.

Modern structured writing tools allow you to partition the content system in ways that significantly enhance the rhetorical quality of content. This includes methods that free writers to focus on quality, to create consistent content, to validate and audit content quality, and to create content structures and products that would not be possible with other methods.

Build quality into the process

The idea that you can build quality into a process by managing the complexity of the design and production process, though accepted in other fields, has not been widely adopted in the content field. To a large extent, process improvements have been in publishing and content management, not rhetoric and content quality. Where structured writing tools have replaced desktop publishing tools, it has largely been in pursuit of process goals. But process and rhetoric are intimately related and treating them holistically can significantly improve both.

Unfortunately, many current content systems not only fail to provide active rhetorical support to writers, they also fail to remove distractions and complexity from the writing environment. This ends up making the writer's life more complex rather than less. If writers end up with more complexity than they can comfortable handle, the result is reduced attention to rhetoric, which means diminished content quality. Sometimes that unhandled complexity also derails attempts to make the process more efficient because the information coming through the complex interface is not reliable enough for the new process to work correctly.

Sadly, the developers of many content systems have never seriously attempted to minimize the amount of complexity they dump on the writer. They treat writing and publishing as separate concerns, as if the publishing process and its demands had no influence on the quality of content that the system produced. The result has been widespread dissatisfaction with both content systems and the resulting output. Lack of attention to where complexity falls in a system, and to the intimate relationship between process and rhetoric, typically results in breakdown and failure of the system. And it gets worse as features are added without adequate thought being given to the impact of the complexity they introduce, resulting in what content strategist Joe Gollner calls the "barnacalization of systems."

Towards a more effective partitioning of the content system

The impact of poor quality content goes straight to the bottom line. To improve content quality on a consistent and economical basis, organizations need an approach to structured writing that recognizes the intimate relationship between process and rhetoric and that partitions the complexity of the content process to the people or processes with the time, skills, and resources to handle it effectively.

Every organization has different process challenges and rhetorical goals. Thus, there is no one structured writing system that is a perfect fit for every organization. The attempt to create a single system to meet all needs has been a large factor in the poor partitioning of many content systems. Rather than prescribe any one partitioning, this book explores structured writing strategies you can use to partition your content system and suggests how you can determine which approach will work best for your organization.

CHAPTER 7
Rhetorical Structure

As described in Chapter 6, correctly partitioning tasks is the key to an efficient process. But process and rhetoric are intimately related, so structures that support the partitioning of process must support (and preferably enhance) rhetoric as well. Fortunately, these aims are much more compatible than it might seem at first. An attention to rhetoric can actually serve process well, since sound rhetoric is the goal of process.

Rhetoric is the way that a piece of content makes its argument or assembles and organizes information to inform, persuade, entertain, or enable the reader to act. The work of the writer is rhetoric and the point of partitioning the content system is, in large part, to allow the writer to focus on rhetoric while still providing the information needed for the other partitions in the content system to operate without dropping complexity.

Achieving an effective rhetoric in the content presented to readers is the ultimate aim of any content system. Any failure to handle complexity is felt largely in the form of impaired rhetoric.[1] When you factor out formatting in the transition to the document domain, you allow a greater focus on rhetoric. When you factor out or enforce rhetorical constraints in the transition to the subject domain, you achieve a higher level of conformance, validity, and repeatability in rhetoric.

The aim of the content system is always to produce better rhetoric, and therefore, structured writing should improve rhetoric, either by removing the complexities that keep writers from focusing on rhetoric or by explicitly supporting the consistent and repeatable development of superior rhetoric.

The rhetorical structure of a piece of content is how it tells its story. For many types of stories, the optimal rhetorical structure is consistent and well known. In other cases, you can determine the best rhetorical structure by carefully considering what needs to be said and by testing with readers. In other words, there is a right way, a best way, to tell a wide range of stories. This best is not necessarily universal. Best for your subject matter and audience may be different from best for mine, just as the best recipe presentation for *Wine Weenie* is different from the best presentation for *The Teetotaler's Trumpet*. However, for you and your readers there is a definable, testable, repeatable best.

---

[1] Impaired ergonomics – for instance, formatting that is hard to read – is a secondary source of difficulty for readers.

Maintaining a consistent, high level of conformance to the optimal rhetorical structure for a particular subject can be complex, especially if you have a variety of contributors. Content quality is greatly enhanced when the rhetorical structure is well defined and followed consistently. Quality suffers if you don't properly manage the complexity of maintaining the optimal rhetorical structure for content. Also, a well-defined rhetorical structure provides an effective baseline against which to compare and measure proposed improvements. Using an explicit, predefined rhetorical structure helps enhance and maintain content quality.

Rhetorical structure consists of information requirements and presentation requirements – what needs to be said and how best to say it. Sometimes information requirements dominate the structure and sometimes presentation requirements dominate. Sometimes there is no regular set of facts to relate across all instances, but there is an approach to presentation that is known to work particularly well.

A recipe contains both information requirements and presentation requirements:

- The information requirements include prep time, servings, number of calories, and publication-specific information, such as wine matches.
- The presentation requirements are the ingredients list, including the precise way that measurements are presented, and the presentation steps.

One of the key presentation requirements for recipes is to present the list of ingredients as a separate section with precise measurements. Yes, this is also an information requirement. But notice that all recipes (or almost all) break out this list and separate it from the preparation steps. The ingredients are mentioned again in the preparation steps, so why not just put the measurements in the steps and omit the ingredients list? This is an important rhetorical decision. It is not like deciding whether to pull out the prep time and number of servings into separate fields or leave them in the introduction. It has a much more important rhetorical purpose.

You separate out the list of ingredients for a recipe because the first step in making a dish is not the first item in the list of preparation steps. The first step is to make sure that you have all the ingredients. One approach to creating the rhetorical structure of a recipe could be to make this an explicit first step, as shown in Figure 7.1.

```
recipe: Hard Boiled Egg
    introduction:
        A hard boiled egg is simple and nutritious.
    preparation:
        1. Make sure you have the following ingredients on hand:
            * 12 eggs
            * 2qt water
        2. Place eggs in pan and cover with water.
        3. Bring water to a boil.
        4. Remove from heat and cover for 12 minutes.
        5. Place eggs in cold water to stop cooking.
        6. Peel and serve.
```

Figure 7.1 – Placing the list of ingredients inside a preparation step.

However, since this is a universal first step for every recipe, it has been implicitly factored out in the typical rhetorical pattern of a recipe and is supported by a separate list.

There are other implicit steps as well. One is to collect and measure your ingredients before you start cooking. Some cooks measure and lay out all their ingredients before they start. Others measure and add as they go. By simply providing a list of ingredients rather than making this an explicit step, a recipe accommodates both behaviors.

In fact, there is a step that is prior to all of these: the step of deciding whether to make the dish or not. This is why most recipes have some form of introduction to the dish and a picture of what the finished dish or a typical serving will look like. Obviously no one is ever going to start writing a recipe with "Step 1. Decide if you want to make this dish." But that decision is actually the first step a cook makes, and the rhetorical structure of the recipe supports that step.

There is an important generality here about what supporting a task means. Tasks are things people do. Procedures are text structures. Supporting a task is a far broader rhetorical problem than writing a procedure. Supporting a task is about the whole experience for users. It is about them deciding to do the task or not. It is about them deciding if doing the task will support their goals or not. It is about their confidence in tackling the task as opposed to hiring someone else to do it for them.

In the recipe case, the reader might be a young man who has asked a young woman to his house for dinner for the first time. Is this the right dish for such an occasion? Will cooking from scratch impress her or lead to a disaster? Does he have the confidence to attempt this recipe with so much at stake? And if he decides to cook the dish, does the recipe talk to him in terms he can understand so that he can successfully prepare the dish and, ultimately, impress the young woman? All these things are at stake in the rhetoric of a recipe. The factual correctness of the steps, while essential, is not nearly enough to fully support the task.

While this is a lot of responsibility for the rhetoric of a recipe to bear, there is also a rhetorical requirement to keep it brief, since the young man is probably going to order out if the recipe you give him is six pages long. So it is also important to look at what a recipe does not do. A typical recipe does not call out a list of the pots and pans and other utensils. Sometimes the instructions will say what kind of pot or spoon to use for a step, but most recipes do not list tools the way they list ingredients, perhaps because they assume all kitchens are similarly equipped. Knitting patterns, which have a similar rhetorical structure to recipes and are frequently used by the same people, tend to identify the specific tools to use. Whatever the reason, not listing tools is as much an established part of the rhetorical structure of recipes as listing ingredients. It is the established way in which recipes present preparation information, and it has been blessed and approved by generations of cooks and young lovers.

But while there is a rhetorical consensus about the general pattern of recipes, individual organizations may require a more specific rhetorical structure. As we saw in Chapter 4, a wine magazine may require every recipe to have a wine match. A health-oriented magazine may require every recipe to contain nutritional information. Other organizations may have specific requirements about how to present recipes, such as requiring ingredients to be presented in a table rather than a list. An organization's specific requirements – its unique constraints – constitute a formal rhetorical structure, a structure you can implement and enforce using structured writing techniques to ensure that your content meets your readers' needs and your organization's goals.

In some cases, rhetorical structures are immediately obvious because they have a visual shape. The components of a recipe look physically different on a page (which is why recipes are the most popular structured writing example). There is the picture; the introduction, which is a block of text; the ingredients, which are in a list; and the steps, which are in a numbered list. The recipe pattern is visually distinctive even without looking at a word of the text.

However, rhetorical structures are not about elements that are visually distinct. They are about the different types of information that are required, the way they are expressed, and the order they are presented in. There may be considerable variation in the second two properties. Whether you would count these variations as options within one rhetorical pattern or as defining different rhetorical structures should probably depend on their effect. If two or more particular organizations and means of expression have the same rhetorical effect, they can reasonably be considered as variations on a single rhetorical structure.

When you look at a page that appears to be just a sequence of paragraphs with perhaps some subheadings thrown it, it is easy to assume that it has no particular rhetorical structure. But this is not necessarily true. If that page presents a consistent set of information for a particular purpose, and you can find (or reasonably imagine) that same set of information being assembled for the same purpose to describe another object of the same type, then you have a repeatable rhetorical structure. Similarly, where there is a deliberate strategy for laying out an argument or demonstrating or supporting a process, you have rhetorical structure. And where you have a repeatable rhetorical structure, you can define a formal rhetorical structure for a specific business purpose.

By formal rhetorical structure, I mean a set of computable structures into which text is inserted and by which text – its creation and interpretation – is constrained. This is not quite the same thing as the mechanical structure of the content, which I look at in Chapter 21. The mechanical structure implements the formal rhetorical structure, but the formal rhetorical structure can be described independent of how it is implemented, and indeed, it can be implemented in more than one way. You can even define a formal rhetorical structure without creating or using a mechanical structure to implement it, though obviously without the mechanical structure, you cannot hand over any part of its validation or processing to algorithms.

Conforming existing content to a type

You should not expect that all your existing content is going to fit into the structures of your formal rhetorical structure without changing a word. The point of structured writing is to improve content, not to faithfully represent its current state. Content that has been written in an unstructured format – even if it obeys a general rhetorical pattern such as a common recipe pattern – may not fit the precise structure you have defined, particularly if you have factored out parts of the presentation when structuring your content in the subject domain.

At this stage of a structured content project, you will generally have turned your formal rhetorical structure into a mechanical structure defined in a markup language. You then start to move existing content into the new structure, either by wrapping the markup language tags around existing content or by cutting and pasting chunks of text into a blank markup document.

When you start this process, you will find that a lot of that content does not fit the formal rhetorical structure particularly well. You will find some instances that omit information that is required in the formal structure, you will find some instances that contain additional information that is not supported by the formal structure, and you will find instances that express information differently from what the formal structure expects.

These discoveries mean one of six things:

- The content you are working with does not have the same rhetoric as your formal rhetorical type, and, therefore, it won't fit into a structure designed for that type. In this case, you have discovered a different type of content, and you need to define a different formal rhetorical model for it.

- The content you are working with has the same rhetoric as your formal rhetorical type but it does not fit with the markup you created to express that model. In this case, the definition of the markup is incorrect. You need to change the markup definition to accommodate all of the examples of the rhetorical model. (And remember when you do this to go back and change the ones you have already marked up. This will help ensure your new markup design really does fit all the relevant examples.)

- The content you are working with has the same rhetorical intent (it is trying to achieve the same end) as your formal rhetorical structure, but it takes a different rhetorical approach (it goes about it a different way). There is often more than one rhetorical model that will achieve the same goals. But in a structured writing environment, such variations get in the way of the validation and processing goals of the system. To keep everything consistent and repeatable, you need to rewrite the content to use your chosen rhetorical approach, at which point it will fit the formal rhetorical model and the markup that expresses it.

- The content you are working with is deficient. It does not accomplish all of the rhetorical goals that you have set for content of this type, and therefore, it does not meet all the requirements of the formal rhetorical structure. This usually means that you need to do new research and new writing, not merely reorder the current content.

- The content you are working with contains extraneous material that does not fit the formal rhetorical model. A formal rhetorical model defines the information and presentation required to meet a specific need for a specific type of user. Content not written to a formal rhetorical model often includes additional material. This can happen because the writer had a different idea of who would read the content or just thought the information should be written down somewhere. In this case, you need to remove the extraneous material, but don't throw it away. That information may be needed somewhere else and may fit well into another rhetorical model.

■ The content you are working with contains valuable information not found elsewhere. This looks a lot like the extraneous content case, except that the extraneous material is actually valuable and should be part of the formal rhetorical model. In this case, you need to update the formal rhetorical model, and the markup that captures it, to include this new information. This also means that all the examples you have already processed are deficient, and you need to go back and add the required information to them.

Interpreting the mismatch between existing content and the formal rhetorical types you have defined can make or break your entire structured writing project. You may be tempted to treat existing text as canonical and to try to shape the formal rhetorical type to fit. However, the purpose of structured writing is not to represent existing texts but to partition and redistribute the complexity of the content system, ultimately resulting in better content for the reader. If your current content processes are so good that all your existing content fits your new structures perfectly, then you are not realizing any gain in content quality and you are wasting your time by adding additional mechanical structure. Finding content that does not fit the model is not a sign that the model is broken; rather, it's a sign that the process is working.

This does not mean that models never need to be changed. However, you should change your models to match the best rhetorical structure for your content to achieve your business goals, not to match your existing content.

This means that applying structure to your existing content is not a trivial or mechanical task. The purpose, after all, is to improve the quality of the existing content, and that is going to mean additional research and writing work to bring the content up to standard.

It is important to note here that not every markup language implements a formal rhetorical structure. DocBook does not implement one at all, nor does Markdown or HTML. Out of the box, DITA implements only the most general rhetorical structures (though it gives you the capacity to define more precise ones through specialization).

People often convert content from one file format to another, including from binary formats to markup formats. This is a mechanical process, though one that may require some cleanup. It does not, in itself, impose any additional constraints on the content. It merely changes the syntax that expresses existing structures. The rhetorical and process improvements that are possible with structured writing do not come about simply by switching the file format of your content from binary to markup. They come through the ability to impose or factor out constraints on content to improve quality and allow algorithms to reliably process the content. These benefits come through the specific structures you use to shape your content and the consistency with which you use those structures. File format conversion does not change the shape or consistency of your content in any way (or, at least, it does not improve it in any way; it may occasionally lose structure that was in the source but which was not recognized or could not be converted by the conversion program).

You can often mechanically convert existing content to general document-domain formats such as DITA and DocBook, even for content that was in the media domain or in very loose document-/media-domain hybrids. This does not mean that the resulting DocBook or DITA output will correctly express the full range of constraints or structures that these formats are capable of.

Your motive for moving to structured writing should be to improve your content process, not to replicate it with a new set of tools just because those tools implement a fashionable set of acronyms like XML or DITA. It is your structures, your conformance to those structures, and the algorithms you implement to process those structures that will improve your rhetoric and process, not the number of acronyms your tools boast. Adding structure to your content is a writing task, not something than can be done mechanically.

Presentation-oriented rhetorical structures

The complete rhetoric of a subject includes both what is said and how it is presented. For instance, in a recipe, the decision to present the ingredients as a separate list is a presentation element that is separate from the purely informational requirement that the ingredients must all be mentioned somewhere in the recipe. But these kinds of presentation decisions – ones that isolate particular chunks of information – can be factored into the subject domain, by making the list of ingredients a set of records that can be presented in many different ways.

This will often be the case. The presentational aspect of a rhetorical structure is about isolating a particular set of facts and organizing them in a particular way. If you can isolate those facts in subject-domain markup, the presentation algorithm can address the presentational aspects of the rhetorical model. In fact, when the rhetorical design consists of a particular arrangement of facts, the presentation algorithm can create any reasonable rhetorical design from subject-domain content. And isolating facts in subject-domain markup makes the data available for other uses.

But not all rhetorical models are reducible to an arrangement of facts. This is clearly true of philosophical essays and even of books like this one. In works of this sort, the rhetorical structure – the course of the argument – cannot easily be reduced to a repeatable structure. But there are also cases where rhetorical structures can be highly repeatable, and yet they do not consist merely of an arrangement of facts. In other words, there are rhetorical models that focus on the optimal presentation of information in certain circumstance, independent of the specifics of the subject matter. A well-known example is the pyramid structure used in newspapers, which clusters the key points of a story at the top.

There is a presentation-oriented rhetorical pattern that is useful in technical communication (and perhaps in other fields) that I call the think-plan-do pattern. Many technical communication tasks simply involve telling a user how to perform specific functions on specific pieces of machinery. But there are cases in which the user's task has complex input conditions and potentially far-reaching consequences. In such cases, the technical communication task goes well beyond telling users how to operate the machine. It is about helping them to correctly plan their actions to achieve the desired business outcome.

You can approach this problem by simply collecting all the relevant facts needed to make a correct decision. But merely listing relevant facts does not help a user who does not fully understand the complexity of the task or the seriousness of its potential consequences. For example, a user may not understand the security implication of a particular configuration option of a computer system. The safety of that option may depend on a variety of factors, such as who has access to the system, what software is running, what data it contains, and how other settings are configured.

A user who does not understand all of this information might either be intimidated by the list of facts and do nothing or simply skip them and go straight to the beginning of the procedure. Either course of action could have negative consequences.

The think-plan-do rhetorical structure addresses this problem by walking users through each decision so they can plan their actions correctly. This can consist of a number of carefully designed discrete questions designed to help users figure out which issues apply to their situation and, if they do apply, how to deal with them.

In other words, the think-plan-do model presents a formal planning methodology in the form of a set of questions which break the planning process into manageable pieces that users can successfully comprehend and act on. It supports cognition by breaking a complex subject down into manageable pieces.

Depending on the material, you may be able to find a common pattern in the subject matter of these questions. For example, you may find that the same questions need to be considered for each configuration setting. However, in many cases, the questions that need to be asked are particular to the individual case. Whether you can find patterns or not, the rhetorical device of breaking the planning process into a set of discrete questions is most important to improving the quality of the content and ensuring that the reader is successful.

So far in this book I have mostly presented document-domain models as rather loose collections of generic document structures mainly used to separate content from formatting or to facilitate content reuse. But this example shows that the document domain can also be used to model a specific rhetorical strategy.

In some cases, you may find that specific subjects require a specific rhetorical strategy but do not lend themselves to the subject-domain approach of breaking out a consistent set of information for each instance of a subject. In these cases, creating a document-domain model that enforces the appropriate rhetorical strategy may be the best approach.

Rhetorical meta-models

There are different ways of thinking about the rhetorical structure of content. Above, I describe the topic pattern of a recipe as consisting of a picture, an introduction, ingredients, and a list of preparation steps.

However, there are a great many other type of information with a similar pattern. For instance, a knitting pattern usually has a picture of the garment, an introduction describing the project, a list of the yarns and needles required, and a list of steps for knitting and assembling the pieces. Lots of other things look similar. Instructions for assembling flat-pack furniture, for example, or planting flowers in your garden.

These are not the same rhetorical structure. You would not confuse a recipe with a knitting pattern. And each of them can have specific information fields that would make no sense for the others. A pot roast will never have washing instructions. A flat-pack bookcase will never have a wine match. Nonetheless, they all have the basic pattern of picture, description, list of stuff you need, steps to complete. I call this the make-thing-out-of-stuff-with-tools pattern.

The make-thing-out-of-stuff-with-tools pattern is a meta-model. It is not based on seeing similarities between texts, but on seeing similarities in the rhetorical models of texts. A meta-model is not intended for creating content directly, but it can provide hints that help you develop individual rhetorical models. That is, the make-thing-out-of-stuff-with-tools meta-model might suggest how to construct the rhetorical model for a fly-tying guide.

Not only are there meta-models for topics, like the make-thing-out-of-stuff-with-tools pattern, there are also meta-models for the different types of information that go into a meta-model, such as the picture, description, list of stuff you need, and steps to complete. These are sometimes called "information types" (a confusing term, since text at any scale expresses information, and therefore the structure of information at any scale is an information type).

Two notable examples of these information type meta-models are found in Information Mapping and DITA. Information Mapping proposes that documents are composed of six information types: procedure, process, principle, concept, structure, and fact.[2] In Information Mapping, every document is composed of some combination of these six information types, where the arrangement is specified by a map.

DITA proposes something similar, but it has just three types: concept, task, and reference,[3] which, confusingly, it calls topic types. Like information mapping, DITA assembles documents out of these topic (information) types using a map.

In the concept/task/reference meta-model, our recipe topic pattern would consist of one concept topic (the introduction), one reference topic (the list of ingredients), and one task topic (the preparation steps). And our make-thing-out-of-stuff-with-tools meta-model would similarly consist of one concept topic (description), one reference topic (list of stuff you need), and one task topic (steps to complete). (DITA's information model does not include pictures. It just provides a mechanism for including them in textual topic types.)

Meta-models vs generic models

Ideally, a meta-model should just be a model of models. You should not be able to use it for anything other than to create concrete models. It should not only suggest those things that each specific model should have in common but also those things that are specific to particular instances of the pattern. For example, the meta-model should in some way suggest that a recipe instance of the make-thing-out-of-stuff-with-tools meta-model might include a wine match. (A thing-goes-with-other-thing relationship, perhaps.)

In practice, though, meta-models often turn out to be simply a list of those things that all instances of the meta-model have in common. In many cases, instead of inventing an entirely new notation

---

[2] http://www.informationmapping.com/fspro2013-tutorial/infotypes/infotype1.html

[3] Or, at least, it originally had these three types. The DITA specification now includes other topic types, some of which are much more concrete than these original three.

for describing meta-models, people just create a model with only the common properties. Thus the expression of the meta-model takes the form of a generic model, which means that it is perfectly possible to write content using that generic model. Thus while DITA's concept, task, and reference topic types are intended as meta-models to be specialized into concrete models, they are implemented as generic models which can be used directly.

A great many DITA users don't specialize at all. They write all of their content using the base task, reference, and concept topic types (or the even more basic "topic" topic type, of which task, reference, and concept are actually specializations). This means that the topic type imposes no specific rhetorical pattern. But at the same time, the generic pattern can be confining. For instance, DITA's default topic model does not allow you to have two procedures in a single task topic.

Are meta-models useful for defining topic patterns? If a concrete topic pattern describes the kinds of information that are needed to help a particular audience perform a particular task, do you arrive at that pattern more easily by derivation from a meta-model or from observation of multiple concrete examples of actual topics?

The obvious problem with the current generation of content meta-models is that none of them alert you that a recipe might need a field for a wine match. It is not impossible to imagine that a meta-model could do this. A meta-model could observe that objects are commonly used with other objects and lead you to ask what other objects is a steak dinner used with. There are obviously multiple aspects of this question. A steak dinner is used with a knife and fork. A steak dinner is used with a table and chair. A steak dinner is used with family and friends. A steak dinner is used with a glass of wine. How do you characterize each of these thing-used-with-thing relationships in a meta-model, and how do you decide which of these types of thing-used-with-thing types is relevant to a recipe?

Perhaps, for instance, you might decide that because a recipe describes a foodstuff, thing-used-with-thing relationships are relevant when the other thing is also a foodstuff. In other words, you might decide that a thing-used-with-like-thing relationship is part of the meta-model.[4] (I am not, by the way, suggesting that this is a useful part of a meta-model, I merely wanted to illustrate the problem of defining a meta-model that would comprehend all the specific models you might care about in the real world.)

---

[4] Rob Hanna's Enterprise Content Metamodel[https://www.oasis-open.org/committees/download.php/41040/Enterprise%20Content%20Metamodel.pptx] does attempt to do something like this for business information, attempting to describe the relationships between pieces of business content based on the business functions they serve as a basis for deriving specific information types.

This is getting complicated enough for me to conclude that, while the ontologists may one day come up with a such a model and a reliable way to derive concrete content models from it, for most writers, information architects, and content strategist, building a concrete topic model from the observation of instances is probably the preferable method.

Then again, it is often more a matter of how the information architect's brain works. A top-down thinker will prefer a top-down approach and a bottom-up thinker will prefer a bottom-up approach. As long as they arrive at the same place, and as long as the meta-model that the top-down thinker uses does not impose unnecessary complexity or abstraction on the final model, it is less important how they arrive at it.

Creating a good structured rhetorical model for content is actually pretty simple when you get down to it. You ask yourself five questions:

1. What elements of information does the reader need to fulfill whatever purpose this unit of content is meant to serve?
2. What is the best way to express each of these elements?
3. What is the best way to organize these elements so that readers can quickly identify the right content and find the information they need?
4. What constraints do I need to apply to ensure that writers create content that satisfies the first three questions, and how will I test conformance to those constraints?
5. What level of detail and precision do I need in the content structures to make sure that the structured writing algorithms run reliably.

You can approach these questions with a meta-model, concrete examples, or a combination of both. Of course, answering these questions is not necessarily easy, but the process itself is simple, and you should not make it any more complex than it needs to be.

Making the rhetorical structure explicit

I noted above that there can be a rhetorical structure in a piece of text that is just a sequence of paragraphs. You can discern the topic pattern in those paragraph, model that pattern in a content type, and still present the output as a sequence of paragraphs. Presumably, in each instance of the topic type, those paragraphs would now be more consistently expressed, with fewer errors and omissions than before, but the presentation itself would be the same.

Alternatively, you may choose to make the rhetorical structure more explicit to the reader as well as to the writer. In this case, the sequence of paragraphs might be replaced with a distinct combination of headings, graphics, tables, lists, pictures, and text sections that would repeat in every topic of that type.

The question, of course, is whether making the rhetorical structure explicit in this way improves the content. In its favor, a more explicit rhetorical structure makes it easier for readers to recognize the type. (As I noted above, you can sometimes recognize a recipe by its shape, without reading a word.) This makes it easier to identify relevant content, which is particularly important on the web. It can also make it easier to scan the content to pick out the parts you need. The argument against this treatment is that it can lead to a noisier page that is harder to read straight through.

Whether you want to make the rhetorical structure of your pages explicit in these ways needs to be decided on a case-by-case basis. But don't fall into the trap of supposing that because you have chosen a plain presentation, that means there is no rhetorical structure and, therefore, no structured type. The rhetorical structure of the content is separate from the presentation of the content, and the aim of structured writing is to improve the rhetorical structure, not just to make the presentation more uniform.

Structure and repeatability

However much success you may have in defining common rhetorical structures, most content does not surrender to the analytical knife entirely. When you define a rhetorical structure, you are essentially defining something repeatable. While you could hypothetically take a piece of exposition and define the structured that it follows, this is only useful in a structured writing sense when that structure is repeatable – that is, when there is another piece of content that can follow the same structure, enabling you to reuse your design and testing work and allowing algorithms to take over parts of the processing of that structure.

Where the rhetorical structure of a piece is unique – specifically, when it is irreducibly unique, not just unique because you have not recognized its similarity to the structure of other pieces – there is no benefit to defining an external structure. Irreducible rhetorical uniqueness can occur at any scale. Sometimes it occurs at the scale of an entire book, such as this one. Sometimes it occurs in a single descriptive paragraph in a reference work that otherwise consists entirely of repeatable key/value pairs.[5]

This variation in the degree of unique vs repeatable rhetorical structure in content is the reason you need different structures for different types of content. If you try to fit all content into one structure, however broad that structure may be, you face two potential problems:

- You fail to model much of the repeatable structure of your content, which means you cannot manage or apply algorithms effectively.
- You squeeze unique rhetorical structures into the mold of a repeatable structure, distorting their rhetoric.

The latter is a common problem with strict information typing systems like DITA's task/concept/reference, which often don't fit the rhetorical pattern the writer is trying to create.

You can usefully consider rhetorical repeatability in three categories:

- **Repeatable data:** The same information is required in each instance. For instance, a recipe requires the name, quantity, and unit for each ingredient of a dish.
- **Repeatable argument:** The way in which the information is conveyed to the reader is the same each time. For instance, the ingredients of a recipe are presented as a table each time.
- **Repeatable relationships:** The way in which the subjects mentioned in one document are related to subjects in the real world and to the documents that treat those subjects. For instance, items in an ingredients list might be related (and linked) to pages describing each ingredient in detail – its history, nutritional characteristics, availability, and possible substitutions.

Remember that the ultimate aim is to partition and redirect complexity in the content system. Part of that complexity is the varying levels of repeatability in content – a problem seldom seen in other forms of data management. If you want to deal effectively with all of the complexity in your content system, then you need to deal with this variation in degrees of repeatability. If you don't deal with it successfully, either the repeatability of repeatable content will be lost or diminished, or the uniqueness of unique content will be compromised. Either way, complexity will fall through to the reader in the form of impaired content quality.

---

[5] This is why so many content types have a "description" section for everything that cannot be easily modeled. It's not that that section contains the only descriptive content – all content is descriptive. It's that it contains content that cannot be reduced – or that the person who designed the content type has chosen not to reduce – to simple structures.

Rhetoric and process

How does capturing the rhetorical model of content help improve the content creation process?

If the aim of structured writing is to partition and redirect complexity in the content system without letting any of that complexity leak out of the process and fall down to the reader, then the correctness and consistency of rhetorical models is a core concern. Poor rhetoric means poor content, and poor content means that the complexity of achieving consistent and correct rhetoric has been dropped somewhere in the content system.

A content system relies for its effectiveness on the ability of its principle writers and occasional contributors to maintain a consistent rhetorical standard. The three tools that it has available to do this are:

- Minimize intrusions into the attention of writers. Any attention given to other matters while writing is attention taken away from rhetoric.
- Guide writers to help them provide the correct rhetoric. In other words, reuse the rhetorical design work that you have done and that you have tested and refined with readers.
- Partition out the rhetorical aspects of a composition by collecting facts in a subject-domain structure that can then be transformed into the appropriate rhetorical form by algorithms.

Content inherently varies in how structured it is, so the more structured of these techniques only work for some parts of your content set. You need all three techniques to provide comprehensive rhetorical support across your content system. In Part III, I examine how rhetorical structures support process at the document level. In Part VI, I show how they support the management and maintenance of the content set as a whole.

CHAPTER 8

Information Architecture

The overall goal of a content process is to create an effective information architecture. Information architecture is the arrangement of useful and effective content so it can be found and navigated. An artful arrangement of awful content helps no one; information architecture is rhetoric writ large.

But the rhetorical nature of information architecture is often neglected. It is all too easy to think of information architecture simply in terms of organizing things. However, this can be a trap. Because institutions with a lot of content need a way for their staff to find content, information architecture often becomes an exercise in organizing content for the convenience of the institution rather than the reader.

It is easy to think of organization is an absolute quality; content is either organized or not. But this is not so. Organization is an orientation of content according to your knowledge and expectations, which is to say that organization is a form of rhetoric, as specific to its audience as any other part of rhetoric. Readers consider things to be organized if the location of those things and the means of retrieving them match their expectations. However, what the institution (and its staff) knows and expects is different from what the reader knows and expects. What the institution sees as organization can be seen as chaos by the reader.

More than this, readers are unlikely to spend much mental energy trying to figure out a complex organizational scheme, even one that was designed with them in mind, unless they are already familiar with the scheme or, at least, it makes intuitive sense to them. Instead, they will forage for information, which, in the web world, means that they will use search and will follow links as long as they believe that the scent of the information they are looking for is growing stronger. For an in-depth treatment of information foraging and its implications for information architecture, see *Every Page is Page One: Topic-based Writing for Technical Communication and the Web*.

It is a mistake, therefore, to think of information architecture simply in terms of organization. A building supply store organizes building materials. An architect takes those materials and constructs a navigable edifice full of useful spaces with efficient passages between them. The organization of materials offered by the building supply store is important to the architect, but it is just a starting point for the unique and useful edifice that the architect will design and build.

And for this reason, rhetoric is indispensable to information architecture. You can have good organization of bad material: you can have a neatly organized junk yard. But you cannot build a good edifice from bad materials. Architects are concerned as much with the quality of their materials as they are with the design that integrates them. One does not work without the other. You can't build a high quality car from junkyard parts. Information architecture is not, therefore, a wrapper around content, it is the rhetoric of the content set as a whole.

Information architecture: past and present

If there is a temptation to regard information architecture as merely a form of organization, its origins may lie in the past. For centuries, the basic unit of information was the book and the architecture of the book was an integral part of the responsibility of the writer and editor. Larger sets of information were created by collecting and organizing books, and that was the responsibility of the librarian or bookseller.

Those larger collections were simply forms of organization. If there was an architecture at that scale, it came from the expertise of librarians and book sellers as they made inferences from their clients' needs and created useful connections.

With the advent of online media, first in the form of large capacity electronic media such as CD-ROMs and then the web, this division of responsibilities was overthrown. Now, the basic unit of information in electronic media is not the book but the page. Thanks to hypertext linking, the relationships between pages in electronic media are much more complex than on paper. Information architecture is a response to this challenge.

Information architect is not merely a new name for librarian but the integration of roles that were formerly divided or, rather, were the result of a different partitioning of responsibilities. Often the process changes required to support rhetoric will partition some content complexity away from writers and direct it to information architects. This doesn't just relieve writers of a burden, it also gives information architects visibility into and control over the rhetoric they need to do their job. This does not transfer all control of rhetoric from writers to information architects, because information architects do not have the intimate knowledge of the subject matter required to define rhetorical best practices for every type of content. Rather, it means that writers and information architects must work together to define and implement those best practices.

Another way in which things are different today is that the architecture of online media has to support the ability to add, modify, and delete individual bits of content at any time. It is possible to think of book or library architectures in largely static terms. It is a serious mistake to think of web architectures as static.

This leads to the development of architectures composed of smaller units that have more complex relationships with a larger, more diverse, and more rapidly changing set of resources. These architectures include not only text and static graphics but active media: videos, animation, dynamic feeds, and information widgets.

Given these factors, the old separation of roles between writer and librarian no longer works. Writers now have to be much more conscious of how their pages interact with other pages in the collection, including those created by others. The scale at which these small pieces of content relate with each other is much greater than the scale at which the pieces of a book related to each other. This constitutes a significant increase in complexity, which calls for a whole new approach to information architecture and for the appearance of a function and a role that had no equivalent in the paper world.

However, information architecture still needs to support paper delivery of content and to ensure that the design of content delivered on paper is as effective as it can be. In many cases, content delivered on paper will be the same content that was delivered on the web and will come from the same repository. However, if the roles of writer and information architect have been partitioned to support web output, responsibility for paper output is also likely to fall to the information architect, and structured writing algorithms such as differential single sourcing are likely to become an important part of the information architect's concerns. See Chapter 12 for more on differential single sourcing.

As noted above, in many cases structured writing is used to transfer complexity from the writer to the information architect. It is the vast increase in the complexity of information architecture that makes this transfer necessary. But it also points out how much information needs to pass from the writer to the information architect, and vice versa, for both roles to do their jobs and not let any complexity slip through the cracks. Thus structured writing, which gives you a way to capture and pass on the required information consistently and reliably, becomes much more important in the current environment. Correct partitioning is essential to success in this area, and it cannot be achieved without the reliable transmission of information between roles that structured writing enables.

Top-down vs. bottom-up information architecture

In looking at information architecture as a form of rhetoric, it is useful to start by making a basic distinction between two types of information architecture: top-down and bottom-up. Top-down information architecture deals with navigational aids and organizing systems that stand apart from the content and point to it. A table of contents or a website menu system is a piece of top-down information architecture. Bottom-up information architecture deals with navigation and organization that exists within the content itself. A website with a consistent approach to hypertext links within its pages is an example of a bottom-up information architecture.

But bottom-up information architecture is not just about linking, it is about the way content is written. A page in a bottom-up information architecture is designed to be entered via search or links from almost anywhere (as opposed to being designed to be entered exclusively from a previous chapter). But it is also designed to help readers with onward navigation and to be a hub of its local subject space, offering readers many onward vectors according to their needs and interests.

This is a very different approach to information design because it makes navigation an integral part of the rhetoric of a document and the content set into which it is integrated. Linking is not an afterthought but part of the rhetorical design of the document. This is no longer an architecture in which navigation takes readers to content and leaves them there. Navigation and rhetoric are intertwined and continuous.

I call this approach to information design Every Page is Page One, and it is described in my book, *Every Page is Page One: Topic-based Writing for Technical Communication and the Web*. One of the key principles of Every Page is Page One is that a topic should follow a well defined rhetorical structure or type. Structured writing, particularly subject-domain structured writing, is very useful in developing Every Page is Page One content.

Bottom-up and top-down information architectures are not incompatible with each other. In fact almost every information architecture has both top-down and bottom-up elements. (Books, for instance, which are principally top-down, based on a table of contents, may also have internal cross references, which are a bottom up mechanism.)

Structured writing can be used to drive both the top-down and bottom-up aspects of information architecture.

CHAPTER 9

Writing

Most humans are bad writers. By that I don't just mean that they use poor grammar or spelling or that they create run-on sentences or use the passive voice too much, though all those things may be true and annoying. I mean something more fundamental: they don't say the right things in the right way for the right audience. They leave out stuff that needs to be said, they weigh their text down with stuff that does not need to be said, or they write in a way that is hard to understand.

We all suffer from a malady called the curse of knowledge,[1] which makes it difficult for us to understand what it is like not to understand something we know. Harvard psychologist Steven Pinker regards the curse of knowledge as single best explanation for bad writing.[2] Because we forget what it is like to not understand, we take shortcuts, we make assumptions, we say things in obscure ways, and we just plain leave stuff out.

This is not a result of mere carelessness. The efficiency of human communication rests on our ability to assume that the person we are communicating with shares a huge collection of experiences, ideas, and vocabulary in common with us.[3] Laboriously stating the obvious is as much a rhetorical fault as omitting the necessary. Yet what is obvious to one reader may be obscure to another. The curse of knowledge is that as soon as something becomes obvious to us, we can no longer imagine it being obscure to someone else.

Thus, much human-to-human communication fails. The rhetoric is inadequate. The recipient of the communication simply does not understand it or does not receive needed information because the writer left it out.

We write better for machines than we do for humans

These days we create a lot of information to be read by machines. The machines are pretty stupid and extremely literal, which forces us to be very careful in how we create and structure information for machines to act on. The computer science community coined the phrase "garbage in, garbage out" very early, because the machines were, and to a large extent still are, too stupid to identify

---

[1] https://en.wikipedia.org/wiki/Curse_of_knowledge

[2] https://www.linkedin.com/pulse/single-reason-why-some-people-cant-write-according-glenn-leibowitz

[3] http://everypageispageone.com/2015/08/04/the-economy-of-language-or-why-we-argue-about-words/

garbage input, and they did not have the capacity to seek clarification or consult other sources, as a human would. They just spit out garbage.

Therefore, we had to improve the quality and precision of the data going in. We worked out precise data structures and implemented elaborate audit mechanisms to make sure that data was complete and correct before we fed it to the machine.

We have never been as diligent in improving the quality of the content we feed human beings. Faced with poor content, human beings do not halt and catch fire; they either lose interest or do more research. Given our adaptability as researchers and our tenacity in pursuing things that really matter to us, we often manage to muddle through bad content, though at considerable economic cost. And writers are often so removed from their readers that they have no notion of what the poor reader is going through. If readers did halt and catch fire, we might put more effort and attention into content quality.

Even today, with many companies implementing enterprise content management and making the store of corporate knowledge available to all employees, most of the emphasis is on making content easier to find, not on making that content more worth finding. (This despite the fact that the best thing you can do to make content easy to find it to make it more worth finding.) People trying to build the semantic web spend a lot of time trying to make the data they prepare for machines correct, precise, and complete. We don't do nearly as much for humans.

Part of the problem is that improving content quality runs up against the curse of knowledge. Both writers and the subject matter experts who review their content suffer from the curse, which makes it difficult to audit written content. Style guides and templates can help, but their requirements are difficult to remember and compliance is hard to audit, meaning there is little feedback for a writer who strays.

The curse of knowledge and the distance separating writers from readers are a major source of complexity. Structured writing provides a way to guide and audit content, specifically, the *rhetoric* of our content: what we say and how we say it to achieve a specific end.

Structured writing enables us to write in the subject domain (wholly or partially), which allows us to guide and audit rhetoric in ways not otherwise possible. It also allows us to factor out many constraints, simplifying the writing task and allowing writers to devote more attention to the quality of their content, whether that content is read by people or robots.

Structure, art, and science

Many writers reject the idea that imposing constraints can improve quality. They see writing as a uniquely human and individual act, an art, not a science and, therefore, immune to the encroachment of algorithms. But I suggest that structures and algorithms do not diminish the human and artistic aspects of writing. Rather, they supplement and enhance them.

We see this pattern in all the arts. Music has always depended on making and perfecting instruments as tools of the musician. Similarly the mathematics of musical theory gave us well-tempered tuning, on which modern Western music is based.

Computer programming is widely regarded as an art[4] among practitioners, but the use of sound structures is an inseparable part of that art. Art lies not in the rejection of structure but in its mature and creative use. As noted computer scientist Donald Knuth observes in his essay, *Computer Programming as an Art*, most fields are neither an art nor a science, but a mixture of both.

> Apparently most authors who examine such a question come to this same conclusion, that their subject is both a science and an art, whatever their subject is. I found a book about elementary photography, written in 1893, which stated that "the development of the photographic image is both an art and a science." In fact, when I first picked up a dictionary in order to study the words "art" and "science," I happened to glance at the editor's preface, which began by saying, "The making of a dictionary is both a science and an art."
>
> —http://dl.acm.org/citation.cfm?id=361612

Writers use structures, patterns, and algorithms as aids to art, just like every other profession. Of course, few writers would claim that there is no structure involved in writing. We have long recognized the importance of grammatical structure and rhetorical structure in enhancing communication. The question is, can the type of structures that structured writing proposes improve your writing, and if so, in what areas? Traditional poetry is highly structured, but using an XML schema won't help you write a better sonnet. On the other hand, following the accepted rhetorical pattern of a recipe will help you write a better cookbook, and using structured writing to create your recipes will help you improve the consistency of your recipes, produce them more efficiently, and exploit them as assets in new ways.

---

[4] http://ruben.verborgh.org/blog/2013/02/21/programming-is-an-art/

The question then becomes: how much of our work is like recipes and how much is like sonnets? That is, how much of business and technical communication would benefit from structured writing? The answer, I believe, is that a great deal of business and technical communication would benefit. Although you may not see any obvious structure in much of that communication, do not take that as evidence that structure is inappropriate. Rather, it means that structure has not been developed and applied to the content.

Contra-structured content

Many writers resist the use of structure writing techniques. Some resist because of an attachment to long-established work patterns. Some resist because of a reluctance to comply with externally defined rhetorical patterns. And some remain attached to desktop publishing – despite its distractions and complexity – because it gives them a sense of ownership over the entire work

Often, writers overcome their reluctance when they have a chance to work with a well-designed and well-managed structured writing system. However, we must also acknowledge that many writers have had a bad experience with structured writing. Often, the structured writing system was not chosen or designed by the writers to enhance their art; it was imposed externally for some other purpose, such as to facilitate the operation of a content management system or to make it easier to reuse content.

Too frequently, such systems move complexity away from some other function and dump it on writers without sufficient thought being given to where that complexity should go. Such systems actively interfere with the writer's task, hindering the production of quality content.

A badly designed structured writing system forces writers to use structures that don't support their tasks. The result is not merely unstructured for these purposes, it is actually *contra-structured*. Such a system enforces structures that actively block writers from doing their best work.

Writers have often shown me page designs and layouts that make no sense, lamenting that the system gives them no alternatives. Content structure is not generic, and you cannot expect to simply install the flavor-of-the-month CMS or structured writing system and get a good outcome. The system you use must at least be compatible with, if not actively supportive of, the rhetoric you need to create.

A large part of the problem of contra-structured content can be blamed on a tendency to install structured writing systems that seek to impose a single format on all content across the organization. As Sarah O'Keefe describes it:

> [W]e have been trying for a world in which all authors work in a single environment (such as DITA XML, the Darwin Information Typing Architecture), and we pushed their content through the same workflow to generate output. We have oceans of posts on "how to get more people into XML" or "how to work with part-time contributors." The premise is that we have to somehow shift everyone into the One True Workflow.
> —https://www.scriptorium.com/2018/04/single-sourcing-dead-long-live-shared-pipes/

The problem with this model is that content, and particularly structured content, is simply not a one-size-fits-all proposition. In practice, such a model does not fit anyone, and you end up with contra-structured content for everyone.

The model that O'Keefe suggests instead is one she calls "shared pipes" in which content can come from many different sources but passes through shared processing stages as appropriate. By allowing each writer to use a model appropriate to their subject matter and audience, the shared pipes model has the potential to allow you to develop structured writing systems that are actually palatable to writers. Subject-domain structured writing is a key enabler of this model.

Constraints versus usability

Most of the complexity of the content process flows through the writer. Before computers, the writer prepared a typescript that was handed off to other people to manipulate before the work was eventually set in type and printed. Today, we generally expect that the writer will be the last person to manipulate content. Everything after that will be handled by an algorithm. The writer must supply everything the publishing process needs. Making that task manageable for writers is key to successfully partitioning your content system.

But the complexity imposed by publishing and content management functions is not the only source of complexity that writers must deal with. They must also develop the best rhetorical strategy for a given subject matter, apply that strategy consistently, and work effectively with other writers to achieve the overall content strategy of the organization. These factors are major sources of complexity, and all that complexity flows through the writer as well.

To relieve writers of as many extended decisions as possible, you need to create structures that direct those decisions away from them. But you can't do this by creating a loose, unconstrained writing environment. If you don't get a high level of conformance to your content structures, your algorithms will not be reliable, partitioning will break down, and complexity will fall back on the writers or down to the readers. The structures you present to writers must combine strict conformance with ease of use.

In media-domain systems, such as word processors and desktop publishing systems, writers must make decisions about formatting while writing. One of the traditional arguments for structured writing is that it relieves writers of the burden of making and executing formatting decisions, so they can focus on writing. This means moving to the document domain. But in the document domain, writers have a new set of structures to think about. Is it easier for writers to think about and manipulate document-domain structures rather than media-domain structures? In some cases, yes. For instance, writing a blog post or a web page in Markdown may be less cumbersome for some writers than using a WYSIWYG HTML editor.

However, Markdown does not contain enough structure – or enough constraints on its structure – to enable many of the structured writing algorithms. If you want to support these algorithms, you need something more structured.

If you want to implement management-intensive algorithms such as content reuse, writers may need to manipulate a content management system and its management policies. Depending on how complex these policies are and how foreign they are to the writer's experience, this can create a far greater burden than formatting content according to a style guide.

You could look at this and say, "Yes, this additional structure makes writing more complex than it was before, but we gain advantages as well, such as reuse and content management." The problem is, attention is a finite resource. Writing takes full attention, and added decision making takes attention away from writing and requires more knowledge and skill. As writing gets more difficult, writers handle each component task less well. In addition, it becomes harder to find writers who have the necessary skills. The more attention writers must spend on structure, the less they have available for writing, and quality suffers.

As writing quality suffers, writers become frustrated with the system, and, in order to focus their attention on content, they start to ignore any structural rules that get in their way. When that happens, the quality of the structure suffers as well. And if the quality of both the writing and the structure declines, your algorithms become less reliable, compromising all of benefits you hoped to obtain. Too much complexity dumped on any one person or process compromises all downstream persons and processes.

Conformance is fundamentally a human activity. It is writers who must conform, and they conform best when you create structures that are easy to conform to. It all begins with writing. Unfortunately, writing is often the last thing people think about when designing content management and structured writing systems. Often, writing is where complexity gets dumped.

One of the most familiar tropes of the content management industry is that problems with content management systems are not technology problems, they are human problems. The solution, this trope suggests, lies in better change management and more training. The presumption here is that the tools work fine if you give them correct input. If the input is not correct, that is the fault of the humans who created the input.

We would not accept this argument for any other kind of system. For any other kind of system we would say, "this system is too hard to use," not "the problem is that everybody needs to be better trained and more accepting of change." The real fault here is poor system design, often the result of the "One True Workflow" approach that O'Keefe describes. If humans cannot conform to the structures the system requires, the fault is in the system design. If the system doesn't partition and distribute complexity correctly, the structures need to be redesigned so humans can conform to them. (I look at this in more detail in Chapter 29.)

How do you design structures that are easy to use but still provide the constraints you need for quality and efficiency? As I've described, moving from the media domain to the document domain allows you to factor out or impose certain structural constraints, but to do so, you often must introduce the management domain to impose content management constraints. This added complexity detracts from ease of writing. However, moving to the subject domain allows you to factor out many of the document-domain and management-domain structures. Designing structures for writing, therefore, often consists of factoring out complex publishing and content management structures using subject-domain structures that are lucid for writers.

One of the most important consequences of this, both for ease of writing and reliability of data, is that in the subject domain you don't ask writers to think, decide, and structure content in terms of algorithms. In this sense, the move to the subject domain not only factors out specific constraints from writers, it factors out the need for writers to think in algorithms at all, leaving them free to think and decide in terms of subject matter. This freedom to focus on content is a property I call *functional lucidity*.

Functional lucidity

Functional lucidity refers to the total intellectual burden that a system imposes on writers: how clearly do they understand what the system requires and how fluidly and naturally can they create structures and content in that system. The functional lucidity of a markup language is not a matter of its size or its complexity. It is a matter of how natural and obvious it is for writers to use and how well the markup language aligns with a writer's core decisions, the decisions that are essential to the task, as opposed to extended decisions, which are external to it.

If you have ever tried to learn a language, you know how painful it is to write even a paragraph in a language you are not fluent in. The effort of finding words and correct grammatical structures takes all of the attention that should be reserved for what you are trying to say. Writing in a markup language where the structures don't make intuitive sense, where they don't seem to fit the thoughts you are trying to express, where they require knowledge and skills external to your core task, is very much like this. Lucidity is essential to avoid having the markup absorb all of the attention that should be focused on the content.

There is a bit of a catch to functional lucidity, however. Many professional writers, particularly technical writers, are so used to making extended decisions that they find it disconcerting, at least at first, to use a system that removes those decisions. Thus, writers currently using a complicated document-domain tool such as FrameMaker may feel more comfortable switching to an equally complicated document-domain markup language like DocBook, since it involves most of the same extended decisions they were making before. On the other hand, writers who are not used to making those decisions will find neither FrameMaker nor DocBook to be functionally lucid.

Writers' initial impressions of the functional lucidity of a tool or markup language are often based on how well the extended decisions required by the new tool align with those required by the old tool, rather than on how well the language aligns with their core decisions. Their old tools have created a rut in the mind that leads them to expect a new tool to present the same extended decisions as the old tool.

However, remember that the aim in adopting structured writing is to handle elements of complexity that are not currently handled well. Thus, you need to assess functional lucidity based on a long-term view of how the new system partitions complexity, not on a short-term view of how similar the structures are to the current system. A well-designed subject-domain language will be functionally lucid for writers who are familiar with the subject matter, even if they have to get used to no longer making the extended decisions required by their old tools. Writers will get over their attachment to the old tools if the new tools produce superior results.

Simplicity and clarity

One of the biggest benefits of subject-domain markup for writers is a much higher degree of functional lucidity compared with a typical document-domain language.

While a document-domain language such as DocBook must have a wide range of document structures, a recipe markup language such as the one developed in Chapter 4 has only a few simple structures. Better still, there are very few permutations of those structures. The decisions writers must make align closely with the core task of expressing the method for preparing a dish.

```
recipe: Hard Boiled Egg
    introduction:
        A hard boiled egg is simple and nutritious.
    ingredients:: ingredient, quantity
        eggs, 12
        water, 2qt
    preparation:
        1. Place eggs in pan and cover with water.
        2. Bring water to a boil.
        3. Remove from heat and cover for 12 minutes.
        4. Place eggs in cold water to stop cooking.
        5. Peel and serve.
    prep-time: 15 minutes
    serves: 6
    wine-match: champagne and orange juice
    beverage-match: orange juice
    nutrition:
        serving: 1 large (50 g)
        calories: 78
        total-fat: 5 g
        saturated-fat: 0.7 g
        polyunsaturated-fat: 0.7 g
        monounsaturated-fat: 2 g
        cholesterol: 186.5 mg
        sodium: 62 mg
        potassium: 63 mg
        total-carbohydrate: 0.6 g
        dietary-fiber: 0 g
        sugar: 0.6 g
        protein: 6 g
```

Figure 9.1 – Subject-domain recipe for hard-boiled eggs

Because subject-domain structures describe the subject matter they contain, they are also much more lucid to writers, who may not understand complex document structures (or, more often, the subtle distinctions between several similar document structures), but who do (we hope) understand their subject matter.

The combination of simplicity and clarity means that writers can often create subject-domain structured content with little or no training. For instance, even if you add some additional fields to the recipe markup, a writer could easily use a sample like the one in Figure 9.1 as a template without any training or special tools.

Of course, the downside is that recipe markup is good for only one thing: recipes. Complexity is never destroyed, only moved somewhere else. So this approach moves complexity away from the writer to the person who has to design and maintain these structures and the algorithms that process them. (Information Architect and Content Engineer are both titles sometimes used for the person with this responsibility.)

This can seem scary because we are not used to partitioning complexity in this way. But then, the way we have partitioned complexity in the past has not been as successful as we would like – it has, in fact, left much of the complexity of content creation unhandled, resulting in impaired rhetoric and all the costs associated with it. A new partitioning of the content creation process requires us to accept that some of these new methods of partitioning will be unfamiliar.

Algorithms

If rhetoric and process are inseparable, how do you structure content to support both at the same time? I have described how structured writing can constrain or factor out many aspects of rhetoric. But how are those structures compatible with the processes you need to run on your content? To answer this, we must look at the principal structured writing algorithms.

Structured writing algorithms basically come down to four operations: separate, combine, relate, and transform. The chapters in this part look at how to use these basic operations to implement the fundamental structured writing algorithms.

However, these are not chapters on programming. Defining an algorithm and writing software to implement that algorithm are two separate tasks. This part introduces the design and capabilities of the structured writing algorithms, but it does not describe how to code software to implement those algorithms.

In this part, I look at algorithms for manipulating content itself. In Part VI, I look at algorithms for managing the content set.

CHAPTER 10

Separating Content from Formatting

If there is one phrase that most people associate with structured writing, it is "separating content from formatting." This basic, well-known structured writing algorithm can be used to achieve a number of process and rhetorical goals. It is a useful place to start the study of algorithms because it illustrates the strategies used for all the other algorithms. Separating content from formatting separates decisions about the appearance of a document from decisions about its content.

Separate out style instructions

Let's start with a piece of text that includes a description of its format. I use CSS syntax to describe the format because CSS is easy to understand and helps illustrate the separation, but don't focus on the syntax, it is just there to illustrate the principle. I also represent certain characters (bullets and tabs) by their names, so you can see exactly where everything is going:

```
{font: 10pt "Open Sans"}The box contains:
{font: 10pt "Open Sans"}[bullet][tab]Sand
{font: 10pt "Open Sans"}[bullet][tab]Eggs
{font: 10pt "Open Sans"}[bullet][tab]Gold
```

This file contains content and formatting, so let's separate the two. Of course, when you remove the formatting, you need to add something in its place so you can add the formatting back later. The simplest method is to replace the formatting with a named style:

```
{style: paragraph}The box contains:
{style: paragraph}[bullet][tab]Sand
{style: paragraph}[bullet][tab]Eggs
{style: paragraph}[bullet][tab]Gold
```

Then, you need to record the style information. You aren't removing that information; you're just separating it from the text:

```
paragraph = {font: 10pt "Open Sans"}
```

Once you separate the style information, you can change the formatting without changing the text; for example, you could choose a different font:

```
paragraph = {font: 12pt "Century Schoolbook"}
```

Separate out formatting characters

Cool, but suppose you need to change the style of the bullets. The bullet style is certainly formatting, but bullets are text characters. To change them you can't just change the font, you have to change the characters themselves. Some characters in your text are part of the content, and some are part of the formatting. Therefore, a style may include characters as well as formatting directives, as shown here:

```
paragraph = {font: 12pt "Century Schoolbook"}
bullet-paragraph = {font: 12pt "Century Schoolbook"}[bullet]
```

Now the content looks like this:

```
{style: paragraph}The box contains:
{style: bullet-paragraph}[tab]Sand
{style: bullet-paragraph}[tab]Eggs
{style: bullet-paragraph}[tab]Gold
```

However, you still must begin the bulleted lines with a tab, which is awkward and error prone, so let's move that character to the style as well:

```
paragraph = {font: 12pt "Century Schoolbook"}
bullet-paragraph = {font: 12pt "Century Schoolbook"}[bullet][tab]
```

Now the content looks like this:

```
{style: paragraph}The box contains:
{style: bullet-paragraph}Sand
{style: bullet-paragraph}Eggs
{style: bullet-paragraph}Gold
```

And now you can change the bullet style:

```
bullet-paragraph = {font: 12pt "Century Schoolbook"}[em dash][tab]
```

Because the style now uses an em-dash, rather than a bullet, the name `bullet-paragraph` may not be the best choice. The content still contains a style named for a particular piece of formatting, which means writers are still making the formatting choice "bullet" when it would be better for them to focus on the content choice "list."

Name your abstractions correctly

The names you choose for styles and other abstractions matter. The wrong name sets up a false expectation, which can lead to writers using a style incorrectly. So the first lesson about separating content from formatting is that naming matters. You are creating an abstraction, so you need to figure out what that abstraction is and name it appropriately.

What is the abstraction here? Sand, Eggs, and Gold are list items. So maybe you do this:

```
{style: paragraph}The box contains:
{style: list-item}Sand
{style: list-item}Eggs
{style: list-item}Gold
```

and

```
list-item = {font: 12pt "Century Schoolbook"}[em dash][tab]
```

Make sure you have the right set of abstractions

But then, of course, you run into lists like the following:

```
{style: paragraph}To wash hair:
{style: list-item}Lather
{style: list-item}Rinse
{style: list-item}Repeat
```

These list items should have numbers, not dashes or bullets. So the abstraction does not cover all list items. If you look at the different kinds of list items, group them into abstract types, and come up with names for those types, you probably come up with ordered-list-item and unordered-list-item. Then you have:

```
{style: paragraph}The box contains:
{style: unordered-list-item}Sand
{style: unordered-list-item}Eggs
{style: unordered-list-item}Gold
```

and

```
{style: paragraph}To wash hair:
{style: ordered-list-item}Lather
{style: ordered-list-item}Rinse
{style: ordered-list-item}Repeat
```

The style for ordered-list-items now looks something like this:

```
ordered-list-item = {font: 12pt "Century Schoolbook"}<count>.[tab]
```

But you need a way to increment the count and reset it to 1 for a new list, which leads to:

```
{style: paragraph}To wash hair:
{style: first-ordered-list-item}Lather
{style: ordered-list-item}Rinse
{style: ordered-list-item}Repeat
```

and

```
first-ordered-list-item =
    {font: 12pt "Century Schoolbook"}<count=1>.[tab]
ordered-list-item =
    {font: 12pt "Century Schoolbook"}<++count>.[tab]
```

(++count here means add one to count and then display it.)

This is pretty much how FrameMaker, Microsoft Word, and many other tools handle lists. The reason for going through this example in such detail is to point out how much is involved in even this simple bit of partitioning. You began by removing formatting commands, followed by characters, which forced you to include characters in the style definitions. And then you had to calculate the value of characters. At each step, you had to consider all possible cases and create abstractions to handle them.

Create containers to provide context

One problem with this approach to creating lists is that you have to apply a different style to the first item of a list. It would be better if you could use the same style for each list item and have the numbering just work. However, this is hard to do because nothing in the content says where one numbered list ends and the next begins. For this you need a new abstraction. So far, you have abstractions for ordered and unordered list items, but you don't have an abstraction for lists themselves.

Up to this point, everything has been purely in the media domain. You replaced direct formatting definitions with indirect definitions through styles. But now it's necessary to venture into the document domain, creating the abstract idea of a list and inserting that abstract idea into your content.

```
paragraph: To wash hair:
list:
    ordered-list-item:Lather
    ordered-list-item:Rinse
    ordered-list-item:Repeat
```

There are two significant changes here. First, the structure is no longer flat. Previously a list was a series of paragraphs with different styles attached. Now there is a container, list, that, as far as the formatting is concerned, never existed in the original.[1] The writer and reader knew that the sequence of bulleted paragraphs formed a list, but that was an interpretation of the formatting. Now that interpretation is recorded explicitly in the content itself.

By creating the idea of a list, you further separate list formatting from the list content. Now, an algorithm can recognize a list and make formatting decisions based on that knowledge.

The second important thing is that the content no longer refers to style names. Instead you have structures. list is a structure and so are paragraph and ordered-list-item. You can give the same structure a different style depending on where it is in the document. The formatting algorithm can determine which ordered-list-item is first and format it accordingly; this is how list formatting works in CSS.[2]

Now, writers no longer apply styles directly to content, even styles with abstract names. Rather they place content in structures and allow the formatting algorithm to apply styles appropriately. This separates the content even more from the formatting.

Move the abstractions to the containers

But what if a writer inadvertently does this:

```
paragraph: To wash hair:
list:
    ordered-list-item:Lather
    unordered-list-item:Rinse
    ordered-list-item:Repeat
```

---

[1] The SAM markup in this example uses indentation to identify hierarchy. The list container contains three instances of ordered-list-item.

[2] https://css-tricks.com/numbering-in-style/

To avoid this type of error, move the abstraction outward. Instead of creating ordered and unordered list items, create ordered and unordered lists:

```
paragraph: To wash hair:
ordered-list:
    list-item:Lather
    list-item:Rinse
    list-item:Repeat
```

and

```
paragraph: The box contains:
unordered-list:
    list-item:Sand
    list-item:Eggs
    list-item:Gold
```

Now, you have a single `list-item` structure that you can use in either an unordered list or an ordered list; the formatting algorithm determines the formatting based on which type of list the list-item belongs to. The name `list-item` describes a role in the document that is entirely separate from formatting.

Moving the abstraction out to the container is an important part of the algorithm for separating content from formatting. It partitions ordered and unordered lists, making the construction of each simpler and more reliable. This helps keep things consistent and reduces the number of things writers have to remember. (Notice that this requires writers to decide whether a list is ordered or unordered, but they had to decide this anyway; the decision is implicit in the rhetoric of the piece.)

Some markup languages take better advantage of containers than others. For example, HTML and Markdown both provide six different heading levels. However, content under an H2 or an H5 heading is not in any container. The content simply comes after the heading. This means that is it perfectly possible and legal to use heading elements in any order you want. Writers must select the heading level and ensure that it fits the structure of their document.

By contrast, DocBook has a `section` structure. Like a list, a section records your interpretation of what you are creating in the document. The `section` structure instantiates the concept of a section. And once you have the instantiation of a section, you don't need six levels of heading. You can have one structure called `title`. Sections can be nested inside other sections, and the formatting algorithm applies the correct style to the title based on context.

This eliminates decisions about which heading element to use; there is just one: `title`. This ensures that headings in output consistently reflect the document's section and subsection structure.[3]

```
section:
    title:
    paragraph:
    section:
        title:
```

Separate out abstract formatting

To separate content from formatting for ordered and unordered lists, you must separate out some of the content as well. Specifically, you must separate out some of the characters. The distinction between content and formatting is not the same as the distinction between characters and the styles applied to them. Sometimes characters are part of the formatting rather than the content. Consider the following labeled-list structure:

Street:	123 Elm Street
Town:	Smallville
Country:	USA
Code:	12345

The generic structure of a labeled list might look like this:

```
labeled-list:
    list-item:
        label: Street
        content: 123 Elm Street
    list-item:
        label: Town
        contents: Smallville
    list-item:
        label: Country
        contents: USA
    list-item:
        label: Code
        contents: 12345
```

---

[3] Not everyone holds to the view that headings in a text must reflect a hierarchy of sections. Instead, headings may be simply signposts along the way, where the size of the sign reflects some quality – size of the town or the importance of the section – other than a strict hierarchy. If that is how you look at document structures, you should choose a different way to separate content from formatting.

But what if you have hundreds of addresses, all with the same labels? Are the labels content or are they presentation? Since the labels don't change from one list to another, you could look at them as being part of the presentation rather than the content. So let's look for a way to separate them from the content.

As always, when you separate something from your content, you have to replace it with something else. In Figure 10.1, named structures reflect the subject matter of the entries, which moves this content into the subject domain.

```
address:
    street: 123 Elm Street
    town: Smallville
    country: USA
    code: 12345
```

Figure 10.1 – Address markup in the subject domain

Here, once again, it is important to distinguish formatting and presentation. *Formatting* refers to the precise details of a text's appearance: the font chosen, the width of the text column, the size of the characters, the spacing between lines, the size and shape of the bullet characters, and so forth. *Presentation* refers to the organization of text.

Deciding to use a list for a certain piece of information is a presentation decision, which is independent of the formatting details applied to lists. When you move content from the media domain to the document domain, you separate the *formatting* of the content from its *presentation*. The decision to present the information as a list remains; the decision about what that list looks like is separated out.

In contrast, moving content from the document domain to the subject domain separates the *information* from the *presentation*. The subject-domain structure in Figure 10.1 is not a labeled list. It is a data record that can be turned into many different forms of presentation.

Turning such structures into a specific form of presentation is the job of the presentation algorithm (see Chapter 19). The presentation algorithm could turn it into a labeled list, a table, a paragraph, or an address label.

In the subject domain – with the content separated from both formatting and presentation – you also gain the ability to query and reorganize the content in various interesting and useful ways (which I explore in later chapters).

This is as far as you can go in separating content from formatting, and you can't separate all content from formatting to quite this extent. Separating content from formatting is not a binary operation. There are various degrees of separation that you can apply for various reasons. It is important to understand exactly which degree of separation best serves your needs.

CHAPTER 11

Processing Structured Text

Chapter 10 looked at separating content from formatting. Separating them is necessary to achieving certain rhetorical and process goals. But of course they cannot stay separated. Content must be formatted before you present it to the reader. So let's look at the algorithms for putting content and formatting back together again.

Understanding how to put things back together is a big part of understanding how to separate them. Indeed, describing how you would put them back together is the only way to know for certain that you didn't lose anything when you did the separation. This is essential to ensuring a clean partitioning that does not allow any complexity to fall through the cracks.

Two into one: reversing the factoring out of invariants

Moving content from the media domain to the document domain or the subject domain involves progressively factoring out invariants in the content. Each step in this process creates two artifacts: the structured content and the invariant piece that was factored out.

Processing structured text is about putting the pieces back together – combining the structured content with the invariants that were factored out. But processing structured text is also about making and/or executing the decisions that were partitioned away from the writer. These decisions may be invariant, such as applying approved corporate style, or they may be variable, such as presenting content differently in different media or for different audiences.

If factoring out the invariants moves content toward the document or subject domains, recombining the content with the invariants moves it in the opposite direction, toward the media domain. This could mean moving the content from the subject domain to the document domain or from the document domain to the media domain, or simply from a more abstract form in the media domain to a more concrete form, which is our first example.

Restoring style information

In Chapter 10, the first example of separating content from formatting factored out the style information from this structure:

```
{font: 10pt "Open Sans"}The box contains:
{font: 10pt "Open Sans"}[bullet][tab]Sand
{font: 10pt "Open Sans"}[bullet][tab]Eggs
{font: 10pt "Open Sans"}[bullet][tab]Gold
```

The style information was replaced with style names:

```
{style: paragraph}The box contains:
{style: bullet-paragraph}Sand
{style: bullet-paragraph}Eggs
{style: bullet-paragraph}Gold
```

And then the styles were defined as follows:

```
paragraph = {font: 10pt "Open Sans"}
bullet-paragraph = {font: 10pt "Open Sans"}[bullet][tab]
```

To unite each paragraph with the correct style, you can write a set of search-and-replace rules:

```
find {style: paragraph}
    replace {font: 10pt "Open Sans"}

find {style: bullet-paragraph}
    replace {font: 10pt "Open Sans"}[bullet][tab]
```

This algorithm combines two sources of information: the structured text and the style definitions. Usually, the style definitions are embedded in the rules, as in this example. In some cases, however, the rules may pull content from a separate file. I show examples of this later.

Applying these rules brings back the original content:

```
{font: 10pt "Open Sans"}The box contains:
{font: 10pt "Open Sans"}[bullet][tab]Sand
{font: 10pt "Open Sans"}[bullet][tab]Eggs
{font: 10pt "Open Sans"}[bullet][tab]Gold
```

If you want to change the styles, you can apply a different set of rules (in other words, make a different set of decisions):

```
find {style: paragraph}
    replace {font: 12pt "Century Schoolbook"}

find {style: bullet-paragraph}
    replace {font: 12pt "Century Schoolbook"}[em dash][tab]
```

Applying these new rules gives you the following:

```
{font: 12pt "Century Schoolbook"}The box contains:
{font: 12pt "Century Schoolbook"}[em dash][tab]Sand
{font: 12pt "Century Schoolbook"}[em dash][tab]Eggs
{font: 12pt "Century Schoolbook"}[em dash][tab]Gold
```

Rules based on structures

The tools that do this sort of processing do not literally use search and replace like this. Rather, they parse the source document to pull out the structures and allow you to specify your processing rules by referring to those structures.

The mechanism by which a processing tool recognizes structure is not the focus here. You just tell the tool what to do when it finds a structure. So, the next set of rules match structures rather than looking for literal strings in the text:

```
match paragraph
    apply style {font: 12pt "Century Schoolbook"}

match bullet-paragraph
    apply style {font: 12pt "Century Schoolbook"}
    output "[em dash][tab]"
```

The result of applying these rules is the same as before:

```
{font: 12pt "Century Schoolbook"}The box contains:
{font: 12pt "Century Schoolbook"}[em dash][tab]Sand
{font: 12pt "Century Schoolbook"}[em dash][tab]Eggs
{font: 12pt "Century Schoolbook"}[em dash][tab]Gold
```

The way I have written these rules is an example of pseudocode. Pseudocode is a means for sketching out an algorithm so you understand what you are trying to do before you write actual code. There is no formal syntax for pseudocode. It is intended for humans, not computers, and you can use whatever approach you like as long as it is clear to your intended audience. However, pseudocode should lay out a set of logical steps for accomplishing something and make it clear how the steps go together.

Writing an algorithm in pseudocode is a great way to make sure you understand the algorithm without worrying about the details of code – or even learning how to code. And pseudocode algorithms are a great way to tell programmers what you need a program to do.

The order of the rules does not matter

You may have noticed that these rules do pretty much exactly what style sheets do in applications such as Word or FrameMaker. If you understand style sheets, you understand a good deal of how structured writing algorithms work.

One important thing to notice is that when you create a style sheet in Word or FrameMaker, you don't specify the order in which styles are applied to the document. The same is true when you create a CSS style sheet. The style sheet is just a flat list of rules. The order in which the rules are applied to the document depends entirely on the order in which the various structures occur in the document.[1]

To put it another way, rather than being a set of steps to follow, style-sheet rules define a set of decisions to be made when certain events occur. If you see X, do Y. The structure of the source document determines the order in which the processor applies the rules; the order of the rules in the style sheet doesn't matter.

This may seem obvious, but it is key to understanding how structured text is usually processed. This can be confusing even to skilled programmers, because style sheets are processed differently from the way most programmers learn to write computer programs.

Things get more complex when you move into processing the nested structures of the document domain and subject domain, but the basic pattern of a set of unordered rules to describe a transformation algorithm still applies.

---

[1] Actually, there is a wrinkle to this. It is possible to have more than one rule that could match a particular structure. In this case the language in which the rules are implemented has to decide which rule to pick. There are various mechanism that a language can use to decide which rule to choose, for example, choosing a more specific rule over a more general one. But in some cases, the selection may be based simply on which rule comes first (or last) in the rule set. But this is a minor detail and does not distract from the idea that rules are fundamentally unordered and rule selection is fundamentally document driven.

Applying rules in the document domain

Suppose you have a piece of document-domain structured text that contains this `title` structure:

```
title: Moby Dick
```

You want to transform this document into HTML. When a rule matches a structure in the source document, the processor outputs the equivalent HTML structure. Here is the pseudocode for this rule (it is in a slightly different format from the pseudocode above):

```
match title
    create h1
        continue
```

This says, when the processor sees a `title` structure in the source, it creates an `h1` structure in the output and then continues applying rules to the content of the title structure.

The `continue` instruction is indented under the `create h1` instruction to indicate that the results will appear inside the `h1` structure. Another way of expressing this would be:

```
match title
    <h1>
        continue
    </h1>
```

Again, it does not matter what form you use for your pseudocode as long as you clearly express the decision to make when the event in question occurs.

In this example, the processor outputs the text content of the structure automatically (as is the case in many tools), so the output of this rule (expressed in HTML) is:

```
<h1>Moby Dick</h1>
```

But suppose there is another structure inside the title? In this case that structure is an annotation of part of the title text:

```
title: Review of {Rio Bravo}(movie)
```

Here the annotated text is set off with curly braces and the annotation itself is in parentheses immediately after it. (This is a feature of the SAM markup syntax that I use for most of the examples in this book.) This annotation says that the words *Rio Bravo* refer to a movie. The annotation is a content structure, just like the title structure, and is nested inside the text of the title.

So what do you need to do to your `title` rule to make it handle `movie` annotations embedded in the title text? Absolutely nothing. Instead, you write a separate rule for handling `movie` annotations no matter where they occur:

```
match movie
    create i
        continue
```

When the processor hits `continue` in the `title` rule, it processes the content of the title structure. In doing so, it encounters the `movie` structure and executes the `movie` rule. The result is output that looks like this:

```
<h1>Review of <i>Rio Bravo</i></h1>
```

The `continue` instruction is all you need to add to your rules to allow them to deal with nested structures. The rules remain an unordered collection, just like a style sheet. (In fact, XSLT, a language that implements this model, calls a set of processing rules a "stylesheet.")

Processing based on context

When you move to the document domain, you use context to reduce the number of structures you need. For example, where HTML has six different heading structures (`H1` through `H6`), DocBook has only one (`title`), which can occur in many different contexts. So how do you apply the right formatting to a title based on its context? You create a different rule for the `title` structure in each context. You express the context by listing the parent structure names separated by slashes:

```
match book/title
    create h1
        continue

match chapter/title
    create h2
        continue

match section/title
    create h3
        continue

match figure/title
    create h4
        continue
```

Now here is the clever bit. You don't have to change the `movie` rule to work with any of these versions of the `title` rule. Suppose your title is the title of a section, like this:

```
section:
    title: Review of {Rio Bravo}(movie)
```

When the content is processed, the processor will execute the `section/title` rule to deal with the title structure and the `movie` rule when the `movie` structure occurs in the course of processing the content of the `title` structure. The results are as follows:

```
<h3>Review of <i>Rio Bravo</i></h3>
```

Here is the basic pattern for most structured writing algorithms:

- For each input structure, create a rule that says how to transform that structure into the new structure you want.
- For each rule, specify the new structure(s) and where to place the content in the new structure(s).
- For each rule, specify where to process any nested structures. In my pseudocode, I use the term `continue` for this function.
- If you want a different rule for a structure that occurs in different contexts, write a separate rule for each context.

Why is it important to understand this pattern? Because when you abstract out invariants to move content to the document domain or subject domain, understanding how those invariants will be factored back in can help you recognize them in your source and give you the confidence to factor them out. Writing pseudocode helps you validate that you have factored out invariants correctly, that the structures are easy to process, and that the processing rules are clear, consistent, and reliable.

Obviously, building a complete processing system is much more complex, and I won't go into all the details here, but let's look at a few common cases.

Processing container structures

When you move content to the document domain or the subject domain, you often create container structures to provide context. These container structures have no media-domain analog, so what do you do with them when you publish? The containers provide context for the rest of your processing rules, but what do you do with the containers themselves?

In the previous example the content was contained in a `section` structure. So how does the `section` structure get processed?

```
match section
    continue
```

Yes, it's that simple. You don't output any new structure in its place. The section container has done its work at this point so you simply discard it. You still want what's inside it, so you use the `continue` instruction to make sure the contents get processed. But the container is just a box; you unpack the contents and discard the box.

Restoring factored-out text

Sometimes when you factor out invariants, you not only factor out styles, you also factor out text. To process the content, you need to restore the text (obviously you can restore different text if you need to, which is why you factored it out in the first place).

As we saw in Chapter 10, a simple example of factoring out text is numbered and bulleted lists, where you factor out the text of the numbers and bullets. Let's look at how you create rules to put them back.

Suppose you have a document that contains these two different kinds of lists:

```
paragraph: To wash hair:
ordered-list:
    list-item:Lather
    list-item:Rinse
    list-item:Repeat

paragraph: The box contains:
unordered-list:
    list-item:Sand
    list-item:Eggs
    list-item:Gold
```

Let's write a set of rules to deal with this document. Converting this to HTML lists isn't very interesting, since HTML handles list numbering and bullets itself. Instead, let's create instructions for printing on paper. Real printing instructions get tediously detailed, so let's use the same style specification shorthand used earlier. The `paragraph` rule is simple enough:

```
match paragraph
    apply style {font: 10pt "Century Schoolbook"}
    continue
```

Now let's deal with the `ordered-list`. The ordered list structure is just a container, so you don't need to create an output structure for it. But because this is an ordered list, you need to start a count to number the items in the list. That means you need a variable to store the current count. The `$` prefix indicates that you are creating a variable:

```
match ordered-list
    $count=1
    continue
```

Then, the rule for each ordered list item outputs the value of the variable `$count` and increments it by one:

```
match ordered-list/list-item
    apply style {font: 12pt "Century Schoolbook"}
    output $count
    output ".[tab]"
    $count=$count+1
    continue
```

Every time the `ordered-list/list-item` rule is executed, the count increases by one, numbering the list items sequentially.

When the processor encounters a new numbered list, it executes the `ordered-list` rule, which resets `$count` to 1.

This rule does not match `list-item` elements that are children of an `unordered-list` element, so you need a separate set of rules for unordered lists:

```
match unordered-list
    continue

match ordered-list/list-item
    apply style {font: 12pt "Century Schoolbook"}
    output "[em dash][tab]"
    continue
```

Applying the combined set of rules produces output like this:

```
{font: 10pt "Century Schoolbook"}To wash hair:
{font: 10pt "Century Schoolbook"}1.[tab]Lather
{font: 10pt "Century Schoolbook"}2.[tab]Rinse
{font: 10pt "Century Schoolbook"}3.[tab]Repeat
{font: 10pt "Century Schoolbook"}The box contains:
{font: 10pt "Century Schoolbook"}[em dash][tab]Sand
{font: 10pt "Century Schoolbook"}[em dash][tab]Eggs
{font: 10pt "Century Schoolbook"}[em dash][tab]Gold
```

This process flattens the structure and removes the document-domain containers. You are back in the media domain, with a flat structure that specifies formatting and text.

Processing in multiple steps

You may not always want to apply final formatting to your content in a single step. When you separated content from formatting, it took several stages. You may want to put them back together in several stages. This not only results in simpler, easier-to-maintain algorithms, it allows you to reuse some of the downstream steps (nearer the media domain). So far, the examples have come from the media and document domains. Here is a subject-domain example based on Figure 10.1.

```
address:
    street: 123 Elm Street
    town: Smallville
    country: USA
    code: 12345
```

Moving this content back to the document domain should make it look like this:

```
labeled-list:
    list-item:
        label: Street
        contents: 123 Elm Street
    list-item:
        label: Town
        contents: Smallville
    list-item:
        label: Country
        contents: USA
    list-item:
        label: Code
        contents: 12345
```

Here is the set of rules to accomplish this transformation:

```
match address
    create labeled-list
        continue

match street
    create list-item
        create label
            output "Street"
        create contents
            continue

match town
    create list-item
        create label
            output "Town"
        create contents
            continue

match country
    create list-item
        create label
            output "Country"
        create contents
            continue

match code
    create list-item
        create label
            output "Code"
        create contents
            continue
```

Notice that the label text, which you factored out when you moved to the subject domain, is being factored back in and is specified in the processing rules. As you moved the content from the media domain to the document domain to the subject domain, you first factored out invariant formatting and then invariant text. Here, you restore the text and the formatting, each at a different processing stage. Of course, you can choose different label text, too.

Processing content in multiple steps can save you a lot of time. The subject-domain `address` structure is specific to a single subject, and you might have similar structures for different subjects in your subject-domain markup. However, you might choose to present those different subjects using the same `labeled-list` structure. Since a labeled list is a document-domain structure that can be used to present all kinds of information and formatted for many different media, you don't have to write any code to format the `address` structure – or any of the similar structures – directly. You can format them all correctly for multiple media using the existing `labeled-list` formatting rules.

Query-based processing

The rule-based approach shown here is not the only way to process structured writing. There is another approach that I call the query-based approach.[2] In this approach, you write a query expression that reaches into the structure of a document and pulls out a structure or a set of structures from the middle of the document.

This is a useful technique if you want to radically rearrange the content of a document or if you want to pull content out of one document to use in another. Any reference in these pages to using an algorithm to go through a content set and pull out certain pieces of data is an example of the query-based approach. You will see more examples of this in future chapters.

This chapter does not by any means describe the full range of available content processing techniques or all the ways in which algorithms can recognize and manipulate structures in content. The point of introducing structured writing algorithms and the basics of content processing is to enable you to think about structures with an eye to how they can be processed. Often when you look at structures in this way you can see how you can use a simpler structure and how that simpler structure can be processed to achieve the same result as a more complex structure.

Don't be afraid to write your algorithms in pseudocode to make sure you have a clear idea of the processing you intend for your structures, and don't feel confined by the algorithms or pseudocode used in this book. Choose a format for pseudocode that is clear to you and that communicates clearly to whoever implements the code. Working with that person can help you improve your ability to create useful algorithms and express them in pseudocode.

Learning to write code can help you write pseudocode, but it is not essential. What is essential is to think about how you can distribute parts of your system to algorithms and to be able to describe them well enough that a programmer can implement them.

---

[2] The rule-based and query-based approaches are often called "push" and "pull" methods respectively, but I sometimes find it hard to remember which is which. I find rule-based and query-based more descriptive.

CHAPTER 12
Single Sourcing

The need to publish the same document in more than one medium – for example, print and web – is a common process problem and source of complexity. Although you might not suspect it, this can also pose a significant and difficult rhetoric problem. The term *single sourcing* is used to describe a variety of different ways to address this need, all of which involve using a single body of content to create output in more than one medium.

Single sourcing adds complexity to the job of separating content from formatting because you are now separating content from not one but two or more sets of formatting. In other words, you are not simply separating content from formatting, you are separating a set of decisions about formatting from your content so you can make different decisions for each medium. If you don't get the partitioning correct, problems will show up when you try to apply two different sets of formatting to the same content.

Many organizations have unhandled complexity in this area – complexity that can result in rhetorical problems. Often, their authoring techniques do not allow writers to sufficiently separate media-specific formatting from content. As a result, formatting for one medium gets applied to another medium, which means the content in that medium does not work as well as it could. By improving how you separate content from formatting, you can significantly reduce the amount of complexity that goes unhandled, resulting in a better experience for your readers.

Basic single sourcing

Basic single-sourcing takes a piece of content in the document domain and processes its document-domain structures into a different media-domain structure for each target medium. To do this you have two or more instances of the algorithm for combining content with formatting that was described in Chapter 11.

Suppose you have recorded a recipe in the document domain (see Figure 12.1).

```
page:
    title: Hard Boiled Eggs

    A hard boiled egg is simple and nutritious.
    Prep time, 15 minutes. Serves 6.

    section: Ingredients
        ul:
            li: 12 eggs
            li: 2qt water

    section: Preparation
        ol:
            li: Place eggs in pan and cover with water.
            li: Bring water to a boil.
            li: Remove from heat and cover for 12 minutes.
            li: Place eggs in cold water to stop cooking.
            li: Peel and serve.
```

Figure 12.1 – Document-domain recipe for hard-boiled eggs

You can output this recipe to two different media by applying two different formatting algorithms. First you output to the web by creating HTML (see Figure 12.2).

```
match page
    create html
        stylesheet www.example.com/style.css
        continue

match title
    create h1
        continue

match p
    copy
        continue

match section
    continue

match section/title
    create h2
        continue

match ul
    copy
        continue
```

```
match ol
    copy
        continue

match li
    copy
        continue
```

Figure 12.2 – Pseudocode to convert document-domain recipe to HTML output

In Figure 12.2, paragraph and list structures have the same names in the source format as they do in the output format (HTML) so you just copy the structures rather than recreating them. This is a common pattern in structured writing algorithms.

The algorithm in Figure 12.2 transforms the source into HTML that looks like Figure 12.3.

```
<html>
    <head>
        <link rel="stylesheet" type="text/css"
        href="//www.apache.org/css/code.css">
    </head>

    <h1>Hard Boiled Eggs</h1>

    <p>A hard boiled egg is simple and nutritious.
    Prep time, 15 minutes. Serves 6.</p>

    <h2>Ingredients</h2>

    <ul>
        <li>12 eggs</li>
        <li>2qt water</li>
    </ul>

    <h2>Preparation</h2>

    <ol>
        <li>Place eggs in pan and cover with water.</li>
        <li>Bring water to a boil.</li>
        <li>Remove from heat and cover for 12 minutes.</li>
        <li>Place eggs in cold water to stop cooking.</li>
        <li>Peel and serve.</li>
    </ol>
</html>
```

Figure 12.3 – Result of running the algorithm in Figure 12.2

Outputting to paper (or to PDF, which is a kind of virtual paper) is more complex. On the web, you output to a screen which is of flexible width and infinite length. The browser wraps lines of text to the screen width (unless formatting commands tell it to do otherwise), and there is no issue with breaking text from one page to another. For paper, however, you have to fit content into a set of fixed-size pages.

This introduces a number of formatting issues, including the following:

- Where to break each line of text so the margins look neat and the text is not cramped or spread out too much.
- How to avoid placing a heading at the bottom of a page or the last line of a paragraph as the first line of a page.
- How to handle a list or a table when you run out of space on the page.
- How to create a cross reference to another page when you don't know the page number yet.

Because of issues like this, you don't write a formatting algorithm for paper the same the way you write an algorithm for HTML. Instead, you use an intermediate typesetting system that already knows how to handle tasks such as inserting page-number references and handling line and page breaks. Rather than handling these things yourself, you give the typesetting system general guidance and then let it do its job.

One such typesetting system is XSL-FO (Extensible Stylesheet Language - Formatting Objects). Other typesetting systems include TeX and newer versions of CSS. XSL-FO is a typesetting language written in XML. To format your content using XSL-FO, you transform your source content into XSL-FO markup, just the way you transform it into HTML for the web. But then you run the XSL-FO markup through an XSL-FO processor to produce your final output, typically PDF.

Here is a small example of XSL-FO markup:

```
<fo:block space-after="4pt">
   <fo:wrapper font-size="14pt" font-weight="bold">
     Hard Boiled Eggs
   </fo:wrapper>
</fo:block>
```

The XSL-FO code contains media-domain instructions for spacing and font choices. There is no division between content structures and formatting structures as there is with HTML and CSS. Also, as a pure media-domain language, XSL-FO does not have document-domain structures such as paragraphs and titles. To XSL-FO, a document is simply a set of blocks with specific formatting properties attached to them.

To limit the amount of detail, the algorithm in Figure 12.4 shows only a portion of the pseudocode for handling recipe markup (The point here is not to teach XSL-FO, but to show how the algorithm works).

```
match title
    output '<fo:block space-after="4pt">'
        output '<fo:wrapper font-size="14pt" font-weight="bold">'
            continue
        output '</fo:wrapper>'
    output '</fo:block>'
```

Figure 12.4 – Portion of the algorithm to generate XSL-FO from a recipe

The rule shown in Figure 12.4 simply wraps the XSL-FO formatting structures around the text of the title. It unpacks the text from a document-domain structure and inserts it into a media-domain structure.

Differential single sourcing

Basic single sourcing outputs the same document presentation to different media. However, each medium is different, and what works well in one medium does not always work as well in another. Thus the rhetoric that is appropriate to one media may not be appropriate to another, This is what I meant when I said at the beginning that single sourcing is a rhetoric problem as well as a process problem.

For example, online media generally support hypertext links, while paper does not. Suppose you have a piece of content that includes a link:

```
In Rio Bravo, {the Duke}(link "http://JohnWayne.com")
plays an ex-Union colonel out for revenge.
```

In SAM syntax, the link markup (link "http://JohnWayne.com") specifies the address to link to. In the algorithm examples below, I refer to the value in quotes (in this case the URL) as the *specifically* attribute using the notation @specifically.

In HTML, you output this as a link using the HTML a element, so the algorithm looks like Figure 12.5.

```
match p
    copy
        continue

match link
    create a
        attribute href = @specifically
        continue
```
Figure 12.5 – Algorithm to generate HTML link from the @specifically attribute

The result of this algorithm is shown in Figure 12.6.

```
<p>In Rio Bravo, <a href="http://JohnWayne.com">The Duke</a>
   plays an ex-Union colonel out for revenge.
</p>
```
Figure 12.6 – Output of the algorithm in Figure 12.5

But suppose you want to output this same content to paper. If you output it to PDF, you can still create a link as with HTML, but when that PDF is printed, all that's left is a slight color change in the text and maybe an underline. Readers cannot follow the link.

You can't print an active link on paper, but you can print the URL, so the reader can type it into a browser. Your algorithm can print the link inline or as a footnote. Figure 12.7 shows the algorithm for printing the URL inline. (I dispense with the detailed XSL-FO syntax this time.)

```
match p
    create fo:block
        continue

match link
    continue
    output " (see: "
    output @specifically
    output ") "
```
Figure 12.7 – Algorithm to display the URL for a link in print

This will produce the output in Figure 12.8:

```
<fo:block>
In Rio Bravo, the Duke (see: http://JohnWayne.com)
plays an ex-Union colonel out for revenge.
</fo:block>
```

Figure 12.8 – Output of Figure 12.8

This works, but notice that the effect is different in print (Figure 12.8) and online (Figure 12.6). Online, the link to JohnWayne.com disambiguates the phrase "The Duke" for those readers who do not recognize it. A simple click on the link will explain who "the Duke" is. But in the paper example, such disambiguation exists only incidentally, because the words "JohnWayne" happen to appear in the URL. This is not how you would disambiguate "The Duke" if you were writing for paper. You would be more likely to do something like this:

> The Duke (John Wayne) plays an ex-Union colonel.

This provides readers with less information, in the sense that it does not give them access to all the information on JohnWayne.com, but it does a better job of disambiguation, and it does so in a more paper-like way. It is a rhetoric appropriate to the media. Losing the reference to JohnWayne.com is probably not an issue here. Following that link by typing it into a browser is a lot more work than simply clicking on a web link. Someone who wants more information on John Wayne is far more likely to type "John Wayne" into Google than type "JohnWayne.com" into the browser address bar.

As written, the source content provides no easy way to produce this preferred form for paper. The content is in the document domain, which specifies the presentation of the content. Therefore, it is not a neutral format between media that require different forms of presentation. The choice to specify a link directly in the markup gives it a strong bias towards the web and online media rather than paper. A document-domain approach that favored paper would similarly lead to a poorer online presentation by omitting the link.

Because different media require a different rhetoric, you need a *differential* approach to single sourcing, one that allows you to differ not only the formatting but the presentation of the content for different media.

One way to accomplish differential single sourcing is to record the content in the subject domain, thus removing the prejudice of the document-domain presentation for one class of media or another. Here is how this markup might look:

```
{The Duke}(actor "John Wayne") plays an ex-Union colonel.
```

In this example, you annotate the phrase "The Duke" with subject-domain markup that clarifies exactly what the text refers to. That annotation says that "the Duke" is the name of an actor, specifically John Wayne.

Our document-domain examples attempted to clarify "the Duke" for readers, but they did so in media-dependent ways. This subject-domain example clarifies the meaning of "The Duke" in a presentation-neutral way; it explicitly, formally, and unambiguously marks up what the phrase refers to. This markup is written for algorithms, not readers. It allows an algorithm to generate an appropriate clarification for the reader in any target medium (see Figure 12.9 and Figure 12.10).

```
match actor
    continue
    output " ("
    output @specifically
    output ") "
```
Figure 12.9 – Algorithm to render paper from subject-domain markup

```
match actor
    create link
        $href = get link for actor named @specifically
        attribute href = $href
        continue
```
Figure 12.10 – Algorithm to web links from subject-domain markup

This assumes the existence of a system that can respond to the `get link` instruction and look up pages to link to based on the type and a name of a subject. I look at such a system in Chapter 18, which revisits this example in more detail.

Differential organization and presentation

Differences in presentation between media can be broader than this. Paper documents sometimes use complex tables and elaborate layouts that don't translate well to online media. Effective table layout depends on knowing the width of the page you have available, and online you don't know that. A table that looks great on paper may be unreadable on a mobile device, for instance.

And this is more than a layout issue. Sometimes the things that paper does in a static way should be done dynamically in online media. For example, airline and train schedules have traditionally been printed as timetables on paper, but you rarely see them presented that way online. Rather, you see an interactive travel planner that lets you choose your starting point, destination, and desired travel times and then presents you with the best schedule, including when and where to make connections. If your single source is a timetable designed for print and PDF, you can't produce the kind of online presentation of your schedule that people expect, and that can have a direct impact on your business.

To single source schedule information for both paper and online media, you can't maintain that content in a document-domain table structure. You need to maintain it in a timetable database structure (which is in the subject domain, even though it looks like a database, not a document).

An algorithm can read such a database to generate a document-domain table for print publication. For the web, however, you can create a web application that queries the same database dynamically to calculate routes and schedules for individual travelers.

In this scenario, you move content to the subject domain to partition the presentation from the information, which allows you to produce a different information design for different media (see Figure 12.11). I examine these scenarios in more depth in Chapter 23.

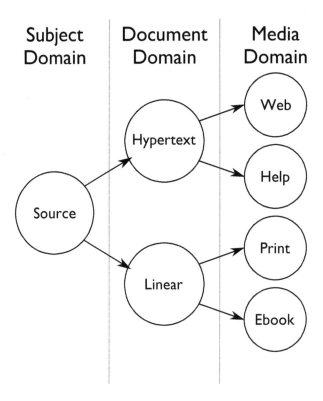

Figure 12.11 – Information design for single sourcing

Conditional differential design

You can also do differential single sourcing by using conditional management-domain structures in the document domain.

For instance, if you are writing a manual that you intend to single source to a help system, you might want to add context setting information to the start of a section when it appears in the help system. The manual may be designed to be read sequentially, meaning that the context of individual sections is established by what came before. But help systems are always accessed randomly, meaning that the context of a particular help topic may not be clear if it was single sourced from a manual. To accommodate this, you could include a context-setting paragraph that is conditionalized to appear only in help output (see Figure 12.12):

```
section: Wrangling left-handed widgets

    ~~~(?help-only)

        Left-handed widgets are used when wrangling
        counter-clockwise.

    To wrangle a left handed widget:

    1. Loosen the doohickey using a medium thingamabob.
    2. Spin three times around under a full moon.
    3. Touch the sky.
```

Figure 12.12 – Differential single source example using management-domain structures

In Figure 12.12, the SAM markup ~~~ creates a fragment structure to which conditional tokens can be applied. Content indented under the fragment marker is part of the fragment.

To output a manual, you suppress the help-only content:

```
match fragment where conditions = help-only
    ignore
```

To output help, you include it:

```
match fragment where conditions = help-only
    continue
```

Primary and secondary media

While differential single sourcing can help you output documents with the appropriate rhetoric in multiple media, there are limits to how far this approach can take you.

Books are a linear medium in which one chapter follows another in order. The web is a hypertext medium in which every page is page one. Readers arrive at a page via search, and they navigate by following hypertext links. The linear and hypertext approaches are fundamentally different ways of writing that invite fundamentally different ways of navigating and using information.[1]

---

[1] For a full examination of the differences, see my book *Every Page is Page One: Topic-based Writing for Technical Communication and the Web.*

Even moving content to the subject domain as much as possible does not entirely factor out these fundamental differences in approach.

When single sourcing content to both linear paper-like media and hypertext web-like media, you generally have to choose a primary medium. Single sourcing that content to other media will be on a best-effort basis. It may be good enough for a particular purpose, but it will never be as good as it could have been had you designed exclusively for that medium.

Many of the tools used for single sourcing have a built-in bias towards one medium or another. Desktop-publishing tools such as FrameMaker, for instance, were designed for linear media. Online collaborative tools such as wikis were designed for hypertext media. It is usually a good idea to pick a tool that was designed for the medium you choose as primary. Tools partition the complexity of content creation to best suit the primary medium they were designed for.

Often, the tools a group uses implicitly determine the primary medium. This usually means that the primary medium is paper, and paper often continues to be primary even after the organization stops producing paper, and their readers have moved to online formats.

Some organizations think they should not switch to tools that are designed primarily for online content until they have entirely abandoned paper and paper-like formats such as PDF. But once online becomes your primary media, your rhetoric should change to a rhetoric appropriate to online media (potentially a very large change, as discussed in Chapter 8). Thus it is appropriate to switch to a rhetoric and tool set that is optimized for online but that can still produce linear media as a secondary output format.

Changing your information design practices from linear, paper-based designs to hypertext designs is not trivial, but such designs better suit the way many people use and access content today. I discuss these design differences in depth in my book, *Every Page is Page One: Topic-based Writing for Technical Communication and the Web*. Don't expect single sourcing, by itself, to successfully turn document-oriented design into effective hypertext. It is usually more effective to adopt an Every Page is Page One approach to information design and use structured writing and single-sourcing techniques as a best-effort method to create linear media for those of your readers who still need paper or paper-like formats – or for those documents where linear organization still makes sense.

Single sourcing vs single format

Finally, adopting a single source for information on a given subject is not the same thing as adopting a single-source format for all content. Often, the best way to achieve sophisticated differential single sourcing is through subject-domain structures that are particular to an individual piece of subject matter. But, unfortunately, most systems designed and sold for doing single sourcing of content are implemented using a single-source format for all content.

It is not hard to see why. With a single-source format – invariable in the document domain with some management-domain structures added – you can develop a single algorithm to transform the common format to multiple output formats. But, such a system cannot do many of the things needed to implement true differential single sourcing.

There is nothing wrong with a single document-domain/management-domain format feeding a set of formatting algorithms for different media, but if you want to do differential single sourcing, it should not be the format writers work in. Instead, it should be used as an intermediate step in a shared pipes[2] architecture. I look at the elements of such an architecture in Chapter 19.

---

[2] https://www.scriptorium.com/2018/04/single-sourcing-dead-long-live-shared-pipes/

CHAPTER 13
Reuse

Another source of complexity in content creation occurs when you want the same information to occur in more than one publication. If there is no coordination between writers, each writer who wants to include that piece of information will research and write it independently. Individual writers who want to use the same information in more than one publication will copy the information from one publication to another. However it happens, you now have two or more instances of the same information that you have to maintain and edit whenever the subject matter changes. If some instances do not get updated, or some of them get updated incorrectly, that is content maintenance complexity that is not being handled, and, as always, it falls through to the reader in the form of inconsistent or incorrect information.

Content reuse is an attempt to handle the complexity associated with using the same information in more than one place. Reuse has become one of the main drivers of structured writing, particularly with the widespread adoption of DITA. Unfortunately, most reuse techniques also introduce a lot of management-domain complexity. A single-minded focus on reuse has sometimes led to the implementation of systems that dump large amounts of complexity on writers.

If you are not careful, reuse can lead to more unhandled complexity than you started with, resulting in both process problems and significant damage to rhetoric. Reuse is also an area in which process and rhetorical goals can easily be at odds; although using the same content for multiple purposes can deliver process benefits, it may not be the best rhetorical strategy.

It is important to remember that content reuse is a means for creating duplication in content. It eliminates duplication on the writing side, but it creates duplication on the output side. It is important to evaluate why you want that duplication at all and see if there are viable alternatives, such as publishing the information once and linking to it from various places. Here are some scenarios where you might want to duplicate information in more than one place:

- You are writing about different products that share common technology, and you want to duplicate information on the common features in the documents for each individual product.
- You are writing about several different releases of a product, and you want information on features that have not changed to be duplicated in the documents for the next release.

- You are writing documents for different audiences (marketing material, technical documentation for various roles, training material, etc.), and you want the same product descriptions in all these documents.
- You are writing general information (e.g., copyright and trademark statements), and you want to duplicate that information in many different publications.

The term *reuse* can suggest that this activity is akin to rummaging through that jar of old nuts and bolts you have in the garage looking for one that is the right size to fix your lawnmower. While you can do it that way, that approach is neither efficient nor reliable. The efficient and reliable approach involves deliberately creating content for duplication in multiple locations. This means that you need to place constraints on the content to be reused and the content that reuses it, and that means you are in the realm of structured writing.

Fitting pieces of content together

To create one piece of content that can be used in many outputs, you have to make sure it fits in each of those outputs. In other words, you have to partition it appropriately for reuse.

If you cut and paste, this is not a concern. You can cut any text you like, paste it in anywhere, and edit it to fit if you need to. But if you reuse a single instance of text that is already used in other places, you can't edit it to fit because that might cause it to no longer fit in the other places. For reuse to work, the content must be written to fit in multiple places. In other words, it has to meet a set of constraints that ensure that it will fit in multiple places. I will look at this in more detail in Chapter 28. In this chapter I focus on the algorithms for fitting the pieces together.

There are seven basic models for fitting pieces of content together:

- Common into variable
- Variable into common
- Variable into variable
- Common with conditions
- Factor out the common
- Factor out the variable
- Assemble from pieces

Common into variable

In the common-into-variable case, you have a common piece of content that occurs in many places. The common content could occur in many documents, in many places in the same document, or both (see Figure 13.1).

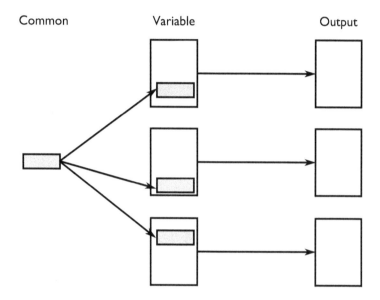

Figure 13.1 – Common-into-variable diagram

There was an example of this in Chapter 5, where a set of dangerous procedures each required a standard warning. Each individual procedure is the variable part and the standard warning is the common part (see Figure 13.2).

```
procedure: Blow stuff up
    >>>(files/shared/admonitions/danger)
    step: Plant dynamite.
    step: Insert detonator.
    step: Run away.
    step: Press the big red button.
```

Figure 13.2 – Common-into-variable markup example

To ensure that the included content will always fit, you need to make sure that there is a clear partitioning of responsibilities between the common content and each place it will be inserted into. In Figure 13.2, the inserted content should be the safety warning, the whole safety warning, and nothing but the safety warning. The procedure structure should describe the steps and only the steps and should insert the reusable warning at the right place.

You can also use the subject-domain approach to the common-into-variable case, which I described in Chapter 5. I will go into more detail on that approach later in this chapter.

Variable into common

With the variable-into-common technique, you have a single document that will be output in many different ways by inserting variable content at certain locations (see Figure 13.3).

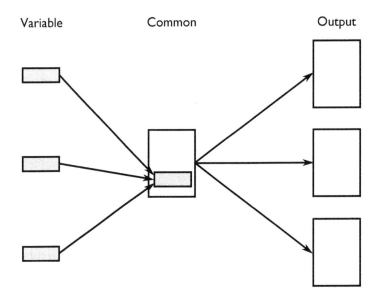

Figure 13.3 – Variable-into-common diagram

For instance, if you are writing a manual to cover a number of car models, you can factor out the number of seats each model has.

```
The vehicle seats >($seats) people.
```

This is the fixed content that will occur in all manuals, with the number of seats pulled in from an external source. Let's say you have a collection of vehicle data that is stored in a structure like Figure 13.4.

```
vehicles:
    vehicle: compact
        seats: four
        colors: red, green, blue, white, black
        transmissions: manual, CVT
        doors: four
        horsepower: 120
        torque: 110 @ 3500 RPM
    vehicle: midsize
        seats: five
        colors: red, green, blue, white, black
        transmissions: CVT
        doors: four
        horsepower: 180
        torque: 160 @ 3500 RPM
```

Figure 13.4 – Variable-into-common data structure example

Figure 13.5 shows an algorithm that selects the correct insert by querying the structure in Figure 13.4.

```
match insert where variable = $seats
    $number_of_seats = vehicles/vehicle[$model]/seats
    output $number_of_seats
```

Figure 13.5 – Pseudocode for the variable-into-common algorithm

Of course, the insert and query mechanism in Figure 13.5 is pseudocode. Exactly how things work and exactly how you delineate, identify, and insert content will vary from system to system.

With the variable-into-common technique, you create a common source by factoring out all the parts of the different outputs that are not common. This is, in some ways, the inverse of the usual pattern of factoring out invariants: you are actually factoring out the variants. But, really, it amounts to the same thing. You are factoring variants from invariants. The only real difference between this and the common-into-variable technique is whether the common parts are embedded in the variable parts or vice versa. Either way, you still end up with two artifacts: the variable piece or pieces and the common piece or pieces.

Variable into variable

The variable-into-variable technique is a variation on common-into-variable reuse in which you select different common elements to pull into a set of variable documents.

For example, suppose you decide to sell your product line in a new market. The new market has different safety regulations, which means you need to insert a different standard warning into your manuals for that market. In this case, you want to swap out the common elements used in your home market and substitute the common elements for the foreign market (see Figure 13.6).

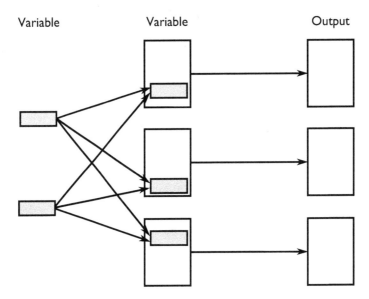

Figure 13.6 – Variable-into-variable diagram

In the common-into-variable example, you inserted the contents of a file that contained a standard warning. But for variable-into-variable reuse, this approach does not work. Variable-into-variable reuse requires you to load a different file for each market, but the content specifies a particular file name, making it difficult to select a different file for the new market.

As always in structured writing, you look for a way to factor out the problematic content. So here, you look for a way to factor out the file name and replace it with something else.

Using IDs

The most basic way to factor out the file name is to give the file an ID. An ID is a management-domain structure used to identify a piece of content in a location-independent way. No matter where the content is stored, it keeps the same ID. Figure 13.7 shows the warning file with the ID #warn_danger added.

```
warning:(#warn_danger)
    title: Danger

    Be very very careful. This could kill you.
```

Figure 13.7 – Danger warning with an ID added

You can then insert the warning into your procedure by referring to that ID (see Figure 13.8).

```
procedure: Blow stuff up
    >>>(#warn_danger)
    step: Plant dynamite.
    step: Insert detonator.
    step: Run away.
    step: Press the big red button.
```

Figure 13.8 – Procedure that refers to a warning using an ID

The decision about which warning to include has been shifted from the writer to the algorithm in Figure 13.9.

```
match insert with ID
    $insert_content = find ID in $content_set
    output $insert_content
```

Figure 13.9 – Pseudocode to locate a warning using its ID

This is a constant pattern in structured writing. When it comes to choosing and locating resources, you want to move that responsibility from the writer to the algorithm, which means moving the identity of the resource out of the content and into the algorithm. This makes it easier to update locations, but it also gives you far more options for storing and managing your content, since algorithms can interact with a variety of systems in sophisticated ways, rather than just storing a static address. It also means you can make wholesale changes in how your content is stored without having to edit the content itself. This is a major win in terms of partitioning complexity so it can be distributed and handled more efficiently.

This method requires that the algorithm have a way to resolve the ID and find the content to include. In many cases, a content management system can resolve the ID. In other cases, you can do something as simple as having the algorithm search through a set of files to find the ID or building a catalog that points to the files that contain content with IDs.

To do variable-into-variable reuse in a system that uses IDs, you maintain a separate set of files that contain the variable content for each market. Each set uses the same IDs but different content. You simply point the algorithm at the set for the market you are generating output for. So if your foreign market requires a different warning, you can create a file that contains the message shown in Figure 13.10.

```
warning:(#warn_danger)
    title: Look out!

    Pay close attention. You could really hurt yourself.
```
Figure 13.10 – Alternate warning message for a different market

By telling the algorithm that builds the foreign market docs to search this file for IDs, rather than the file with the domestic market warning, you automatically get the foreign warning rather than the domestic one.

Using keys

Another way to do this is with another management-domain structure called a *key*. A key is similar to an ID, but it is not directly tied to a resource. Instead, the same key can point to different resources at different times. Since a key does not represent any one concrete resource, you don't assign the key to a resource. Instead, you use an intermediate lookup table to resolve keys to particular resources for a particular purpose.

Suppose you have the warning in a file called `files/shared/admonitions/domestic/danger` with the content shown in Figure 13.11 (no ID):

```
warning:
    title: Danger

    Be very very careful. This could kill you.
```
Figure 13.11 – Warning content in a key-based system

Figure 13.12 shows a procedure that includes the warning using the key %warn_danger.

```
procedure: Blow stuff up
    >>>(%warn_danger)
    step: Plant dynamite.
    step: Insert detonator.
    step: Run away.
    step: Press the big red button.
```

Figure 13.12 – Procedure that includes a warning using key-based linking

(These examples use # to denote IDs and % to denote keys. This is the notation that SAM uses for IDs and keys, but it is purely arbitrary and has nothing to do with how they work. Different systems denote IDs and keys in different ways.)

To connect the key to the warning file, you create a key lookup table (see Figure 13.13).

```
keys:
    key:
        name: warn_danger
        resource: files/shared/admonitions/domestic/danger
```

Figure 13.13 – Key lookup table

When the algorithm in Figure 13.14 processes the procedure, it sees the key reference %warn_danger and looks it up in the key lookup table. The key lookup table tells the algorithm that the key resolves to the resource files/shared/admonitions/domestic/danger. The algorithm then loads that file and inserts the contents into the output.

```
match insert with key
    $resource = find key in lookup-table
    output $resource
```

Figure 13.14 – Pseudocode for the key lookup algorithm

To output your content for the foreign market, you prepare a new key lookup table (see Figure 13.15) and tell the algorithm to use this table instead.

```
keys:
    key:
        name: warn_danger
        resource: files/shared/admonitions/foreign/danger
```

Figure 13.15 – Key lookup table for a new market

Using keys is not necessarily better than using IDs. What it comes down to is that you need some kind of bridge between the citation of an identifier in the source file and the location of a resource with that identifier in the content store. This bridge can be created by a key lookup table, by remapping file URLs, or by modifying a query to a content repository.

One advantage of keys is that, because you don't attach the key directly to the content, you can use a key to identify resources that don't have IDs, including resources you don't control.

One downside of keys is that, by themselves, they can only point to a whole resource. This can force you to keep your reusable units in separate files. To avoid this, you can combine keys with IDs. Figure 13.16 combines both danger warnings into one file and gives each an ID.

```
warnings:
    warning:(#warn_danger_domestic)
        title: Danger

        Be very very careful. This could kill you.

    warning:(#warn_danger_foreign)
        title: Look out!

        Pay close attention. You could really hurt yourself.
```

Figure 13.16 – Warnings file with different versions identified by IDs

Now you can rewrite your key lookup tables to use IDs to pull the right warning out of the common file. Then you can use a key lookup table like Figure 13.17 for the domestic build and Figure 13.18 for the foreign build.

```
keys:
    key:
        name: warn_danger
        resource: files/shared/warnings#warn_danger_domestic
```

Figure 13.17 – Key lookup table with ID for domestic market

```
keys:
    key:
        name: warn_danger
        resource: files/shared/warnings#warn_danger_foreign
```

Figure 13.18 – Key lookup table with ID for foreign market

This method partitions the warnings into separate files and also partitions the location of those files from the writer. Using keys as a bridge between two partitions can be a convenient way to manage content relationships without having to update source files every time the relationships change.

The downside is that keys introduce an abstract element into the writer's world, and abstractions are a form of complexity that can be difficult to deal with. Instead of deciding which warning to use, writers have to decide which key or ID to use. That decision is not necessarily easier to make, and it requires more system knowledge and a more abstract way of thinking than the original decision that was factored out. Keys and IDs abstract out the actual act of inclusion, but that is the least complex part of ensuring that the standard warning is used.

Common with conditions

In some implementations of the variable-into-common approach, the variant pieces are not factored out into a separate file. Rather, each of the possible alternatives is included in the file conditionally (see Figure 13.19).

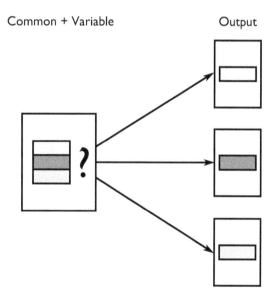

Figure 13.19 – Common-with-conditions diagram

For instance, in content for a car manual you might have conditional text for the number of people the car seats.

```
The vehicle seats {four}(?compact){five}(?midsize){seven}(?van).
```

Here the main text is the fixed piece and the variable pieces are the words "four," "five," and "seven." Which of these will be included in the output depends on which condition is applied during the build. If the condition midsize is applied, then the output text will be "five" and the other alternatives will be suppressed. Figure 13.20 shows this algorithm.

```
match phrase with condition
    if condition in $build_conditions
        continue
    else
        ignore
```

Figure 13.20 – Pseudocode for the common-with-conditions algorithm

The upside of the conditional approach is that it keeps all the variants in one file, so your algorithm does not have to know where to go to find the external content. But there are downsides to this approach:

■ It gets cumbersome to read the source if you apply many different conditions.
■ When the subject matter changes, you have to find all the places where conditions occur and update them.
■ If you refer to the same data point (the number of seats) in many different documents, that information is still being duplicated all over the content, which makes it hard to maintain, verify, or change if, for example, the compact seats five in the next model year.

Common with conditions is not limited to cases where there are alternate values, however. In some cases, content may simply be inserted or omitted for certain outputs (see Figure 13.21).

```
The main features of the car are:

ol:
    li: Wheels
    li: Steering wheel
    li:(?deluxe) Leather seats
    li: Mud flaps
```

Figure 13.21 – Common with conditions used for optional content

In this case, the list item "Leather seats" is only published if you specify the condition `deluxe` in the build. It would be omitted for all other builds. In cases like this, it is hard to avoid using conditionals as a reuse mechanism.

This approach to reuse is often called filtering or profiling. Some systems have more elaborate ways of specifying filtering or profiling. The net effect is the same as the simple condition tokens shown in this section, but such systems may allow for more sophisticated conditions.

Because common with conditions is essentially a form of variable into common where the variable content is contained inside the common source, it can technically be replaced by a variable-into-common approach in all cases. In practice, conditions are typically used in the following cases:

- The number of variations is small and thought to be fixed or to change infrequently.
- The variable pieces are eccentric or contextually dependent.
- The writer or organization wishes to avoid managing multiple files.
- The current tools don't support variable into common.

How successful a common-with-conditions approach will be also depends on what you choose for your conditional expressions. Generally, subject-domain conditions will be more stable and manageable than document-domain conditions. For instance, conditions that relate to different vehicles (subject domain) are based in the real world and are, therefore, objectively true as long as the subject matter remains the same. Conditions that relate to different publications or different media, on the other hand, are not objectively true and can't be verified independently. The only way to verify them is to build the different documents or media and see if you get the content you expected. This makes maintaining such conditions cumbersome and error prone – an indication that complexity is not being distributed in an optimal way.

Factor out the common

In Chapter 5, I noted that the subject-domain alternative to using an insertion instruction for the warning text was to specify which procedures were dangerous, thus factoring out the constraint that the warning must appear (see Figure 13.22).

```
procedure: Blow stuff up
    is-it-dangerous: yes
    step: Plant dynamite.
    step: Insert detonator.
    step: Run away.
    step: Press the big red button.
```

Figure 13.22 – Procedure that factors out explicit inclusions

In this case, writers do not have to identify the material to be included, either directly by file name or indirectly through an ID or a key. Instead, it is up to the algorithm to include it (see Figure 13.23).

```
match procedure/is-it-dangerous
    if is-it-dangerous = 'yes'
        output files/shared/warnings#warn_danger_domestic
```

Figure 13.23 – Pseudocode for the factor-out-the-common algorithm

To produce the foreign market version of the documentation, you simply edit the rule, as shown in Figure 13.24.

```
match procedure/is-it-dangerous
    if is-it-dangerous = 'yes'
        output files/shared/warnings#warn_danger_foreign
```

Figure 13.24 – Pseudocode for an alternative warning using the factor-out-the-common algorithm

Or, to further partition complexity in the code, you can use keys, as shown in Figure 13.25.

```
match procedure/is-it-dangerous
    if is-it-dangerous = 'yes'
        $resource = find key '%warn_danger' in lookup-table
        output $resource
```

Figure 13.25 – Pseudocode for the factor-out-the-common algorithm using keys

The beauty of this approach is that the content is entirely neutral as to what kind of reuse is going on or how dangerous procedures are treated. Because the content contains only objective information about the procedure itself, you can implement any publish or reuse algorithm you like – in any way you like and at any time. Because the content does not specify any form of reuse or any reuse mechanism, you have made it more reusable and partitioned the complexity of reuse much more neatly and reliably.

This approach also makes the content much easier to write, since it does not require writers to know how the reuse mechanism works, how to identify reusable content, or even that reuse is occurring at all. All they have to do is answer a simple question about the content that they should know the answer to: is the procedure dangerous or not.

This approach partitions the entire reuse mechanism away from writers. This is a big win because the biggest problem with most reuse techniques is the amount of complexity they add for writers, which directly compromises their finite and valuable attention.

This is important from the point of view of both complexity and cost. Whenever writers are asked to consciously reuse content, they have to look for that content every time the potential for reuse occurs. This cost is incurred whether or not they find reusable content, whereas any savings from reuse are realized only when reusable content is found.

The factor-out-the-common approach relieves writers of all responsibility for the reusable content. Locating reusable content is the job of an algorithm. If the content does not exist, the algorithm will report that it is missing, and it will be somebody's job to create it, after which the algorithm will locate it automatically every time it is needed. This is far more efficient than having writers look for reusable content over and over and over.

The downside of this approach is that it is not as general. The `is-it-dangerous` metadata applies only to dangerous procedures. It does not address the inclusion of reusable content in other places. You would need to factor out other interesting reuse cases in a similar way to create a complete subject-domain solution. Again, we see that complexity always has to go somewhere. But as we have also seen, if writers cannot fully handle the complexity thrust on them, that complexity goes unhandled, with consequences for both process and rhetoric.

Factor out the variable

You can also factor out the variable content. In the case of the different models of a car, rather than conditionalizing the list of features in the document, you can maintain the list in a database. The organization probably already has a database of features for each vehicle, so you don't need to create anything new. Instead, you simply query the existing database. After all, reusing what already exists is what reuse is all about.

For example, instead of including the feature list, as shown in Figure 13.26, you can factor out the list entirely, as shown in Figure 13.27.

```
The main features of the car are:

ol:
    li: Wheels
    li: Steering wheel
    li:(?deluxe) Leather seats
    li: Mud flaps
```

Figure 13.26 – Conditionalized feature list

```
The main features of the car are:

>>>(%main_features)
```

Figure 13.27 – Factored-out feature list

Now, your algorithm looks like Figure 13.28.

```
match insert with key
    $resource = lookup key in lookup-table
    output $resource
```

Figure 13.28 – Pseudocode for the factor-out-the-variable approach

You then have a key lookup table where the resource is identified by a query on the database (Figure 13.29).

```
keys:
    key:
        name: %main-features
        resource: from vehicles select features where model = $model
```

Figure 13.29 – Key lookup table for the factor-out-the-variable approach

This query retrieves a different set of features from the database depending on how the variable $model is defined for the build. Launch the build with $model = 'compact' and you get the feature set for the compact model. Launch the build with $model = 'van' and you get the feature set for the van model.

Naturally, this leaves out a lot of detail about how this query gets executed and how the results get structured into a document-domain list structure. But those are implementation details.

Assemble from pieces

In the assemble-from-pieces approach, there is no common vs. variable distinction and no single source document into which you insert reused content or apply conditions. Instead, you assemble a set of content units to form a finished document (see Figure 13.30).

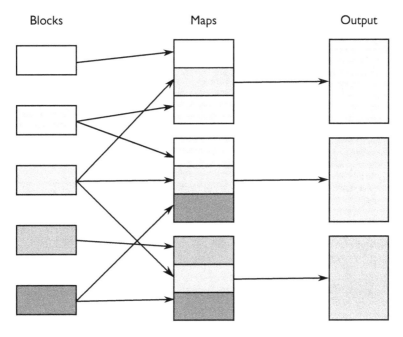

Figure 13.30 – Assemble-from-pieces diagram

For example, if you have a range of products with common features, you might assemble the documentation for those products using a common introduction followed by a piece representing each feature of each model. This piece could be a flat list, or it could be a tree structure. For instance, you might assemble a chapter of a manual with an introductory piece and then several sections below it in the tree, as shown in Figure 13.31.

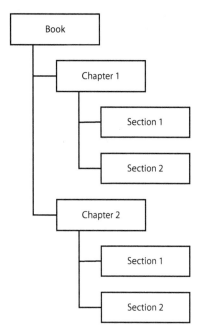

Figure 13.31 – Assemble-from-pieces tree structure

The assembly approach requires a structure to describe how the units are assembled. This structure is often called map. (It is called a map in DITA, for instance.) Some applications refer to it as a table of contents. Figure 13.32 is an example of a map.

```
map: Widget Wrangler Deluxe User Manual
    unit: units/ww/deluxe/intro
        unit: units/ww/shared/basic_features
        unit: units/ww/deluxe/deluxe_features
    unit: units/ww/shared/install/intro
        unit: units/ww/shared/requirements
        unit: units/ww/deluxe/requirements
        unit: units/ww/shared/install
        unit: units/ww/deluxe/install_options
```

Figure 13.32 – Assemble-from-pieces map

Rather than using a map, you can allow the units themselves to pull in other units, which may in turn pull in other units. So the Widget Wrangler Deluxe install introduction unit might look like the example in Figure 13.33.

```
unit: Installing the Widget Wrangler Deluxe

    You should be very careful when installing the
    Widget Wrangler Deluxe. Follow these steps carefully:

    >>>(unit units/ww/shared/requirements)
    >>>(unit units/ww/deluxe/requirements)
    >>>(unit units/ww/shared/install)
    >>>(unit units/ww/deluxe/install_options)
```

Figure 13.33 – Alternate structure for the assemble-from-pieces approach

This method avoids the need for a map, but it can make the units less reusable. In Figure 13.33, you would need a separate introduction unit for the regular Widget Wrangler, since the introduction file imports all the requirements and procedural units. By assembling units with the map, you can use a shared install introduction, which increases the amount of reuse you can do.

Combining multiple techniques

There is one problem with using a common install introduction for both the regular and the deluxe Widget Wrangler. The introduction mentions the name of the product. To solve this problem without requiring two different units, you can use the variable-into-common or common-with-conditions reuse techniques. Figure 13.34 shows an example using variable into common:

```
unit: Installing the >($product_name)

    You should be very careful when installing
    the >($product_name). Follow these steps carefully:

    >>>(unit unit/ww/shared/requirements)
    >>>(unit unit/ww/deluxe/requirements)
    >>>(unit unit/ww/shared/install)
    >>>(unit unit/ww/deluxe/install_options)
```

Figure 13.34 – Combination of the assemble-from-pieces and variable-into-common approaches

There are a number of ways to mix and match the basic reuse patterns to achieve an overall reuse strategy. Most systems designed to support reuse allow you to use all of these approaches and combine them as needed. However, don't lose sight of the amount of complexity you introduce into your content and your writers' workload when you rely on a complex set of reuse techniques.

Content reuse is not a panacea

Content reuse can seem like an easy win, and in some cases it can return substantial benefits, but there are pitfalls to be aware of. You need to plan carefully to make sure that you avoid the traps that await the unwary.

Many reuse techniques introduce a lot of complexity into the writer's job and into the content itself. These techniques may address major sources of content complexity, but they come with a high cost in terms of new complexity introduced. Even if a reuse tool looks easy to use, it can introduce large amounts of complexity.

Rhetoric traps

Designing content to be reused can harm its rhetoric, either by distracting writers with complexity or by motivating them to prefer reusability over rhetorical quality. There are three main rhetoric traps with content reuse:

- Making content too generic
- Losing the narrative flow
- Failing to address the audience appropriately

Many works on content reuse casually recommend making content more generic or more abstract as a means to making it more reusable, without saying anything about the potential downside. This is very dangerous and can do serious harm to the rhetorical quality of your content. Specific, concrete statements are easier to understand, and they communicate better than generic or abstract statements. Replacing specific and concrete statements with generic or abstract statements will reduce the effectiveness of your content significantly. This is a classic case of one process directing complexity away from itself without regard for the effect on the whole.

Unfortunately, as noted in Chapter 9, human beings suffer from the curse of knowledge. The curse of knowledge is a cognitive bias that makes it very hard for people who understand an idea to appreciate the difficulties that idea presents to people who do not understand it. The curse of knowledge makes a generic or abstract statement of an idea appear to the writer to be as clear as a concrete and specific statement – and perhaps more succinct and precise. This pulls writers away from the kind of specific and concrete statements that make ideas easier to comprehend. The desire to make content reusable reinforces this temptation.

Replacing the specific and concrete with the generic and abstract always reduces content quality and effectiveness. You may decide that the economic benefits of content reuse outweigh the economic costs of less effective content, but you should at least be aware that there are real economic consequences to this choice.

Another potential quality problem comes with the loss of narrative flow. Not all content has or needs a lengthy narrative flow, but if you start breaking your content into reusable units and putting those units back together in different ways, the narrative flow can easily be lost. In some cases you can avoid this problem by making your topics more self-contained, using an Every Page is Page One information design. But don't assume that you have an effective Every Page is Page One design just because you have broken your content into reusable units. If that content was written in a way that assumed a narrative flow, it won't work when reused in a way that breaks that flow.

Finally, reuse can encourage you to come up with one way of telling our story that you present to all your audiences. But not all audiences are alike, and the way you tell your story to one audience may not work for another audience. Good rhetoric tells a good story to a particular audience. Two different tellings of the same story do not constitute redundant content if they address different audiences.

Cost traps

It is easy to see content reuse as a method to achieve big cost savings. Reusing content means you do not have to write the same content over and over again. It is easy to add up the cost of all that redundant writing and regard that number as pure cost savings from a content reuse strategy.

But all of the reuse techniques create multiple artifacts – including content and algorithms – that need to be managed. You need a mechanism to make sure that your content obeys the constraints required to make the pieces of content fit together reliably. You need a mechanism to make sure that your reuse processes produce the documents you need. The cost of such management can be non-trivial, and the consequences if the management breaks down can be significant.

While reuse is supposed to reduce the cost of modifying content when the subject matter changes, there are hidden cost traps here as well. It is often not until the subject matter changes that you find out if the content you have treated as common is really common. If not, you have to sort out what is common and what isn't, which can involve complex edits that have to be tested and verified.

If you get everything right, you can realize major savings when it comes time to modify your content, but if you get it wrong, your costs can multiply. Not only that, you may find that the money you spent factoring duplicate content out in anticipation of a change that did not happen is wasted as well. Reacting to changes that actually happen is sometimes cheaper than preparing for multiple changes that never happen.

If you do not audit and validate your content collection regularly, it can become chaotic over time and lose cohesion. This can make adding new content or changing existing content increasingly difficult and expensive.

Depending on the techniques you use, content reuse strategies can complicate the lives of writers, which may reduce the pool of writers you can use or reduce their productivity.

With many reuse techniques, writers have to look for reusable content frequently. Not only does this take time and add complexity, writers have to pay that cost every time they look for content to reuse, whether they find it or not. Indeed, if writers must look for reusable content before they write anything, much of their time may be taken up with unsuccessful content queries. Reducing the expense of content queries, therefore, has to be a major component of any general reuse strategy. I look at this in more detail in Chapter 38.

Localized and constrained forms of reuse can avoid this cost trap. For instance, if your reuse is focused on producing manuals for different versions of a product, you can train writers on techniques specifically for this purpose, thus avoiding the need for them to make frequent, broad-based queries to find reusable content.

Some content reuse techniques are easy to use in non-structured ways, and early in a project, it may seem like a non-structured approach to reuse speeds things up by allowing writers to reuse content wherever they find it. Over time, however, this approach can lead to a rat's nest of dependencies and relationships that makes it hard to update the content with any confidence.

Once the cohesion and discipline of a content set starts to break down, the decline can accelerate. As it becomes harder to find content to reuse, more duplication occurs, which further complicates the search for reusable content, creating a vicious cycle. As links and other content relationships break down, people tend to form ad hoc links and relationships to get a job finished, further tangling the existing rat's nest. Under the gun, it is almost always easier to get the next document out by ignoring the discipline of the content set structure, but the effects of this are corrosive. Without consistent discipline, even in the face of deadlines, a reuse system can fail over time.

All of these issues can be managed successfully with the right techniques and the right tools, but they all introduce costs as well, both up-front costs and ongoing costs. Those costs have to be calculated and subtracted from the projected cost savings before you can determine if a content reuse strategy is going to save money.

One final trap: there may not be as much potential reuse as you thought. Some organizations have plunged into reuse strategies in hopes of a big payoff only to find that they had far less reusable content than they first thought. The cost of systems and the complexity added to the process by complex reuse techniques require a high level of reuse to pay off. If you can't achieve that level of reuse, your system will end up costing you more than it saves.

Alternatives to reuse

Here are three alternatives to content reuse:

- **Duplicate but label:** Allow information to be stored in more than one place but clearly label the information so you can easily find it when you need to do an update. If this sounds like an inferior solution, consider that reuse techniques cannot possibly eliminate all repeated information without adding more complexity than they eliminate. Therefore, no matter how much reuse you do, when you have a major change in subject matter, you will still have to search your content for varying statements of the same fact that need to be updated.

- **Reduce duplication:** Content reuse is always content duplication. Reuse is predicated on the idea that you actually want to duplicate information in multiple publications. This idea is out of date. In the paper world, you had to give each customer a book that contained all the information needed, even if the same information occurred in many different books. But in an online environment, you don't need to duplicate information to make it available to every user who needs it. Maintaining the information once and linking to it whenever it is needed is often a much less complex solution. To do this efficiently may require some of the techniques described in Chapter 18.

- **Combine and differentiate:** Another approach to reducing duplication in your content is to combine information on multiple products into a single publication and differentiate them in the text. There are limits to this, obviously, and some organization got into content reuse specifically so that they could remove the complexity of handling multiple versions in a single manual. Still, there are times when it is an appropriate solution.

CHAPTER 14
Generated Content

I have mentioned that one of the advantages of the subject domain is that it allows you to generate different types of rhetoric from a base of subject-domain data. Here, I look at the content generation algorithm in greater depth.

There is nothing new about generating content. Word processors and desktop publishing programs can generate indexes and tables of contents, for instance, and the content reuse algorithms in Chapter 13 generate content by combining smaller pieces of content or data to form larger units of content.

The fundamentals of content generation are pretty simple. You take separate pieces of content and data and combine them to form new content. Thus, when I spoke earlier about storing the ingredients of a recipe as a data structure and using an algorithm to present them as either a table or a list, that was an example of content generation.

Figure 14.1 shows an algorithm to turn that data into a list (again, in pseudocode).

```
match ingredients
    create ul
        continue

match ingredients/record
    create li
        output name
        output [tab]
        output quantity
        if unit is not 'each'
            output [space]
            output unit
```

Figure 14.1 – Algorithm to turn a recipe data structure into a list

Figure 14.2 shows an algorithm for interpolating the serving and preptime fields into the introduction of a recipe.

```
match description
    continue
    output 'Preparation time is '
    output /recipe/preptime
    output '. Serves '
    output /recipe/serves
    output '.'

match preptime
    ignore

match serves
    ignore
```

Figure 14.2 – Algorithm to interpolate serving and preptime into a recipe

But this kind of content generation does not have to confine itself to working within a single document. It can pull content from several files or assemble different collections of content to serve a common rhetorical purpose, such as compiling a low-calorie cook book based on nutrition information in a collection of recipes. And it can generate many of the elements required to build a top-down or bottom-up information architecture.

Categorization

One of the key elements of top-down information architecture is categorization. An information architect develops categories of content and develops an organizational schema (such as a table of contents) based on those categories. This may include levels of subcategories forming a hierarchical categorization scheme.

Not all categorization is hierarchical, though. In some cases content can be classified on several independent axes, allowing for the development of what is called *faceted navigation*. The easiest place to see faceted navigation in action is on a used-car site, where you can narrow down your selection using any set of criteria that matter to you, such as selecting blue convertibles or all-wheel drive vehicles with manual transmissions.

Categorization of content requires metadata to identify which category it belongs to. (Even if you just sort papers into piles, as soon as you put a label on each pile, you are adding metadata, and if you don't add a label, you will soon forget which pile is which.) Categorization may involve the addition of new metadata or it may rely on metadata that is already attached to the content. This effectively means that your categories are expressed as query statements, and those queries do not have to operate on a single piece of metadata. A query can create a category out of the conjunction of several pieces of metadata. For example, you could create a category of heart-healthy recipes by writing a query that looks at the salt, fat, and calorie metadata of a collection of recipes.

For content in the subject domain, the metadata required to assign a piece of content to a category may be inherent in its subject-domain markup. It is the nature of the subject domain to describe the subject matter. Therefore, any markup that describes the subject matter may already contain the fields you need for categorization. This is one of the attractions of the subject domain: the markup can serve many purposes, which simplifies both markup design and content authoring and often means that you don't need to create additional structures to support a new algorithm.

Relying on the subject-domain metadata already in the content, rather than creating a separate metadata record, can be a tremendous advantage, because it makes submission of content to a repository so much easier for writers. But, in some cases, it can also avoid the need for a costly content management system, since it allows the publishing algorithm to categorize content at build time without needing a separate metadata store or a separate system to manage categorization. I look more at the role of the content management system in Chapter 36.

Tables of contents

If you are creating a top-down information architecture, your structured writing system needs to generate tables of contents just as a word processor or desktop publishing application does.

Tables of contents can serve different purposes, depending on the nature of the content and the form of the output. Some describe a linear reading order, some provide a classification scheme for random access to the content, and some simply provide a list of chapters that does not necessarily imply an intended reading order.

A table of contents may seem like a document-domain structure, but it is really more of a media-domain structure, for two reasons. First, it contains specific links to specific resources at specific addresses, or specific page numbers in a paper or a virtual paper format such as PDF. Second, it is virtually always factored out in document-domain markup languages. Tables of contents are not written, they are generated.

From a structured writing point of view, what matters is how they are generated. In DocBook, for instance, it is typical to write each chapter of a book in a separate `chapter` file and then pull them together into a book using a `book` file. The order of the table of contents is then determined by the order in which the chapters are listed in the book file. The table of contents is generated by extracting chapter and section headings from the chapter files in the order they appear in the book file.

In DITA, the normal process is to assemble a book using a map file. A map file may assemble a book out of DITA topics or other maps, and this may include assembling the chapters from topics as well. In the end, though, the table of contents is generated in the same way, by traversing the document assembled by the map.

In both cases, the order of the TOC is specified by hand by the person who creates the book or map file. But there are other ways to determine the order of content in a TOC. For instance, a reference work such as an API reference may be organized by listing each library in order by name and each function in alphabetical order by name within its library, creating a table of content with two levels. You don't need to write a map or book file to create this table of contents. There is an algorithm for creating this table of contents. In fact, it is the algorithm stated earlier in this paragraph: "listing each library in order by name and each function in alphabetical order by name within its library." Figure 14.3 shows this algorithm in pseudo code.

```
create toc
    for each library sorted alphabetically
        create toc-entry library name
        for each function in library sorted alphabetically
            create toc-entry function name
```

Figure 14.3 – Generate a TOC for an API reference

Tables of contents serve different purposes. Some describe a curriculum, a designed reading order. Others are simply a means of navigation, a way to select one topic out of a collection of many. If your content is written in the subject domain, the chances are that it already contains the structures on which such classifications could be based, and again, the TOC can be generated based on the metadata already in the content.

One advantage of this approach is that if an algorithm assembles the TOC based on metadata, new content is automatically included in the TOC the next time you generate output. This simplifies the task of adding new content by avoiding the need to update multiple files or systems. This makes life easier for writers because they do not need to know how the TOC is constructed. They only have to create individual pieces of conforming content and submit them to the right location. This also avoids duplication, since the metadata the TOC generation algorithm uses is stored only in one place.

Lists

A major feature of a bottom-up information architecture is the list. Like TOCs, lists are a catalog of resources. But while a TOC is a list of resources defined by their container (contents = things in a container), a list may have any principle of organization or inclusion.

For instance, in a collection of movie reviews, you might want to include a filmography for each actor. Such a list is not only a useful piece of information, it is also an important aid for navigating a site. Maintaining such a list by hand would be laborious and error prone, especially with new movies being added to the collection all the time.

If you have your movie reviews in a structured format that lists the actors in the movie in a format accessible to algorithms, like this:

```
movie: Rio Bravo
    starring:: actor
        John Wayne
        Dean Martin
        Ricky Nelson
        Angie Dickinson
        Walter Brennan
```

you can generate the filmographies for all your actors, like this:

```
create-filmographies
    for each unique actor in movie/starring/actor
        create filmography named actor with link to actor
        for each movie where starring/actor = actor
            create entry named movie with link to movie
```

Tables of contents are a top-down information architecture device. You expect to find them at the top of the information set. Lists are a bottom-up information architecture device. You expect to find them as independent pages or as elements within a page. Thus, if your collection includes the biographies of actors, and you want each biography to include the filmography, you can omit the filmography from the subject domain version of the biography and have the publishing algorithm generate it with the output, as shown in Figure 14.4.

```
match actor-bio
    create html
        create h1 "Biography: " + actor-name
        continue
        create h2 "Filmography"
        for each movie-review where starring/actor = actor-name
            create li
                create a with attribute href
                = address of movie-review
                    output movie-name
```

Figure 14.4 – Algorithm to generate a filmography

Note the close relationship between rhetoric and navigation in Figure 14.4. The generated filmography is both content and navigation, both part of the individual topic and part of the overall navigation scheme – an example of how information architecture unites rhetoric and navigation.

Collections and selections

One of the applications of subject-domain markup that I mentioned in the recipe example is that it can be used to select content for a collection. Thus, if you capture calories and preparation time in your recipe markup, you can use that information to assemble a cookbook with a title like "Diet-Friendly Dishes You Can Make in 30 Minutes or Less." If you store seasonal information, you can create "Diet-Friendly Christmas Treats" or "Summer Suppers in 20 Minutes."

One of the most important aspects of creating collections based on subject-domain markup is that you did not have to think of those collections while the recipes were being written. Nothing ties the recipes to these publications. The recipes simply record certain significant facts about the dishes that may matter to readers. A content strategist can then dream up all kinds of collections, and, because the recipes record significant truths about the dishes in a form that is accessible to algorithms, chances are that you will be able to assemble those collections quickly and get them to press while the market demand is hot.

The subject domain is the gift that keeps on giving. You don't have to anticipate all of the possible uses for your subject-domain data, and collecting that data is relatively inexpensive since you are simply asking writers to enter information they already know in fields with concrete, specific names that are easy to understand.

Content queries

If you know what subjects the phrases in your content refer to, you can use that information to form queries to pull in additional information from other sources. For instance, let's say that you are writing about novels, and you annotate the titles of novels that are mentioned in your text:

```
{War and Peace}(book "ISBN:1400079985") is a very long book.
```

Here, the title is marked up as a book title, and, to make things more precise, an ISBN number is provided. An ISBN number is the key to a large amount of data about a published book. If you have the ISBN number, you can look up all sorts of other information. For instance, you can use the ISBN to look up publication details using a web service like ISBNdb (http://isbndb.com).

Most web services return information in XML. A hypothetical ISBN web service might return an XML document that looked like Figure 14.5.[1]

```
<book>
    <isbn>1400079985</isbn>
    <title>War and Peace</title>
    <author>Leo Tolstoy</author>
    <publisher>Vintage</publisher>
    <publication-year>2008</publication-year>
    <page-count>1296</page-count>
    ...
</book>
```

Figure 14.5 – Information returned by an ISBN web service

You could then pull pieces from that XML document and add them to your own content, thus allowing you to produce output like this:

War and Peace (Leo Tolstoy, Vintage, 2008, 1296 pages) is a very long book.

---

[1] This is not what ISBNdb returns, just a simplified example.

The algorithm to do this looks something like Figure 14.6.

```
match p/book
    $isbn = @specifically

    $book-info = get 'http://example.com/isbn/lookup?' + $isbn

    create i
        continue

    output " ("
    output $book-info/book/author
    output ", "
    output $book-info/book/publisher"/>
    output ", "
    output $book-info/book/publication-year"/>
    output ", "
    output $book-info/book/page-count"/>
    output " pages"
    output )
```

Figure 14.6 – Algorithm to look up book information using the ISBN

This basic technique opens all kinds of doors. The power of structured writing as a tool to merge information from different sources is enormous. Here are just a few of the tricks you could pull using information retrieved using the ISBN number:

- Pull in a picture of the book cover.
- Create a link to an article about *War and Peace* on your website.
- Create a link to an online bookstore where readers can buy the book. If you belong to an affiliate program for the online bookstore, you can earn a commission each time a reader follows your link and buys the book.

You can also realize major process efficiencies by capturing this kind of metadata in your content. If you can use metadata keys to pull information from external sources, writers don't have to look up that information when they write. And writers don't have to decide which book details will appear in the final output. That decision is made separately by editing the algorithm, and you can change that decision for all your existing content simply by changing the algorithm.

Having writers enter the ISBN number in the content makes writing the algorithm straightforward, and sometimes it is appropriate because you are referring to a particular edition of a book and the ISBN number is the most reliable identifier of a specific edition. But in many case it is actually too specific, and it complicates life if writers have to look up the ISBN when all they want to refer to is the novel itself, regardless of the edition. This distinction can be important. There are many other editions of *War and Peace*, in many languages. *War and Peace* is a very long book in all those editions and all those languages. The paragraph does not refer specifically to the Vintage Edition of 2008. It refers to *War and Peace* as a novel generally.

```
{War and Peace}(novel) is a very long book.
```

Here, I have replaced the `book` annotation with the more specific `novel` annotation. If you are concerned that there might be other novels named *War and Peace* by other writers, you could make the annotation more specific:

```
{War and Peace}(novel (Leo Tolstoy)) is a very long book.
```

In SAM, a phrase in parentheses inside an annotation is a namespace identifier. A namespace is a context in which a set of names is guaranteed to be unique. No author publishes more than one novel with the same name, so the name of the author is generally an adequate namespace identifier for the name of a novel. This markup is obviously easier for writers to create than an ISBN. It asks them only for the things they already know, so they won't have to stop to look anything up. That is an important part of functional lucidity.

However, without an ISBN number, can you still get the book data you want? You can, but you have to use a different query to extract it:

```
match p/novel

    $title = #content
    $author = @namespace

    $book-info = get 'http://example.com/isbn/lookup?category=novel&title='
                     + $title + '&author=' + $author
```

The only thing different about the results you will get from this query is that it may return records for more than one book (actually, for *War and Peace*, you will certainly get multiple records, since there are many editions in print). So the code that adds the book info to the content must pick one of the alternatives based on some relevant piece of publication data, such as the most recent publication date.

I could have chosen an example that did not have this kind of ambiguity (title vs. ISBN as identifier of a novel), but you will often come across issues like this in the real world. This is one of those issues that forces you to make a decision about how to correctly partition complexity in your system. You have a choice between an approach that uses a simple, easy-to-write algorithm but requires effort and research from writers versus an approach that requires more thought and effort to write the algorithm but provides greater functional lucidity for writers.

Stated like this it seems obvious which choice you should make, but in practice these decisions are often made by the people developing the algorithms, and they often choose to make their lives easier at the expense of the writers. This may seem like a detail, but when correct partitioning and distribution is at stake, decisions should not be left to one partition to make. These decisions require input from all sides and the attention of the project owners.

Personalized content

A key feature of modern web architecture is personalized content, which means content that is generated in response to either what the site already knows about you – from your account information or a transaction token such as a cookie – or the selections or entries that you make on the page.

For example, when you log into Amazon, the first page you see is crafted for you based on everything Amazon knows about your browsing and purchasing history. As you make selections, such as adding an item to your shopping cart or wish list, that information is used to generate the next page you see.

If you browse a used car site like Autotrader.com, you can select features that you are interested in (red convertibles with manual transmission under $20,000, for instance), and the next page will be generated based on that input.

The ability of a site to personalize pages depends on whether it can identify content that is relevant – based on everything it knows about the reader – and assemble that content to form a page. For this to work, the content must be easy to identify unambiguously and must fit together easily.

As we have seen, these properties are maximized when you store content in the subject domain, both because the subject domain makes the relevant metadata available and because working in the subject domain helps writers produce more consistent content that works better with personalization algorithms.

The consistency of your content is most important in personalized content applications. Because output is assembled in real time by an algorithm based on a combination of unique things you know about the reader and your content, there is no opportunity for a writer or editor to inspect the output of a personalized content publication before the reader sees it. This requires total confidence that:

- the content conforms to its constraints
- the markup expresses those constraints completely and correctly
- the algorithm correctly processes and delivers the content

All three of these requirements depend on the soundness and simplicity of the markup design. They require precise content structures with few alternatives, clear guidance for writers, and good audit capability. Without these properties, your content and its markup will be inconsistent, and it will be hard to write and test reliable algorithms because of the wide variety of markup combinations they may encounter.

Most personalized content applications model their content in relational database tables for these very reasons. However, with the correct markup design, almost certainly in the subject domain, there is no reason why you cannot use markup-based tools alone or in concert with database tools and solutions to achieve the same kind of result.

Audit reports

Finally, you can use content generation algorithms to generate things other than content to be published. You can use it to generate reports about your content itself, which you can use to audit and manage your collection. I look at this in more detail in Chapter 39, but it is worth looking at the basics here because it is just another application of the capacity for content generation that you gain when you move content into the subject domain.

For instance, suppose your content strategist establishes an editorial calendar that says that you are going to put out a Christmas-themed diet cookbook every October as people are starting to prepare for Christmas. Do you have enough Christmas seasonal recipes under 300 calories? An algorithm can quickly go through your subject-domain recipes and create a list of all the recipes that meet that criteria.

Or suppose that you want to make sure that your book-related site has reviews and shopping links for every book that has been mentioned on the site this year. An algorithm can go through your content collection looking for the `book` or `novel` annotations in your articles, compile a list, sort it, eliminate duplicates, compare it to the list of reviews you currently have, and create a list of every book that is mentioned but not reviewed.

Reuse versus generation

As you have probably noticed, the methods used to generate content have a lot in common with those used to reuse content. That is because content reuse is just a form of content generation. Content reuse means generating more than one form of output content from the same collection of input content. Technically, the mechanisms and algorithms are the same; the differences have more to do with how you think about the problem.

The biggest difference is that when you think in content generation terms, you are automatically thinking in subject-domain terms. Content generation starts by treating content as data and then generates content from that data. When people think in content reuse terms, however, their thoughts often go straight to the management domain, to conditional content and pulling in content by reference. Thinking in content reuse terms also tends to give first use primacy over secondary uses. The first instance is created to serve a specific purpose. You then serendipitously discover another potential use for the content. Thinking in content generation terms does not give primacy to any one use. The content is created as data with many potential uses. There is no first use or secondary uses. Every use is a production from the source data, and it really does not matter how many such uses there are.

As I noted in Chapter 13, there are certain kinds of reuse that are only feasible using management-domain constructs, but a great deal can be accomplished using the subject-domain approach. Creating conditional markup in your recipe to provide a different beverage match for *Wine Weenie* and *The Teetotaler's Trumpet* is a case of reuse thinking. Creating structure for alcoholic and non-alcoholic beverage matches to achieve the same goal (and potentially many more) is a case of content generation thinking. In many cases, you will create a far more valuable content resource that is is easier to write for and is easier to maintain by thinking of your content set as a data source from which many kinds of content can be generated rather than a collection of re-usable content components.

CHAPTER 15

Extract

A great deal of content, particularly technical and business content, is essentially a report in human language on the features of a product, process, or data set. Much of the data that defines those things is contained in some kind of formal data set, such as a database or software source code.

In the traditional approach, writers treat these data sets as a research source. They look up information and then write their content. This is complex and requires a large amount of work, which is ongoing because writers must keep their content in sync with the original data set as it changes.

The source data for those systems is, in effect, subject-domain content for those products, processes, and data sets. Rather than researching and recreating that content, you can use structured writing techniques to extract information from those sources and create or validate content. This is the extract algorithm. It partitions and redirects the complexity of dealing with these sources away from writers and towards information architects and content engineers who create algorithms that extract the data and generate subject-domain content from it.

Tapping external sources of content

I have talked throughout this book about moving content from the media domain to the document domain and from the document domain to the subject domain. I have described the advantages of creating content in the subject domain and looked at the processing algorithms that use subject-domain content to produce various kinds of publications in different media.

Subject-domain content is created and annotated in structures that are based on the subject matter rather than on the structure of documents or media. Subject-domain structures tell you what the content is about rather than how it should be published. You can, therefore, write algorithms that process content based on what it is about rather than how it is presented. This allows you to transfer and delay decisions about presentation and formatting and to have algorithms make those decisions based on the subject-domain markup.

Any data source that is contained in subject-domain structures is a source of subject-domain content, regardless of whether or not that source was created with the expectation that it would be used to produce content. This includes virtually all databases and quite a bit of software code. It also includes all authored content (under an appropriate license) that contains usable and accessible subject-domain structures or annotations. You can potentially extract subject-domain content from all of this material. This means you can use the extract algorithm to feed content to any of the structured writing algorithms that work with subject-domain content.

As a source of subject-domain content, the extract algorithm naturally separates content from formatting and contributes to the differential single sourcing algorithm.

By tapping existing information to build content, the extract algorithm works hand in hand with the content reuse algorithm. In fact, it is really the highest expression of reuse, since it reuses content not only in the content system but also from the organization at large or even beyond the organization, which further reduces duplication within the organization.

Because the extract algorithm taps directly into external sources of information, it is a great source of information for content auditing (which I discuss in Chapter 39). At one level, it provides a canonical source of information to validate existing content against. At another level, it factors out part of the conformance problem from the authoring function. It transfers the entire responsibility for maintaining information to the creators of the source you are extracting content from, which is a responsibility they already have.

Extracting information created for other purposes

There is nothing new, of course, about generating content from database records. Database reporting is an important and sophisticated field in its own right, and it would be entirely correct to characterize it as a type of structured writing. What sets it apart from other structured writing practices is that the databases it reports on serve other business purposes besides being sources of content. An insurance company policy database, for instance, may be used to publish custom benefit booklets for plan participants, but it is also used for processing claims. The design of the structures and data entry interfaces of these systems tends to fall outside the realm (and the notice) of writers and authoring system designers.

This is a pity, because organizations often develop separate processes, tools, and repositories for content creation that duplicate information that is already contained in databases that are researched, validated, recorded, and managed independently. Rather than treating code and databases as sources of content, writers treat them as research sources. They look up information in these sources, write content to describe the information from those sources, and then store that content in a separate repository.

The essence of the problem is that many content organizations choose to work in the media domain or the document domain and have neither the tools not the expertise to bridge the gap to all this material already available in the subject domain. But even when content organizations extend their efforts into the subject domain, they often don't realize that the subject-domain content they plan to create already exists in the systems of another department.

Another drawback is that content produced from these other systems – for instance, from a database reporting system – exists in isolation from the rest of the content produced by the organization. Such content can often be quite sophisticated and beautifully formatted and published. But it is the product of an entire structured publishing chain that has to be separately developed and maintained.

In the field of software documentation you can see the same pattern with programming language API documentation. Much of the material of an API reference guide is a description of each function, what information is required as input (its parameters or arguments), the information it produces as output, and the errors or exceptions it can generate. All of this information already exists in the code that implements the function.

API documentation tools such as JavaDoc or Sphinx extract this information from code and comments and turn it into publishable content. This is an application of subject-domain structured writing, and these API documentation tools implement an entire structured publishing system that produces final output, often in multiple formats.

And here you see the same problems again:

- An entire publishing chain is maintained separately from the main content publishing chain.
- The content produced from this publishing chain is isolated from all the other content produced by the organization.
- Much of the same information gets re-created in the form of programming guides or knowledge base articles and maintained in a separate repository using a separate tool chain.

There are other cases of entirely separate publishing chains producing information that is isolated from the rest of an organization's content. Technical support organizations create knowledge bases to answer commonly asked questions. The material in these knowledge bases is technical communication, plain and simple, yet it usually exists in isolation from the product documentation set, even to the extent that users may not be able to search both the documentation and the knowledge base from the same search box. Most users, however, have no way of guessing whether the answers they are looking for are in the documentation, the support knowledge base, or a users' forum (yet another independent publishing system).

You can address these redundancies and the isolation that goes with them in two ways. One approach is to attempt to unify all content authoring and production in a single enterprise-wide system, often with a single set of content structures intended for use across all departments. However, this is expensive and disruptive, and it tends to create interfaces and structures that are less usable and less specific to your business functions than the ones they replace.

This approach also ignores the local complexity of individual groups and subject matter, which means that complexity gets shoved downstream, eventually, to the reader. And many of these systems exist for purposes other than generating content, which means they have subject-domain structures specific to, and required for, the database or software functions they were built for.

The second approach is to leave the subject-domain systems in place (and perhaps create more of them) and feed their output into a common document-domain publishing tool chain through shared pipes. The publishing algorithm normally passes subject-domain content through the document domain on its way to media-domain publication. Subject-domain content, by its nature, is not strongly tied to a particular document-domain structure, so integrating many sources of subject-domain content into a single publishing chain is not particularly onerous. The management-domain features of some tool chains make things more complicated, but since the subject domain factors out a lot of the management domain, this is not an insurmountable problem.

Most enterprise-wide content systems are based on document-domain languages. After all, there is no way to create a single enterprise-wide subject-domain system, since an enterprise creates content on many subjects for many audiences. In principle, a document-domain system should be capable of integrating content from domain-specific subject-domain systems. Unfortunately, neither subject-domain systems nor enterprise content systems are commonly designed with this kind of integration in mind.

Because of this, you sometimes have to find ways to extract content from these sources and feed them into a unified publishing chain.

A common example of the extraction algorithm is found in API documentation tools such as JavaDoc. These tools parse application source code to pull out information such as the names of functions, parameters, and return values, which they use to create reference documentation (or at least an outline). These tools generate a human language translation of what the computer language code says.

How the extraction algorithm works depends on how the source data is structured, but it usually creates output in the subject domain that clearly labels the pieces of information it has culled from the source. For instance, a Java function definition is a piece of structured content in which the role and meaning of each element can be determined from the grammar of the Java language (see Figure 15.1).

```
boolean isValidMove(int theFromFile,
                    int theFromRank,
                    int theToFile,
                    int theToRank) {
        // ...body
    }
```

Figure 15.1 – The syntax of a Java function definition

The same information can be extracted by an algorithm that knows the grammar of Java to produce something that looks more like subject-domain content (see Figure 15.2):

```
java-function:
    name: isValidMove
    return-type: boolean
    parameters:: type, name
        int, theFromFile
        int, theFromRank
        int, theToFile
        int, theToRank
```

Figure 15.2 – Subject-domain rendering of a Java function definition

Figure 15.2 contains the same information as Figure 15.1, but in a different structure. In this structure, however, the information can be easily accessed and processed through the publishing tool chain like any other content.

Any structured data source that consistently expresses the semantics of its data can be a source of subject-domain content. You just need to find a way to get at it.

The diversity of sources

When it comes to drawing content from diverse sources, the term single sourcing can be misleading, because it can lead you to think that all source content is kept in a single place. Some vendors of content management systems would like to encourage this interpretation, but a better definition is that each piece of information comes from a single source. That is, there may be many sources, each containing different pieces of information, but each piece of information comes from only one source.

Ensuring that you store content only once means making sure that both the information and the repository that contains it meet an appropriate set of constraints. The constraints that establish the uniqueness of a piece of content are different for different types of content and different subject matter, so such constraints are best expressed and enforced by systems that are designed for a particular type of content and subject matter.

This is not to say that consolidation and standardization of sources is never appropriate. Often two departments will store substantially the same information with trivial differences in how it is structured and expressed. Combining the information from both departments into one source, or at least standardizing the way each department structures and expresses its data, makes data exchange simpler and more reliable. Most organizations have all kinds of isolated and ad hoc information that could potentially be managed much more efficiently and be more easily accessible with a degree of rationalization and centralization. But it does not follow that absolute centralization of all data into a single system or a single data model is appropriate, useful, or even possible.

The best way to ensure that information is stored only once is to store each type of information in a system with the right constraints and the right processes for the people who create and manage that information. By tightly constraining each source, you prevent it from accidentally accepting information that should be stored in another system. By contrast, a single central system that is loosely constrained may end up accepting many copies of the same information because neither the system nor the users can detect the duplication.

Preserving many tightly constrained information systems that serve the needs of different departments may mean that an integrated publishing system will have to draw from diverse sources of information and content. Therefore, the ability to extract content from these sources and merge it with other content for publication is central to effectively partitioning complexity and eliminating duplication and error.

Of course, a system that relies on drawing content from many different sources, even if all those sources are well-constrained and well-managed introduces a lot of integration and maintenance complexity. Remember that the point of the exercise is to minimize the overall amount of unmanaged complexity in the system. Integrated systems can manage complexity in sophisticated ways, but they also introduce complexity that has to be factored into the calculation. Sometimes it may be less complex and less costly to allow some duplication of information between systems rather than try to manage the complex relationships required to eliminate all duplication. Sometimes the optimal solution is less than total integration.

CHAPTER 16

Merge

Although the extract algorithm gives you access to new sources of subject-domain content, it does not always give you everything you need for a complete document. In order to present a complete document, you sometimes need to combine extracted content with content you have written. This is a job for the merge algorithm.

You can extract useful information about an API function from the code that implements it, but there is not enough detail to build a complete API reference. A good reference entry also requires an explanation of the purpose of the function, a little more detail about its parameters, and possibly a code sample illustrating its use. To address these needs, you can merge authored content covering these topics with content you have extracted from the source code.

API documentation tools often allow you to include authored content in the source code files. This content is contained in code comments and is often written in small subject-domain markup languages that are specific to the tool. Of course, as with all subject-domain structures, another tool can read them if it wants to.

```
/**
 * Validates a chess move.
 *
 * Use {@link #doMove(int theFromFile,
 *                    int theFromRank,
 *                    int theToFile,
 *                    int theToRank)} to move a piece.
 *
 * @param theFromFile file from which a piece is being moved
 * @param theFromRank rank from which a piece is being moved
 * @param theToFile   file to which a piece is being moved
 * @param theToRank   rank to which a piece is being moved
 * @return            true if the move is valid, otherwise false
 */
boolean isValidMove(int theFromFile,
                    int theFromRank,
                    int theToFile,
                    int theToRank) {
    // ...body
}
```

Figure 16.1 – API documentation included in source code using JavaDoc

Figure 16.1 is an example of authored content combined with source code in JavaDoc.[1] In this example, everything between the opening /* and the closing */ is a comment (as far as Java is concerned), and the rest is a Java function definition. However, JavaDoc sees the comment block as structured text marked up using a style of markup specific to JavaDoc.

The JavaDoc processor extracts information from the function definition (the extract algorithm) and then merges it with information from the authored structured content (the merge algorithm). In doing so, it can validate the authored content, for instance, it can ensure that the names of parameters in the authored content match those in the function definition. This ability to validate authored content against extracted data is an important aid to conformance and auditing.

However, the merge algorithm does not require that the authored content and the extracted content be part of the same file. You can just as easily place the authored content in a separate file. In many cases, this will be the only available approach, since some source formats do not provide any place to include documentation. To merge the two, you need two things:

1. Sufficient structure in both the extracted data and the written content that you can identify the pieces you are selecting to merge (this is required for composability).
2. An unambiguous term or combination of terms in the extracted data that you can use as a key to connect the extracted data with the authored content by entering the same terms as a field or fields in the authored content.

For example, on one project I worked on, we needed to create a reference for a large body of operating system components. The components were defined in a data file that allowed for a one-line description of the component, which was intended for display in a GUI configuration editor. We needed to supply much more extensive documentation for each component, so we created an extract algorithm to pull information from the data file that defined the components and save it as XML. We then developed a subject-domain XML format to capture the additional information we needed to document each component. We then merged the two sources using the name of the component as the key, as shown in Figure 16.2. (And note the similarity of this diagram to Figure 2.3)

---

[1] https://en.wikipedia.org/wiki/Javadoc#Example

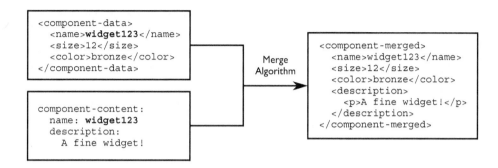

Figure 16.2 – Merging extracted and authored content

This produced an integrated subject-domain document, which we then processed into the document-domain format we used for the whole documentation system. We then processed that document to produce the formatted reference. Because the operating system components we were documenting supplied additional APIs to the operating system, we also merged in information from the API reference and created links from the component reference to the API reference, all with algorithms.

One of the downsides of API documentation tools like JavaDoc is that they tend to be tightly coupled systems that directly produce media-domain output such as formatted HTML, thus providing little or no control over presentation or formatting. This is a problem because it means that your API reference content will not look like the rest of your content. And worse, your API reference won't be integrated with or linked to the rest of your content. This means any mention of an API routine in your programmer's guide won't link to the documentation for that routine in the API reference.

It would be much better to generate subject-domain content from the API documentation tool and then process it with the rest of your content. Many API documentation tools make this possible by allowing you to export XML, which may be either subject domain or document domain. Even if the output is document-domain XML, it may be regular enough that you can extract subject-domain structures reasonably easily. You can then skip the API documentation tool's built-in publishing chain and run the XML output through your regular publishing chain.

CHAPTER 17
Modeling

As noted in Chapter 15, it is possible to extract content from existing data sources, which can then be used as input for generating content (as described in Chapter 14). Some of these sources were not intended to be used as content, but in many cases you can turn them into content by using algorithms to wrap the data into sentences or field values in a document. Such sources essentially present a model of some object in the real world, and when you use them as a source of content, you create content from that model.

Even when an external data source does not exist, you can apply the same techniques to creating content by creating your own model of a real-world object using structured writing techniques. You can then extract content from that model just as you would from an external data source. This can be an extremely powerful technique for managing complexity in cases where content frequently traverses the same ground, such as describing the steps a user takes when navigating a menu system, or where content must express complex relationships that are easier to model and control in a formalized system.

An entire book could be written about how to model various types of subjects and how to derive content from those models. Here, I present one simplified model to demonstrate the principle at work and point to some of its key benefits.

Many technical documents contain numerous procedures for navigating an application interface. Writers can spend a lot of time writing and maintaining these procedures. However, these procedures are often dictated entirely by the interface and could be generated by an algorithm if the right data was available. The modeling algorithm is about creating a data set you can generate content from.

The navigational structure of a typical desktop application is essentially a tree consisting of a well-defined set of objects: menus, menu items, dialog boxes, field values, buttons, and assorted other controls. The entire navigation system of an application can thus be modeled as a tree, and, indeed, there are languages designed for modeling interfaces in just this way for programming purposes.[1] However, rather than using one of those languages, I will use a simpler model that is just complex

---

[1] https://en.wikipedia.org/wiki/User_interface_modeling

enough to demonstrate the technique. Figure 17.1 shows a model for an interface that creates a new document.

```
application: MyApp
    menu-set:
        menu: File
            menu-item: New
                dialog-box: New document
                    fields:: name, purpose
                        Name, the name of the new document
                        Type, the type of the new document
                    confirm: Create
                    cancel: Cancel
                    screen-shot:
                        windows: graphics/windows/new.png
                        mac: graphics/mac/new.png
                        linux: graphics/linux/new.png
```

Figure 17.1 – Simple structure to model a user interface

Most of what is contained in a procedure for using a GUI follows a formula. With a model of the UI like Figure 17.1, you can write an algorithm to follow that formula for all procedures.

Figure 17.2 shows how you might write a procedure for using this menu using conventional techniques.

```
procedure: Create a new document
    step:
        From the {File}(menu) menu, select {New}(menu-item).
        The {New document}(dialog) appears.

        >>>(%image.new_dialog)
    step:
        Enter the name of the new document into the {Name}(field) field.
    step:
        Enter the type of the new document into the {Type}(field) field.
    step:
        Press {Create}(button) or {Cancel}(button) to cancel.
```

Figure 17.2 – Steps for writing a procedure for a GUI operation

Using the model in Figure 17.1, however, you can reduce all of this to:

```
procedure: Create a new document
    use-dialog: New document
```

Every piece of text used in Figure 17.2 can be derived from the model by an algorithm. The `use-dialog` field in the reduced version triggers an algorithm that uses the "New document" `dialog-box` in Figure 17.1 as its source and generates the full text of the procedure, including the correct screen shot.

Clearly, you won't always want to generate the full procedure with a single line. There will be cases where you need to say something specific about a particular case. For example, if this procedure is part of a tutorial, you might want to specify a particular file name. In that case, you could use the model to generate only part of the procedure (see Figure 17.3).

```
procedure: Create a document for testing
    navigate-to: New document
    step:
        Enter "test.abc" into the {Name}(field) field.
    step:
        Enter "normal" into the {Type}(field) field.
    confirm: New document
```

Figure 17.3 – Partially customized procedure

In Figure 17.3, the fields `navigate-to` and `confirm` tell an algorithm to generate parts of the procedure from the model (the navigation to the dialog box and the button presses to confirm the action) while allowing the writer to insert custom instructions in the middle.

Procedures contain huge amounts of repeated text, none of which can be effectively factored out by standard variable substitution techniques. But by creating a model of the interface, you can factor out most of the text for most procedures, leaving it to an algorithm to perform the tedious task of typing out all those instructions over and over again.

There are a number of potential benefits to factoring out content into a model:

- You reduce the amount of content that needs to be written, and you remove the necessity for writers to trace carefully through every procedure to make sure every detail is correct. The details come from the model, so all you have to do is ensure that the details are correct in the model. And if you find a flaw in the model, fixing it in the model automatically fixes it in all the content derived from it.
- When changes occur in the subject matter, you only have to change the model, not any of the documents that contain text derived from the model.

- The model can be maintained by one person who is an expert on the structure of the subject matter, thus removing the need for multiple writers to become experts.
- If there is a change to the subject matter that could affect the way the model is called from the content, this will be immediately obvious because the names used to identify parts of the model will no longer match, triggering a referential integrity error. This means it is almost impossible for a change in content to be missed when the subject matter changes.
- It allows you to change the way that the subject matter is described independent of the existing content. If you want to shorten the navigation instruction in the procedure example to `File>New`, you only need to change the algorithm, not hundreds of places in the content.
- Because you can change the way a subject is expressed by simply changing the algorithm, you can test different forms of expression to see which work best.
- The ability to easily change the output also allows you to personalize content for different groups. You can even allow individuals to choose how they want content to be expressed. Using active content techniques, you can even change how content is expressed at read time based on the information you have about your users, or the way you see them interact with the content (showing screen shots automatically if the users frequently click on a screen shot link, for instance).
- The fully-written-out example of the procedure in Figure 17.2 uses keys to factor out the difference between screen shots for different platforms, a technique described in Chapter 13. But while keys are a powerful tool, they are also an abstract concept and writers have to learn to use them correctly. Using the modeling algorithm, the choice of screenshots is moved from the content to the model. This factors out the complexity of using keys. It also allows you to make choices about how, where, and if screen shots are displayed. And you can test which works best or even allow users to select whether they prefer to see screen shots or not.

As with every other algorithm described in this book, it is important not to get carried away with enthusiasm of modeling in isolation. Although it is technically satisfying to build a model and derive a variety of content from it, the model and its algorithms are also sources of complexity.

As I have noted, there is nothing wrong in principle with introducing new complexity into the content system. All tools do this. The point is to ensure that all of the complexity in the content system is handled by a person or process with the skills, time, and resources to handle it properly. Modeling partitions and redistributes a lot of complexity, which means it can create substantial change to the overall content system. It is important to make sure that the effect on your overall content system is positive.

CHAPTER 18
Linking

A large part of the complexity of rhetoric and information architecture arises because readers do not read in a straight line. Numerous studies[1] have shown that readers often read opportunistically, looking for one piece of information or the answer to a particular question. Their behavior can best be described as information foraging. They are sniffing for the scent of the information they want and will follow its trail as long as the scent gets stronger and the trail is easy to follow.

The best way to support this information foraging behavior is with a bottom-up information architecture (as described in Chapter 8). Linking is key to building a bottom-up information architecture. But as I discussed in Chapter 12, different media often require different rhetoric, and linking strategy is a big part of that difference. This makes linking a complex problem.

Linking is at the heart of a bottom-up information architecture. In a bottom-up architecture, a page is not simply a leaf on a tree – the prize you find at the end of the search. It is a junction point in the exploration of an information space and the quest to understand a subject.

For example, while reading your content, readers may discover new subjects they need to understand and new options they need to consider. Readers may discover that what they thought they knew is wrong or that what they thought they wanted to do was not the right choice. They may find that their search or navigation has led them to the wrong place, or they may discover new worlds to explore. On a more mundane level, they may discover that they need additional information, such as reference data, to complete a task.

These are all pointers to some next page that a reader may need. However, even the most prescient writer cannot make every possibility the next page in a linear narrative. To serve readers, writers need to pave all of the possible paths, and the way to do that is with hypertext links.

This means that linking is not something that happens at arbitrary points where the writer feels like adding a link. Linking must be planned for as part of the information architecture. Whether you specify explicit links in the media domain or the document domain, manage links with keys in the management domain, or generate them from subject annotations in the subject domain, links should be created in a disciplined and consistent manner according to a deliberate plan.

---

[1] For a discussion of these studies and their implications for how we write, see my book *Every Page is Page One: Topic-based Writing for Technical Communication and the Web.*

The trail readers follows through an information set has junction points, places where they can decide to keep going straight through the current document or turn aside to look at a different document. While readers can decide to switch documents at any point and for any reason, the most common reason for switching is because they want more information about, or are simply more interested in, a subject mentioned in the current document.

Subject affinities

Even if each document describes just one subject, it still mentions many other subjects in the course of describing its own subject. Thus, a movie review mentions actors, directors, and other movies; a recipe mentions cooking techniques such as whisking and grilling; and a programming topic mentions functions, libraries, and data structures.

I call the points where a document mentions related subjects *subject affinities*. Subject affinities are the junction points where the subjects of different pieces of content intersect – the points where the reader may choose to turn or to go straight on.

Suppose you are reading a recipe and come across the instruction "sweat the onions," but you don't know how to do that. You can't continue without that knowledge, so you turn aside in search of it. The words "sweat the onions" are a point of subject affinity between the recipe and the cooking technique of sweating vegetables. Your decision to turn in search of additional information at this point is not arbitrary. It follows from the relationship between the task of making the dish and a cooking technique used in that task. Writers can anticipate that some readers will need to make this turn at this point of subject affinity.

This particular subject affinity, between a recipe and a cooking technique, is neither arbitrary nor unique. The subject affinity between recipes and basic cooking techniques applies universally and can be modeled using structured writing techniques.

The need to seek more information on a subject or task is part of the complexity of information seeking. The goal of information delivery is to make information seeking simpler. If you don't make it possible for your readers to find this information when they need it, you are dumping the complexity of navigating the subject affinity onto them. Managing subject affinities, therefore, is an important part of managing the overall complexity of the content system. Making sure that every part of the complexity of the content system is handled by a person or process that has the skill, time, and resources to handle it extends to the reader. You don't want to give readers any navigational complexity that they are not fully equipped to handle. Indeed, you want to make it as easy as you can for them to navigate subject affinities.

The most obvious, and generally the most powerful, way of handling subject affinities is to create links from the point in the text where the affinity occurs to a suitable resource that provides the needed information. If you force readers to use search, go to a table of contents, or consult an index when you could easily provide a direct link, you are dumping complexity on them.

But linking is not the only way to handle subject affinities, nor is linking an option in all media. You can use footnotes, cross references, sidebars, or parenthetical statements to provide additional information. You can even attempt to anticipate and forestall a reader's information need by using data about that individual to dynamically reorder the content. A discussion of linking, therefore, needs to consider other ways of handling subject affinities that may be appropriate in particular circumstances.

If you single-source content between different media, you might want to handle subject affinities differently between paper and hypertext outputs. For example, you might want to include a chunk of explanatory content in a sidebar in paper documents but link to a single copy of it in hypertext. Linking, in other words, is a kind of reuse that avoids duplication by referencing content rather than copying it.

Thus, you should not think solely in terms of managing links. Instead, you should think about managing subject affinities and supporting them in a way that works best with each of your output formats.

One of the biggest problems with link management is that it is complex and often time-consuming. As a result, many organizations do not link their content adequately, nor do they provide alternative ways to handle subject affinities. They just dump the complexity of finding the next piece of information right back on the reader.

However, there are ways to partition the complexity of link management and subject affinity management that make it much easier and more economical for writers to provide good subject affinity support. To understand how to do this, it is useful to look at how link and subject affinity management works in each of the structured writing domains.

Subject affinities in the media domain

In the media domain, you simply record the various devices used to express subject affinities: cross references, tables, links, etc. For example, in HTML a link specifies a page to load:

```
<p>In Rio Bravo,
<a href="https://en.wikipedia.org/wiki/John_Wayne">the Duke</a>
plays an ex-Union colonel out for revenge.</p>
```

The phrase "the Duke" is a subject affinity. The reader may not know who "the Duke" is or may want more information about him. The link allows the reader to navigate the subject affinity and find the information needed.

But if the page is printed, the link is lost. The phrase "the Duke" is still a subject affinity, and the reader may still want more information. The reader can still get more information by doing a search for "the Duke" or asking a friend what the phrase means. But the printed version doesn't support the reader's need. It dumps the complexity of the search back on the reader.

If the content had been written for paper, the subject affinity might be supported in a different way. For example, it might be supported by adding an explanation in parentheses:

> In Rio Bravo, the Duke (John Wayne) plays an ex-Union colonel out for revenge.

Or it might be handled with a footnote:

```
In Rio Bravo, the Duke* plays
an ex-Union colonel out for revenge.

...

* "The Duke" is the nickname of the actor John Wayne.
```

Clearly, this is a case where you would like to do differential single sourcing and handle the subject affinity differently in different media. To accomplish this, you need to move the content out of the media domain.

Subject affinities in the document domain

Moving to the document domain is about factoring out the formatting specific structures of the media domain. But a link is not really a piece of formatting, so conventional refactoring into abstract document structures doesn't apply. For this reason, people working in the document domain often enter hypertext links exactly the way they would in the media domain – by specifying a URL. Thus, in DITA you might enter a link as:

```
<p>In Rio Bravo,
  <xref href="https://en.wikipedia.org/wiki/John_Wayne"
        format="html">The Duke</xref>
  plays an ex-Union colonel out for revenge.
</p>
```

The difference from HTML is slight. The link element is `<xref>` rather than `<a>`. But the meaning of `<xref>` is bit more general. The HTML `<a>` element says, "create a hypertext link to this address." The DITA `<xref>` element says, "create some sort of reference to this resource." (As you will see in a moment, DITA is capable of linking to things other than HTML pages, which is why it requires the `format` attribute to specify that in this case the target is an HTML page.) This gives you a little more leeway in processing. From this markup, you can create usable print output that looks like this:

> In Rio Bravo, the Duke (see: https://en.wikipedia.org/wiki/John_Wayne) plays an ex-Union colonel out for revenge.

This is not the way you would handle the subject affinity if you were designing for paper, but it is a definite improvement from a differential single sourcing point of view. At least the link is now visible to readers.[2]

Fundamentally, though, this is not a satisfactory differential single sourcing solution. Unless there is no alternative (such as when you are citing a specific source), you would not normally direct someone reading paper to the web or vice versa. Linking to an already published file, such as an HTML page, means linking to the address where the published file resides. It commits you to a particular format for the link target. If, instead of linking to the published address, you link to the address of the unpublished source file, or to an identifier for that file, you gain the freedom to link to any format of that content that you choose to publish.

---

[2] Technically you could do this from the HTML markup as well, but that would be cheating. The HTML markup doesn't really give you permission to do this. It says create a hypertext link and nothing else. The problem with cheating is that you are basing your algorithm on constraints that are not promised or enforced, and this can fail in ways you may not expect or catch. Some cheats are more reliable than others, but you probably don't want to get into the habit. It's better to create content in a format that doesn't require cheating to get the output you want.

In DITA, you can link to another DITA file (this is the default format, so you don't need to use the `format` attribute):

```
<p>In Rio Bravo, <xref href="John_Wayne.dita">The Duke</xref>
plays an ex-Union colonel out for revenge.</p>
```

You don't yet know if that content will be published to paper or the web, what the address of the published topic will be, or if that topic will stand alone or be assembled into a larger page or document for publication. This means that the publishing system is taking on responsibility for both ends of the link. It has to make sure that the target page is published in a way the source page can link to and that the source page links to the right address. But transferring this responsibility to an algorithm gives you the leeway to publish this link as you see fit.

If you publish as a printed book and the target resource ends up as part of a chapter in that book, you can render the `xref` as a cross reference to the page that resource appears on. And you can format that cross reference in line or as a footnote. These are all legitimate interpretations of the `xref`'s instruction to create a reference to a resource.

If you publish to a help system and the target resource ends up as a topic in the same help system, you could render the `xref` as a hypertext link to that topic.

This is a big step forward, but it still does not let you do this:

> In Rio Bravo, the Duke (John Wayne) plays an ex-Union colonel out for revenge.

In other words, you can render the `xref` as a cross reference, a link, or a footnote, but you can only handle the subject affinity as a reference to that specific resource. You can't decide to link to a different resource or render it as a sidebar instead. If you want those capabilities, you need to turn to the management domain.

Subject affinities in the management domain

Linking to a source file rather than to an address gives you more latitude about how the link or cross reference is published, but you are still linking to the same resource. If you are doing content reuse, this is a problem, because you do not know if the same resource will be available everywhere you reuse your topic. You need to be able to link to different resources when your topic is used in different places.

To accommodate this, you can factor out the file name and replace it with an ID or a key. IDs and keys are management-domain structures that I discussed in Chapter 13. They allow you to refer to resources indirectly. Using IDs lets you use an abstract identifier rather than a file name to identify a resource. Using keys lets you remap the resources you point to, which is a more efficient way to address this problem. So instead of referring to a specific resource on John Wayne, you refer to the key John_Wayne. In DITA this would look like the following:

```
<p>In Rio Bravo, <xref keyref="John_Wayne">The Duke</xref>
plays an ex-Union colonel out for revenge.</p>
```

Somewhere in the DITA map for each publication, the key John_Wayne points to a topic. Publications link the keyref to the resource pointed to by that key in each of their DITA maps. This allows you to link to different resources in each publication.

The problem with IDs and keys

However, there is still a problem with linking based on IDs and keys. Keys let you vary which resource a keyref resolves to, but what happens when the resource doesn't exist? The xref must link to a resource, but there is no resource to link to. You have a broken link and fixing it is not easy. You can't simply go in and remove the xref from the source for one publication, because it defeats the purpose of content reuse if you have to edit the content every time you reuse it. Removing the key reference would fix the broken link in one publication, but that would result in the link being removed from all publications, even where the resource does exist and the link ought to be created.

Relationship tables

One solution to this problem is to use a relationship table. In a conventional linking approach, the source page contains an embedded link that points to the target page. The source knows it is pointing to the target, but the target does not know it is being pointed to (see Figure 18.1).

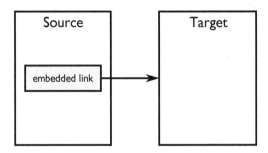

Figure 18.1 – A simple link

Because the target resource does not know it is being pointed to, it does not have to do anything for other resources to point to it. This is fundamental to the rapid growth of the web. If the target resource had to participate in the link process, every link would require negotiation between page owners, and the owner of the target resource would have to edit that resource to accept the link. It would have been impossible for the web to grow as explosively and organically as it has under those conditions.

A relationship table takes this one step further. When you create a link using a relationship table, you factor the link out of the source document and place it in a separate table. The relationship table in Figure 18.2 says resource A links to resource B, but neither resource A nor resource B knows anything about it.

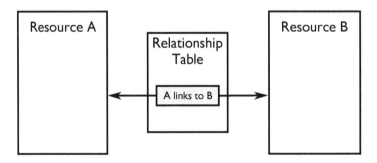

Figure 18.2 – A relationship table

Once the links are factored out of a piece of content, you can reuse it anywhere you like. If there is a suitable resource available to link to, you enter it in a relationship table for that build and have the presentation algorithm create the link at build time. If no suitable resource is available for a different publication, no entry is made in the relationship table for that publication, and the presentation algorithm does not create a link.

The problem with relationship tables

The problem with relationship tables is that they separate the link from the subject affinity it supports. The link was created to serve a subject affinity in the content – a point where readers might encounter something they don't understand or might need more information – but the relationship table does not record the subject affinity, it simply says that there is some connection between topic A and topic B. And it does not say where in topic A the subject affinity is located. Therefore, there is no way to insert a link at the point where the subject affinity occurs. Links generated by relationship tables end up in a block, usually at the end of the page.[3]

The fundamental problem here is that links exist as a way to service subject affinities in content. If we lose sight of this and think only in terms of managing links between resources, it makes sense to pull links out of the content and manage them separately. However, this ignores the reason links exist in the first place, which is to provide readers with support at a point of subject affinity. Using relationship tables is a classic case of partitioning that focuses on one part of the problem without adequate regard for the complexity that is being redirected. Moving the link away from the subject affinity simplifies link management but makes it harder for readers to find supporting information – a classic case of sacrificing rhetoric in the name of process efficiency.

The other problem with the relationship table approach is that it is time consuming. You have to rewrite the links for each content set, and because the subject affinities are not recorded in the content source, you have to figure out the appropriate links each time. This goes against the spirit of recording something once and using it many times. A mechanism intended to help you reuse content ends up forcing you to rework links for each publication you create. In other words, this mechanism does not partition the complexity of link management very well.

---

[3] It is not impossible to imagine a markup system in which you could markup the source of the subject affinity in topic A and then have the relationship table reference that point by ID, thus allowing you to insert a link at the subject affinity. But this would be cumbersome for writers, and it would complicate the management task. I don't know of any systems that work this way.

Conditional linking

Before leaving the management domain, it is worth mentioning a management-domain approach that could address the differential single sourcing problem and handle subject affinities for both online and print publishing. This approach uses conditional structures to include content for both options in the source file. With a specialization to support `media` as a conditional attribute, you could do this in DITA as shown in Figure 18.3.

```
<p>In Rio Bravo,
<ph media="online"><xref keyref="John_Wayne">The Duke</xref></ph>
<ph media="print">The Duke (John Wayne)</ph>
plays an ex-Union colonel out for revenge.</p>
```

Figure 18.3 – Conditional linking in DITA

In Figure 18.3, the DITA `<ph>` element delineates a phrase in the content that you want to apply management-domain attributes to. There are two versions of the phrase "the Duke," one for print that uses a parenthetical expression and one for online that uses a link. The `media` attribute identifies the version. The synthesis algorithm chooses the appropriate version of the phrase for each publication based on the conditions set for the build.

There are obvious problems with this approach. It doubles the writer's job and it doubles the cost of maintenance. This approach also flies in the face of creating formatting-independent content.

Unfortunately, in a general purpose document-domain markup language with management-domain support, it is nearly impossible to support differential single sourcing and content reuse without forcing writers to do things like this. And, in practice, writers do end up using conditional markup like this, which can lead to tangles of conditions that are hard to maintain and debug.

The management-domain approach does a poor job of partitioning the complexity of links and handling subject affinities:

- Management-domain link structures are artificial. They don't correspond to things in the writer's everyday world, which makes them harder to learn and use. They introduce complexity into the writer's world, which leads writers to severely limit the links they create.
- You can't link to a key or an ID that does not exist. This means that as you develop a set of content, the first pages have few other pages to link to. You cannot enter links to content that has not been written yet. This introduces the complexity of system dependencies into the writer's world.

- In reuse scenarios, the use of IDs and keys does not solve the whole problem because you cannot guarantee that the resource that an ID or key refers to will be present in the final publication. You can use relationship tables to address this problem, but they create additional complexity for writers and don't correctly handle subject affinities in the middle of the content.

- Unless you resort to ugly conditional structures, you can't use differential single sourcing to handle subject affinities differently for different media.

Subject affinities in the subject domain

As described in Chapter 13, you can often eliminate the need for management-domain structures by moving content to the subject domain. The same is true with subject affinities. In the subject domain, you can mark up subject affinities as subject affinities. This means that you can move away from managing links and manage subject affinities instead. Figure 18.4 and Figure 18.5 show what this looks like in XML and SAM.

```
<p>In <movie>Rio Bravo</movie>,
<actor name="John Wayne">the Duke</actor>
plays an ex-Union colonel out for revenge.</p>
```

Figure 18.4 – Subject-domain subject affinity markup in XML

```
In {Rio Bravo}(movie),
{the Duke}(actor "John Wayne")
plays an ex-Union colonel out for revenge.
```

Figure 18.5 – Subject-domain subject affinity markup in SAM

This markup clarifies that "the Duke" (a subject affinity) refers to the actor John Wayne (its subject). It specifies both the type of the subject (actor) and its value (John Wayne).

This is subject annotation, not link markup. Unlike document-domain `xref` markup, it does not insist that a reference should be created nor does it specify any resource to link to. Instead, it simply states what the subject of the text is. Specifically, it clarifies that the phrase "the Duke" refers to the actor named John Wayne (and not the Duke of Wellington or the Duke of Earl) and that the phrase "Rio Bravo" refers to the movie (and not the city in Texas or the nature reserve in Belize).

Given this markup, you can easily render the subject affinity in a manner that works for print by having the presentation algorithm take the value of the `specifically` attribute (as this annotation is called in SAM) and output it between parentheses, as shown in Figure 18.6.

```
<p>In Rio Bravo, The Duke (John Wayne) plays an
ex-Union colonel out for revenge.</p>
```

Figure 18.6 – Subject affinity rendered for print

Marking up a phrase as a subject affinity does not oblige the publishing algorithm to create a link. If you decide to have the publishing algorithm create a link on the web and a cross reference on paper, nothing in the markup obliges you to use any particular formatting or target any particular resource. I'll describe how to create a link from a subject annotation later in this chapter.

In all the previous examples, mentions of *Rio Bravo* were not marked up, even though the names of movies are clearly subject affinities in a movie review. This reflects the writer's decision not to create a link to support this subject affinity. But what if you want to make a different choice later? By marking up *Rio Bravo* with a subject annotation, you keep your options open by transferring the information needed to make the decision to a different partition. You can tell the presentation algorithm to create links on the names of movies if you want to, but you don't have to.

There are additional reasons to annotate *Rio Bravo* as a subject, because that annotation can be used for other purposes as well.

- The subject annotation says that *Rio Bravo* is the title of a movie. In the media domain, the titles of movies are commonly printed in italics. You can use the subject-domain `movie` tags to generate media-domain italic styling.
- You could use this subject annotation to generate document-domain index markers, so you can automatically build an index of all mentions of movies in a work.

Subject annotation thus serves multiple purposes and, correspondingly, reduces the amount of markup required to support all these publishing functions. This is an important characteristic of subject-domain markup. Because subject-domain markup is not directly tied to specific document-domain or media-domain structures, you can generate multiple document-domain and media-domain structures from the same subject-domain markup.

```
<para>
    In
    <indexterm>
        <primary>Rio Bravo</primary>
    </indexterm>
    <indexterm>
        <primary>Movies</primary>
        <secondary>Rio Bravo</secondary>
    </indexterm>
    <citetitle pubwork="movie">Rio Bravo</citetitle>,
    <indexterm>
        <primary>John Wayne</primary>
    </indexterm>
    <indexterm>
        <primary>Actors</primary>
        <secondary>John Wayne</secondary>
    </indexterm>
    <ulink url="https://en.wikipedia.org/wiki/John_Wayne">
        The Duke
    </ulink>
    plays an ex-Union colonel out for revenge.
</para>
```

Figure 18.7 – DocBook markup generated from subject-domain markup

For example, you could generate the DocBook document-domain markup shown in Figure 18.7 from subject-domain markup. Figure 18.7 contains index markers, movie title formatting, and links on the actor's name, all generated from subject annotations. It should be clear how much less work it takes to create the subject-domain version of this content than the DocBook version, how many fewer decisions writers have to make, and how much less knowledge and skill they have to possess. Yet the same publishing capabilities are supported by both versions.

Generating links from subject annotations has a number of other advantages:

- In a reuse scenario, you never have to worry about broken links or creating relationship tables. The presentation algorithm generates appropriate links to whatever topics are available.
- In a differential single sourcing scenario, you are never tied to one mechanism for handling the subject affinity. You can generate any mechanism you like in whatever media you like.
- You don't have to worry about maintaining links in your content because your source content does not contain any links. The subject annotations in your content are objective statements about your subject matter, so they don't change. All the links in the published content are generated by the linking algorithm, so no management is required.

- Writers can create subject annotations to content that has not been written yet. The subject annotation refers to the subject matter, not a resource. Links to content that is written later will appear once that content becomes available to link to.
- Writers do not have to find content to link to or manage complex link tables or keys. They just create subject annotations when the text mentions a significant subject. This requires no knowledge of the publishing or content management system. It does not even require knowledge of other resources in the content set. It only requires knowledge of the subject matter, which the writer already has, and the format of subject annotations, which is easy to learn.

Finding resources to link to

The subject-domain approach represents a radically different partitioning of the complexity of linking and handling subject affinities. Most notably, it partitions and redirects away from writers the responsibility for finding content to link to.

Partitioning always introduces new complexity to replace what has been partitioned out. In this case, the new complexity is the effort required to create subject-domain annotations. However, this partitioning is particularly effective both because annotating subject affinities requires only knowledge the writer already has and because the same annotation is useful for so many other algorithms. This is one of the best features of the subject domain – its structures can serve multiple algorithms rather than just one, which is, in itself, a highly effective partitioning of complexity.

However, although the responsibility of finding resources to link to has been partitioned away from writers, you still need to consider the people and processes that are now required to locate appropriate resources. They do this by looking up resources based on the subject information (type and term) captured by the subject annotation. For this you need content that is annotated at the document level using those types and terms (or their semantic equivalents). You can annotate your documents to describe their subjects by adding fields at the document level. (This is metadata.) If you have a page on John Wayne, you can annotate its subjects as shown in Figure 18.8.

```
topic:
    title: Biography of John Wayne
    subjects:: type, term
        actor, John Wayne
    body:
        John Wayne was an American actor known for westerns.
```

Figure 18.8 – Subject annotation for John Wayne biography

Now the linking algorithm looks like Figure 18.9.

```
match actor
    $target = find href of topic
              where in subjects
              type = actor
              and term = @name
    create xref
        attribute href = $target
        continue
```

Figure 18.9 – Linking algorithm for subject-annotated content

However, content stored in the subject domain may already be annotated effectively enough by its inherent subject-domain structures. Suppose your content collection includes the subject-domain actor biography shown in Figure 18.10.

```
actor:
    name: John Wayne
    bio:
        John Wayne was an American actor known for westerns.
    filmography:
        film: Rio Bravo
        film: The Shootist
```

Figure 18.10 – Subject-domain actor biography for John Wayne

Here the topic type is actor, and the name field specifies the name of the actor in question. This is all you need to identify this topic as a source of information on the actor John Wayne.

Only minor changes to the linking algorithm are required to use this bio (see Figure 18.11).

```
match actor
    $target = find href of actor topic where name = @name
    create xref
        attribute href = $target
        continue
```

Figure 18.11 – Linking algorithm modified to use biography

There is a lot more to how this mechanism works in practice, including what you do about imperfect matches and what happens when the query returns multiple resources. But that goes into the specifics of individual systems, which is more detail than needed for present purposes.

You can also annotate the subjects of topics using a content management system, in which case the linking algorithm would query the CMS to find topics to link to.

A useful feature of this approach is that you can have the publishing algorithm fall back to creating a link to an external resource if an internal one is not available. If a search of the annotations in your own content fails, you can search indexes of external content. You can build such an index yourself, but some external sites provide indexes, APIs, or search facilities that you can use to locate appropriate pages to link to.

Of course, building these linking algorithms adds complexity to your content system. The subject domain partitions complexity away from writers and distributes it to information architects and content engineers.

Different domain, different algorithm

The linking algorithm – or, to be more precise, the subject-affinity handling algorithm – is perhaps the algorithm that most clearly illustrates how moving from one domain to another leads to a significantly different partitioning of complexity. While the purpose of the linking algorithm remains the same, the way it achieves that purpose differs significantly in each domain.

In the document domain, the data structures tend to have a one-to-one correspondence with their algorithms. As system designers determine they need a particular algorithm, they create structures to support that algorithm. Thus, document-domain languages have separate structures for linking, reuse, indexing, and single sourcing. (Some of these may be management-domain structures, of course.)

In the subject domain, however, the data structures reflect the subject matter. You won't find a one-to-one correspondence between a structure and the algorithm it supports. Thus, you will not find markup for links, reuse, indexing, or single sourcing in the subject domain. You will find subject annotations that clarify and delineate the subject matter of the content. Algorithms that work on subject-domain content must interpret subject annotations and use them as the basis for creating whatever kind of document or media-domain structure you need for publishing.

System designers still must decide which algorithms they need, so they can ensure that the information that drives those algorithms is present in the content. However, since every subject structure can drive many publishing algorithms, you will often find that your subject-domain content already supports any new algorithm you want to apply.

Therefore, moving from the document domain to the subject domain is not a matter of looking for the subject-domain equivalent of a document-domain structure. Instead, it's a matter of asking what information in the subject domain drives the creation of document-domain structures. Subject-domain content can look very different from its document-domain counterpart, and it will often be starkly simpler and easier to understand. This represents a better partitioning of complexity. But as noted before, all complexity has to go somewhere, and the use of the subject domain tends to transfer more of the complexity to the information architect or content engineer.

CHAPTER 19

Publishing

All structured writing must eventually be published. Publishing structured content means transforming it from the domain in which it was created (subject domain, document domain, or the abstract end of the media domain) to the most concrete end of the media-domain spectrum: dots on paper or screen.

Publishing is complex, particularly if you are using single sourcing, content generation, or any of the structured writing algorithms that combine content from different sources in a "shared pipes" model. The best way to handle that complexity is to partition and distribute it to appropriate people and processes.

In this chapter, I describe a partitioning of the publishing process into four basic algorithms: the synthesis, presentation, formatting, and encoding algorithms. These four stages are formalized in the SPFE architecture, which I discuss later, but I think they are a fair representation of what goes on in most publishing tool chains, even if those tool chains don't divide responsibilities exactly as I describe them here or separate them as clearly.

I have noted that successful partitioning depends on the ability to communicate everything needed for each partition to do its job and that one hallmark of a successful partitioning is how clear and functional the required communication is. In the partitioning of the publication process I describe here, each step deals with a different concern, and each step produces an output: a structured document or set of documents that is structured in terms of the concerns it deals with. This output becomes the input to the next step. The structures that these steps produce and consume are stages along the path from the subject to document to media domains. In other words, the publication process moves content through the domains, just as described in Chapter 1.

In describing the publishing process, I am going to start at the media domain and work backward to the subject domain, just as I did in describing the domains themselves. Since the input of each stage is the output of the previous stage, it is easier to understand the process if you look at it in reverse order. This way you will see what each stage needs to produce in order to meet the needs of the following stage.

The rendering algorithm

There is actually a fifth algorithm in the publishing chain, which I call the rendering algorithm. The rendering algorithm is responsible for placing the right dots on the right surface, be that paper, screen, or a printing plate. However, this is a low-level device-specific algorithm, and no one in the structured writing business is likely to be involved in writing rendering algorithms. The closest you ever get is the next step up, the encoding algorithm.

The rendering algorithm requires some form of input to tell it where to place the dots. This usually comes in the form of a page description language. Like it sounds, this is a language for describing what goes where on a page, but it does so in higher-level terms than describing where each dot of ink or pixel of light is placed. A page description language deals in things like lines, circles, gradients, margins, and fonts (see Figure 19.1).

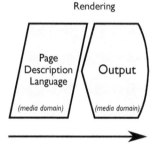

Figure 19.1 – The rendering algorithm

One example of a page description language is PostScript. Figure 19.2 shows the PostScript code for drawing a circle.

```
100 100 50 0 360 arc closepath
stroke
```

Figure 19.2 – PostScript code for drawing a circle.

This code moves a virtual pen over a virtual output device; the equivalent of a hand guiding a pen over paper. But it is a much lower-level operation than you need to worry about in structured writing.

The encoding algorithm

Since most writers don't write directly in a page description language, the page descriptions for your publication are almost certainly going to be created by an algorithm. I call this the encoding algorithm because it encodes your content in the page description language.

While it is possible that a highly specialized publishing tool chain may require a custom-made encoding algorithm, most encoding algorithms are implemented with existing tools that translate formatting languages into page description languages (see Figure 19.3).

Several formatting languages are used in content processing. They are often called typesetting languages. XSL-FO (XSL Formatting Objects) is one of the more commonly used languages in structured writing projects. TeX is another.

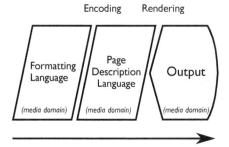

Figure 19.3 – Adding the encoding algorithm

Figure 19.4 shows the XSL-FO sample described in Chapter 12.

```
<fo:block space-after="4pt">
   <fo:wrapper font-size="14pt" font-weight="bold">
     Hard Boiled Eggs
   </fo:wrapper>
</fo:block>
```

Figure 19.4 – Example of XSL-FO

You process XSL-FO using an XSL-FO processor such as Apache FOP. The XSL-FO processor produces a page description language, such as PostScript or PDF, as output.

Writers are not likely to write in XSL-FO directly, though it is not impossible to do so. In fact some boilerplate content such as front matter for a book does sometimes get written and recorded directly in XSL-FO. (I did this on one project.) But when you are constructing a publishing tool chain, you will need to select and integrate the appropriate encoding tools as part of your process.

The job of the encoding algorithm is to take a high level description of a page or a set of pages, their content, and their formatting and turn it into a page description language that lays out each page precisely. For publication on paper, or any other fixed-sized media, this involves a process called pagination: figuring out exactly what goes on each page, where each line breaks, and when lines should be bumped to the next page.

For example, the pagination function figures out how to honor the keep-with-next formatting in an application like Word or FrameMaker. It also has to work out how to deal with complex figures such as tables: how to wrap text in each column, how to break a table across pages, and how to repeat the header rows when a table breaks to a new page. Finally, it has to figure out how to number each page and then fill in the right numbers for any references that include a particular page number.

These are complex and exacting operations, and you need to select a formatting language that can render your content the way you want it.

You also have to consider how automatic you want all of this to be. In a high-volume publication environment you want it to be fully automatic, but this could involve accepting some compromises. For example, in book publishing it is not uncommon for writers and editors to make slight edits to the actual text of a document in order to make pagination work better. This is very easy to do when you are working in the media domain in an application like Microsoft Word or FrameMaker. If you end up with the last two words of a chapter at the top of a page all by itself, for instance, it is usually possible to find a way to edit the final paragraph to reduce the word count just enough to pull the end of the chapter back to the preceding page.

This gets much harder to do when you are writing in the document domain or the subject domain, particularly if you are single sourcing content to more than one publication or reusing content in many places. An edit that fixes one pagination problem could cause another, and requiring this type of editing dumps back on writers the kind of formatting complexity that the document and subject domains are meant to remove.

For web browsers and similar dynamic media viewers, such as ebook readers or help systems, the pagination process takes place dynamically when content is loaded into the view port, and pagination can be redone on the fly if the reader resizes a browser window or rotates a tablet. This means publishers have little opportunity to tweak the pagination process. They can guide it by providing rules, such as the CSS keep-together instruction, but they obviously cannot hand tweak the text to make it fit better each time the view port is resized.

The formatting language for dynamic media viewers is typically Cascading Style Sheets (CSS).

The formatting algorithm

The formatting algorithm generates the formatting language that drives the encoding and pagination process. The formatting algorithm produces a media-domain representation of the content from content in the document domain or a more abstract media-domain format (see Figure 19.5).

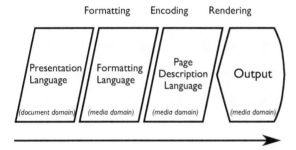

Figure 19.5 – Adding the formatting algorithm

In the case of HTML output, the formatting algorithm generates HTML (with connections to the relevant CSS, JavaScript, and other formatting resources). This is the end of the publishing process for the web, since the browser will perform the encoding algorithm internally and the computer operating system will likely take care of the rendering. In the case of paper output, the formatting algorithm generates a formatting language such as TeX or XSL-FO which is then fed to the encoding algorithm as implemented by a TeX or XSL-FO processor.

In some cases, organizations use word processing or desktop publishing applications to tweak the formatting of the output by having the formatting algorithm generate the input format of those applications (typically RTF for Word and MIF for FrameMaker). This allows them to exercise manual control over pagination, but with an obvious loss in process efficiency. In particular, any tweaks made in these applications are not routed back to the source content, so they will have to be done again by hand the next time the content is published.

This algorithm is usually the province of publication designers. One of the most elementary structured writing algorithms is to separate content from formatting (Chapter 10) which means removing formatting as one of the writer's concerns. Almost every structured writing implement-ation will require you to write formatting algorithms. Even if you use an off-the-shelf language like DITA or DocBook, you will have to customize their formatting algorithms to get the formatting you want. I looked at some examples of basic formatting algorithms in Chapter 11.

The presentation algorithm

The presentation algorithm determines how content is going to be organized as a document. A pure document-domain document represents the presentation of the content. The job of the presentation algorithm is to produce a pure document-domain version of the content. This may mean producing the entire presentation from purely subject-domain content or handling the occasional subject-domain structure in a largely document-domain file.

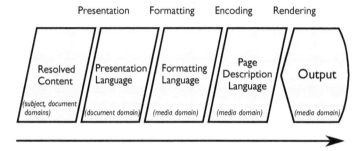

Figure 19.6 – Adding the presentation algorithm

Organizing content involves several things:

- **Ordering:** At some level, content forms a simple sequence in which one piece of information follows another. Writers working in the document domain typically order content as they write, but if they are working in the subject domain, the information architect can choose how to order subject-domain information in the conversion to the document domain.
- **Grouping:** At a higher level, content is often organized into groups. This may be groups on a page or groups of pages. Grouping includes breaking content into sections, inserting sub-heads, inserting tables and graphics, and inserting information as labeled fields. Writers working in the document domain typically create these groupings as they write, but if writers

are working in the subject domain, the information architect may have choices about how to group subject-domain information in the document domain.

- **Blocking:** On a page, groups may be organized sequentially or laid out in some form of block pattern. Exactly how blocks are laid out on the displayed page is a media-domain question that may even be done dynamically. However, to lay out blocks in the media domain the document-domain markup must clearly delineate the types of blocks in a document in a way that allows the formatting algorithm to interpret and act reliably.

- **Labeling:** Any grouping of content requires labels to identify the groups. This includes titles on blocks and labels on data fields. Labels are typically created by writers in the document domain, but are almost always factored out when writers work in the subject domain. Most document-domain labels indicate the place of content in the subject domain (which is why the label you factor out is often similar to the name of the subject-domain structure you insert when you refactor content to the subject domain). Inserting these labels in the presentation algorithm is therefore a necessary part of reversing the factoring out of labels that occurs when you move writing to the subject domain.

- **Relating:** Ordering, grouping, blocking, and labeling cover organization on a two-dimensional page or screen. But information is related to other information in complex ways that you can express by creating non-linear relationships between pieces of content using mechanisms such as hypertext links and cross references.

Differential presentation algorithms

As I described in Chapter 12, you cannot ignore differences in media when organizing content in the document domain. Although the fact that a relationship exists is a pure document-domain issue, how that relationship is expressed, and even whether it is expressed or not, is affected by the media and its capabilities.

Following links in online media is cheap. Following references to other works in the paper world is expensive, so document design for paper tends to favor linear relationships while document design for the web favors hypertext relationships. Therefore, you should expect to implement differential single sourcing and use different presentation algorithms for different media (see Figure 19.7).

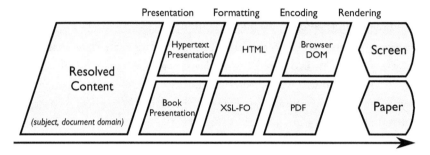

Figure 19.7 – Adding differential presentation algorithms

Presentation sub-algorithms

Many other structured writing algorithms, are executed as part of the presentation algorithm, including the following:

The linking algorithm

How content is linked or cross-referenced is a key part of how it is organized in different media and a key part of differential single sourcing. I looked at the linking algorithm in detail in Chapter 18.

The content generation algorithm

Part of the presentation of a document or document set is generating the table of contents, index, and other navigation aids. Generating these is part of the presentation process. Because these algorithms create new resources by extracting information from the rest of the content, it is often easier to run these algorithms in sequence after the main presentation has run. This also makes it easier to change the way you generate a TOC or index without affecting your other algorithms. For more on the content generation algorithm, see Chapter 14.

The synthesis algorithm

The synthesis algorithm determines exactly what content will be part of a content set. It passes a complete set of content to the presentation algorithm, which turns it into one or more document presentations. If you are pulling in content from multiple sources and multiple source formats, the synthesis algorithm is where you collect and sanitize content to make sure it is ready for the later processing stages.

Among other things, the synthesis algorithm resolves all management-domain structures in the content (unless some are to be retained for downstream post-publication algorithms to work with). This means that it processes all inclusions and evaluates all conditions. The result is document-domain or subject-domain content with the management-domain structures removed and replaced with the appropriate document- or subject-domain structures.

In the case of document-domain content, processing the management-domain structures yields a document-domain structure the presentation algorithm may be able to pass through unchanged (that is, the document-domain markup may already express the desired presentation).

In the case of subject-domain content, processing management-domain structures yields a definitive set of subject-domain structures that can be passed to the presentation algorithm for processing to the document domain (see Figure 19.8).

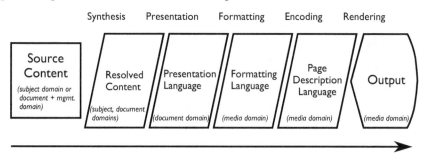

Figure 19.8 – Adding the synthesis algorithm

Differential synthesis

I noted above that you can use differential presentation to do differential single sourcing when you have two publications that contain the same content but are organized differently. If you want two publications in different media to have differences in their content, you can do this by doing differential synthesis and including different content in each publication (see Figure 19.9).

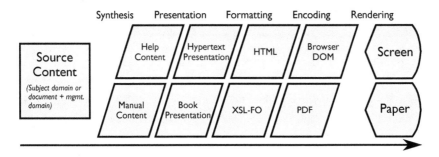

Figure 19.9 – Adding differential synthesis algorithms

Synthesis sub-algorithms

A number of structured writing algorithms are executed at the synthesis stage. In order to keep things well-partitioned, it is often advisable to execute each one as a separate sub step in the synthesis stage.

The reuse algorithm

Pulling in reused content is part of the synthesis process. I discussed the reuse algorithm in Chapter 13.

The content-generation algorithm

As noted in Chapter 14 the content-generation algorithm presents an alternative approach to creating different forms of output from a single source of data. It is run as part of the synthesis algorithm when creating raw content and as part of the presentation algorithm when creating presentation artifacts, such as a table of contents. The content generation algorithm can be used in both the synthesis and presentation stages, since can generate either synthesis or presentation artifacts.

The extract algorithm

In some cases you may wish to extract information from external sources to create content. This can include data created for other purposes, such as application code, or data created and maintained as a canonical source of information, such as a database of features for different models of a car. Extraction is part of the synthesis process. See Chapter 15.

The merge algorithm

In one sense, every structured writing algorithm is a merge algorithm. As described in Chapter 11, most algorithms process content by factoring in information or metadata that was factored out as the point of recording moved from the media domain to the document domain and the subject domain. But it is also possible, and often useful, to combine information from different sources to create a new set of content. Merging content is done at the synthesis stage. See Chapter 16.

The modeling algorithm

Extracting content from models also occurs in the synthesis process (see Chapter 17).

Deferred synthesis

For static presentation, all synthesis happens before the material is presented. But if you are presenting content on the web, you can defer parts of the synthesis algorithm to the browser, which can synthesize and present content by making calls to web services or back-end data sources or by making a request to code running on the server. I look at this in Chapter 20.

Combining algorithms

As described in (Chapter 11), structured writing algorithms are usually implemented as sets of rules that operate on structures as they are encountered in the flow of the content. Since each algorithm is implemented as a set of rules, you can run two algorithms in parallel by combining the two sets of rules to create a single set of rules that implements both algorithms at once.

Obviously, you need to avoid clashes between the two sets of rules. If two sets of rules act on the same structure, you have to ensure that they work together.

In other cases, one algorithm needs to work with the output of a previous algorithm. In that case, you need to run them in sequence.

In most cases, the major publishing algorithms (synthesis, presentation, formatting, encoding, and rendering) must be run in sequence, since they transform an entire content set from one domain to another (or from one part of a domain to another). In many cases the sub-algorithms of these major algorithms can be run in parallel by combining their rule sets, since they operate on different content structures.

Architecture of a publishing tool chain

The architecture of a publishing tool chain should facilitate the partitioning and distribution of complexity in the content system. The structured writing algorithms I have looked at in this part are executed at some point in the publishing process. The management algorithms described in Part VI are generally executed separately.

Your partitioning needs to ensure that any necessary information is passed from one partition to another, whether the task is executed by a person or an algorithm. This means that a partition must be executed after the information it needs becomes available and before the information it produces is needed.

The SPFE architecture – Synthesis, Presentation, Formatting, Encoding – provides a general framework into which the various algorithms and their timing fit reasonably well. However, don't get hung up on either the names or the order of operations that this architecture describes. The goal is to partition and distribute the complexity of your content system so that none of the complexity goes unhandled and no one person or process is asked to handle more of the complexity than they have the time and resources to manage. Any architecture that accomplishes that for your system is a good architecture.

The keys to partitioning the complexity of code are to keep code units small and simple and to reuse code whenever you can. Each piece of code should do one thing and one thing only. Processing content written in the subject domain or the document domain, particularly with management-domain structures added in, is complex and requires the execution of many algorithms. Writing a single program to execute all of those algorithms at once would be complex and would violate the principle of simplicity and leave little room for code reuse.

Every step in a content publishing chain reads in one or more structured content files and produces one or more structured content files in a different domain or, at least, nearer to the media domain and dots on a page than the input file.

In Chapter 11, I presented algorithms that moved subject-domain content to the document domain, and other algorithms that moved that same content from the document domain to the media domain for publishing. This is the template for all publishing processes. In the real world, though, you may have more than one intermediate stage in the journey. This type of architecture is sometimes referred to as a publishing pipeline. Each step in the publishing process is kept simple by doing just one job and then passing the content to the next step in the pipeline.

A publishing pipeline requires individual programs to be simple and straightforward, which makes them easier to design, write, debug, and maintain. A pipeline with many small self-contained programs is usually cheaper to create and maintain and more robust than a single monolithic program that tries to do everything in one step.

The pipeline approach also allows for a great deal of code reuse. For example, if your content is written in the subject domain, you need to get it all the way to the media domain for publication. But along the road to the media domain, it passes through the document domain. There are many robust document-domain publishing tool chains available, including DocBook, DITA, reStructuredText, and LaTeX. You could write a presentation algorithm to transform each of your subject-domain structures into one of these formats and then simply use that format's existing tool chain for the rest of your publishing pipeline.

Other opportunities exist for code reuse at a more granular level. For instance, in addition to subject-specific structures and subject annotations, your subject-domain content will need paragraphs, lists, and other basic text structures. You can use the same definitions for basic text structures and subject annotations across your entire content set and write one rule set for those text structures and subject annotations that you can use everywhere, greatly reducing the amount of code you have to write.

CHAPTER 20
Active Content

When you publish to electronic media, you can create active content, that is, content that has behavior as well as formatting. Here are some examples:

- **Personalized content:** You can select and arrange content for individual readers based on the things you know about them. For instance, if you are logged into Amazon, it customizes elements of every page based on your previous purchases, your wish list, and things you previously browsed.
- **Dynamic arrangement:** Part of the presentation algorithm is arranging content on the page or screen, but with online media you can allow readers to arrange the content. For instance, you can publish tables that they can sort for themselves.
- **Adaptive content:** Similarly, you can create content that adapts itself dynamically to the view port by, for instance, displaying in multiple columns on a wide view port and in a single column on a narrow one.
- **Progressive disclosure:** You can initially display only part of the content and then reveal more when a reader clicks on a link or takes another action. For instance, you might show a high-level summary of a procedure and provide a link that opens detailed steps for those who need them. This is a way to cater to audiences with different levels of knowledge or skill.
- **Feeds and dynamic sources:** You can include content that comes from an independent external source, such as a feed or web service.
- **Interactive media:** You can include apps, widgets, and media that readers can interact with.

Active content simply postpones execution of one or more of the structured writing algorithms until the point at which the content is read.

For example, for personalized content, the synthesis algorithm is executed when you request a page, using personal details to select content for you. When you resize your browser or rotate your phone, the presentation algorithm is executed in the browser, changing the layout of the page. If the browser allows you to change the font size, the formatting algorithm is executed in the browser. If you have a page that pulls in a live stock quote, the synthesis algorithm is executed by the web server or browser when the page is served/displayed.

In principle, therefore, you support active content the same way you support these algorithms in any form of structured writing: you create content in the appropriate domain and with the appropriate structures to reliably support the algorithms you need.

Reliability is key. If you create static books or web pages, you have the opportunity to review the output and even hand tweak it to fix any issues, but you can't do that with active content. You need to trust the algorithm 100%, which means you need to trust the content it is working on 100%. This makes conformance crucial to an active-content project, and, as we have seen, different domains support conformance with varying degrees of reliability.

Another factor to consider is that while active content essentially postpones the execution of one or more algorithms until the content is in the user's hands, you don't necessarily want to postpone running those algorithms on the whole page. You usually postpone execution only for those parts of the page that you want to make active.

For instance, if you want real-time stock quotes embedded in your business stories whenever a company is mentioned, there is no reason to postpone running the synthesis, presentation, and formatting algorithms for the entire page; you need to postpone only the synthesis of the stock quote itself. And if you want that quote to continue to update in real time, while the page is displayed, you certainly don't want to redraw the whole page every time the price changes.

Displaying stock quotes depends on subject-domain metadata that unambiguously identifies the company mentioned in the content. For example:

```
{Microsoft}(company "NASDAQ:MSFT") is a large software company.
```

In a normal, static publishing flow, this subject-domain information would be resolved into a static piece of content by the presentation algorithm:

Microsoft US$60.29 +0.34 (+0.57%) is a large software company.

However, for active content the main presentation algorithm needs to pass the subject-domain information about Microsoft to the browser, so it can retrieve the stock price in real time when the content is displayed.

This means that only those parts of the content that you want displayed as active content need to be captured in the domain and structures required to reliably run the applicable algorithms. The rest of the content can be in a different domain. In practice, this usually means that you embed subject-domain structures to support the active content in document-domain structures. Of course, you can have all your content in the subject domain, in which case you would run the presentation algorithm on the content you want to deliver statically, and pass the subject-domain representation of the active content to the system that will run the active-content algorithms.

However, this is not likely to be a straightforward pass through. The downstream system that will execute the active-content algorithm – for example, a JavaScript app running in the web browser – may expect or require the subject-domain information to be in a particular format. For instance, you might have to embed the ticker symbol in a function call that returns stock quotes:

```
<p>Microsoft (<span onload="getStockQuote(NASDAQ:MSFT)"/>)
is a large software company.</p>
```
Figure 20.1 – HTML markup to get a stock quote

In this case, your server side presentation algorithm needs to transform your subject-domain annotation into the format shown in Figure 20.1 using a rule such as that shown in Figure 20.2.

```
match company
    continue
    $fncall = 'getStockQuote(' + @specifically + ')'
    output ' ('
    create span
        create attribute onload = $fncall
    output ') '
```
Figure 20.2 – Pseudocode to create the HTML markup

The document domain represents the desired presentation of a piece of content. Therefore, the only active-content algorithms that you can execute on ordinary document-domain structures are those that leave the presentation alone and only affect the formatting. For example, you can allow readers to select the font size of a web page, which is a purely formatting concern.

If you want to do active content that dynamically changes the presentation of the content or that changes the text of the content (as in our stock price example), then you need either subject-domain structures or document-domain structures that are explicitly designed to support active-content algorithms.

For example, a generic table structure does not allow the reader to sort on any column. Sorting by column only makes sense if the content is inherently sortable.

item	legs	price
table	4	$400
stool	3	$20
shooting stick	1	$75
chair	4	$60

Figure 20.3 – Table with sortable columns

For example, in Figure 20.3 the columns are sortable. The reader could choose to sort the table on the item name, the number of legs, or the price, all of which might be useful arrangements.

But consider what might happen if you sorted the table in Figure 20.4 on the second column.

1. Don protective clothing.

2. Clear the area.

3. Block all entrances.

4. Activate the destruct sequence.

Figure 20.4 – Table with columns that shouldn't be sorted

Thus, you can't just make all tables sortable by column. Unless your document-domain structure explicitly states which columns are sortable and which are not, you can't implement this kind of active content.

Therefore, to implement column sorting at the document-domain level, you need a table structure which ensures that the sorting behavior is applied only to columns where it makes sense in tables where it makes sense.

Although the document domain typically requires different markup for different algorithms, the subject domain typically does not. In the subject domain you capture the semantics of the subject matter, which are the same no matter what algorithms you apply.

In the subject domain, the product list would be a structured data set with known semantics and would probably be maintained in a separate database, as shown in Figure 20.5.

```
products:: item, legs, price
    table, 4, $400
    stool, 3, $20
    shooting stick, 1, $75
    chair , 4, $60
```

Figure 20.5 – Product list marked up as a structured data set

Knowing what the semantics are, you can determine whether the data set is sortable and, if so, which columns can be used as sort keys, thus letting you know whether it can be presented as a sortable table.

Creating your content in the subject domain gives you the greatest flexibility to generate active content in ways that are appropriate to the subject matter and the device. And because the subject domain does not require different structures to support active content, your writers don't have to understand or even think about how the active content might work. This partitions the complexity of active content from your writers. As we have seen before, partitioning content from its presentation lets you experiment with different forms of presentation. This allows you to test whether active content is working and to change the form of active content you use, all without involving writers or changing your content.

This does not mean that active content is a free benefit of using the subject domain. You still have to design and implement the content behavior, and you have to make sure that you capture the subject-domain semantics needed to drive the active-content behaviors you want. Those semantics may not be different in kind from other subject-domain semantics, but you may need to provide greater detail than you would for other algorithms.

You also need to ensure a high degree of conformance from your writers, because it is difficult to validate the correct operation of active-content algorithms at run time. The success of your active-content strategy depends on the quality and consistency of your input data.

Structures

Now that we have examined the structured writing algorithms, it is time to look at structures in more depth. Once you decide to adopt structured writing, all of your content has to be expressed in structures. This creates issues with representation and functional lucidity, which I examine in this section, looking at how the approach to structure for each domain differs, and the impact those differences make to process, rhetoric, and functional lucidity.

CHAPTER 21
Content as Data

Structured writing represents content as data in order to make it accessible to algorithms. Conventional computer data structures such as relational database tables do not work well for content because they are too regular to fit the shape of content. Creating structures that are regular enough for algorithms to deal with yet irregular enough to fit the rhetorical patterns of written language is a challenging problem.

Fundamentally, a document is just a stream of characters. However, that stream of characters contains rhetorical structures such as headings, bibliographical entries, bold text, chapters, ingredient lists, links, wine matches, tables, function signatures, and labeled lists. In addition, you may need to annotate subject affinities and attach management metadata. The question is, how do you express these various structures within the raw stream of characters?

The most common answer is to divide the text into a series of blocks, divide those blocks into smaller blocks, and continue dividing until all the structures you need are contained in blocks. This is not the only way to do it. Some file formats, such as WordPerfect, use independent stop and start markers to delineate structures, meaning that the boundaries of structures can overlap. But while this kind of structure can work in the media domain, it is difficult for structured writing algorithms to work with such structures. Therefore, most structured writing today uses the nested block approach.[1]

There are cases where non-overlapping block structure is inadequate for some rhetorical structures. This is more of an issue for the academic study of texts than for structured writing, but it can occur in both fields. However, allowing overlapping fields adds complexity to many of the structured writing algorithms, and it is rarely a worthwhile approach to partitioning content.

For our purposes, therefore, I am going to deal with the mechanical structure of content strictly in terms of nested, non-overlapping blocks, such as shown in Figure 21.1.

---

[1] It is technically possible to implement independent stop and start markers, even in languages that are mainly block based. You simply define empty blocks for the start and stop markers. Both DITA and DocBook do this to delineate arbitrary bits of content for reuse and to define arbitrary spans of content for indexing. I recommend against this practice in content that has a lifespan beyond its first publication.

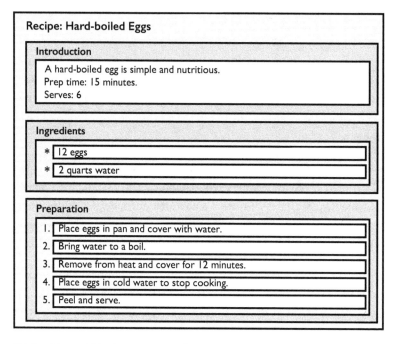

Figure 21.1 – Nested block structure of a recipe

Flat versus hierarchical structures

Even using a strictly nested blocks approach requires some fundamental choices about mechanical structure. The first is flat versus hierarchical structure.

As noted in Chapter 3, HTML has six levels of heading (h1 through h6), whereas DocBook has only title. DocBook lets you divide a document into sections and nest sections inside sections. You can then print the titles of sections inside sections in a smaller font than the titles of first-level sections. You can render headings in different sizes without having six different tags.

But the DocBook model assumes that the real structure of a document is a hierarchy of nested sections and that the size of titles announces the steps up and down that hierarchical tree. HTML makes no such assumption. It will let you put an <h4> immediately after an <h1> if you want to. It treats documents as essentially flat structures punctuated by headings of various sizes.

Which model of a document is correct? Which corresponds best to the rhetorical structure of the document? You can think of a document as being organized hierarchically, with major ideas expressed in sections, sub-ideas supporting the major ideas in subsections, etc. There are many documents that fit that model. But you can also think of documents as being more like a journey in which headings function like road signs. A city gets a big sign, a hamlet a small sign, and a town a medium sign. But the town is not inside the city, the hamlet is not inside the town, and there is no guarantee that on leaving the city you will come to a town before you come to a hamlet.

Studies by Peter Flynn indicate that most writers think of documents much more in terms of a punctuated linear model (a journey) than a hierarchical model.

> The classical theory, derived from computer science and graph theory, is that the document is a hierarchical tree (actually inverted: a root-system) and that all necessary actions can be seen in terms of navigation around the tree, and of insertion into and withdrawal from the the nodes which form the branches and leaves.
>
> The conventional writer, however — and we expressly exclude the markup expert, as well as the experienced technical writers who responded to the survey — is by repute probably only marginally aware of this tree; but we have been unable to measure this at present. In this view, the document is seen as a continuous linear narrative, broken into successive divisions along semantic lines, and interspersed with explanatory material in the form of figures, tables, lists, and their derivatives.
> —Flynn, Peter. "Why writers don't use XML: The usability of editing software for structured documents." Presented at Balisage: The Markup Conference 2009, Montréal, Canada, August 11 - 14, 2009. In Proceedings of Balisage: The Markup Conference 2009. Balisage Series on Markup Technologies, vol. 3 (2009). doi:10.4242/BalisageVol3.Flynn01.

As Flynn's research illustrates, there is a good chance that writers don't think of their arguments as a strict hierarchy of points. Rather, they think of them as a sequence of points with headings occasionally inserted to break up the text or signal a change in emphasis or subject matter.

Nevertheless, to clearly define the rhetorical structure of a document, you have to divide it into blocks such as introduction, ingredients, and preparation. And, as previously noted, treating these parts of the rhetorical structure as blocks is essential to establishing a context where blocks can be identified and processed by algorithms.

Some blocks are naturally more hierarchical than others. Even so, structured writing forces writers to think more hierarchically than they might otherwise, which means imposing a little bit of complexity on them to support all the partitioning and redirection of complexity needed for the content system overall. However, there is no need to do any more of this than the partitioning of the system demands.

If the structured writing constraints that you need to express demand hierarchy, while functional lucidity demands a more punctuated linear model, how do you design a markup language that reconciles these two opposing requirements?

This is of greatest concern in the design of document-domain languages. The structure of media-domain languages follows the shape of the media they are modeling, which is largely flat. In the subject domain, content has been abstracted out of strict document order, which means that hierarchy in the subject domain tends to match the hierarchy of relationships in the subject matter itself.[2] In the document domain, however, this is a real concern. The document domain consists of abstractions of document structures and the nature of their relationship to the structure of thought in the text is not obvious.

Outside of the media domain, the options available are:

- Use a flat document-domain language. Examples are HTML and Markdown. The problem here is that a flat language imposes few constraints, and the lack of context-setting hierarchy makes it hard to model different types of document structures without creating hundreds of tags, which negates any functional lucidity that you gain by keeping the language flat.
- Use a hierarchical document-domain language that has a permissive structure, so you can put blocks inside blocks in lots of different ways. An example of this is DocBook. The problem here is that the permutations make writing algorithms difficult, and you often need to impose additional writing constraints that go beyond those expressed or enforced by the markup language itself. This again diminishes functional lucidity and compromises conformance. (An interesting property of this approach is that the flexibility of the language means that writers can create documents that are very hierarchical or very flat. However, this is not really a virtue, because it is not clear how this choice contributes to improved content quality.)
- Create a smaller, stricter document-domain language that fits the types of documents you want to write. The main difficulty with this approach is that you have to design your own language, which many organizations try to avoid. (In some cases you can reduce this effort

---

[2] This is not universal. For example, addresses, which are based on hierarchical locations, are typically modeled as flat ordered lists. The order reflects the hierarchy, but the structure does not nest city inside country and street inside city.

by using a restricted subset of an existing language such as DocBook, something we will look at in Chapter 34.)

■ Move content creation to the subject domain. This also involves creating your own languages, but it can provide support for rhetoric and process while improving functional lucidity.

Annotating blocks

When treating content as data, you annotate every block to identify – to people and algorithms – what constraints it meets. The most basic annotation is the name of the block, which describes its type. The name of a table block is `table`, which tells us that the block is a table. Blocks may also have other annotations that either refine the type or provide additional information.

These additional annotations don't have to be in the same domain as the block name. Figure 21.2 repeats an example of this from Chapter 4.

```
<section publication="Wine Weenie">
    <title>Wine match</title>
    <p>Pinot Noir</p>
</section>
<section publication="The Teetotaler's Trumpet">
    <title>Suggested beverage</title>
    <p>Lemonade</p>
</section>
```

Figure 21.2 – Conditional text markup (XML)

In Figure 21.2, the `section` element defines a block in the document domain, and the `publication` attribute adds an annotation in the management domain.

Agreeing on names

For structured writing to work, everyone involved must understand and agree on what the names and annotations mean. Annotations tell people and processes what constraints each block obeys. If the names and annotations don't mean what we have all agreed to, communication between the partitions of the content system breaks down and both process and rhetoric suffer.

Confusion and disagreement about the meaning of names and annotations are not uncommon. Large document-domain languages such as DocBook and DITA have large vocabularies, and many of the names they offer are quite abstract. Questions about the right way to tag certain passages are common in the communities around these languages, and opinions can vary considerably in some cases.

These disagreements don't affect just low-level structures. In DITA, for example, it is common to debate whether a topic that contains operational information, but is not procedural, is a concept or a task. And some writers use only generic topics because they don't think DITA's task, concept, and reference topics fit their content.

Therefore, you must define terms precisely when developing a structured writing language. But it is equally important that the language be functionally lucid. Functional lucidity requires you to create structures that do not detract from writing. You can't meet this requirement if writers must puzzle out or debate how to mark something up. You must both define what the names mean and choose structures that are easy to name correctly. As described in Chapter 29, the best way to achieve conformance is to design structures that are easy to conform to.

Different rules for intermediate languages

Of course, functional lucidity only matters for the formats that writers actually write in. As we have seen, the publishing algorithm typically consists of multiple steps, and each one of those steps can create a format that is closer to the media domain than the one before it. It is perfectly possible to design a document-domain structure just to serve as a step in the publishing chain. Separate authoring formats are created for writers to actually write in (perhaps subject-domain formats or simplified highly constrained ad-hoc document-domain formats). Content is transformed from these formats to the document-domain format by the presentation algorithm and then the document-domain format is translated into various different media-domain languages by the formatting software. An arrangement like this eliminates the need to compromise between different demands in designing a single language, generally making each language in the chain simpler and more constrained, which in turn makes each one easier to validate and process.

Secondary structures of interpretation

Under normal circumstance, the structures that constrain content also constrain the interpretation of that content. In other words, the markup that tells the writer what to write also tells the algorithm (or other downstream user) what the content means. But there are cases where annotations are added to document-domain structures to tell downstream processes what the content means.

These structures don't constrain writers, which means they don't guide them either. Writers must know exactly how to create this kind of annotation without any guidance or prompting.

Consider these examples of HTML microformats from Wikipedia.[3] Figure 21.3 shows an address formatted as a list.

```
<ul>
   <li>Joe Doe</li>
   <li>The Example Company</li>
   <li>604-555-1234</li>
   <li>
     <a href="http://example.com/">http://example.com/</a>
   </li>
</ul>
```

Figure 21.3 – HTML markup for an address

In Figure 21.3, the phrase "The Example Company" is contained in `li` tags. This is part of a list structure delineated by `ul` tags, so the markup is largely structural in the document domain. The `li` tag does not tell you what the content itself is about. It does not tell you anything useful beyond what document structure it belongs to. The second example, Figure 21.4, adds vCard microformat markup:

```
<ul class="vcard">
  <li class="fn">Joe Doe</li>
  <li class="org">The Example Company</li>
  <li class="tel">604-555-1234</li>
  <li>
    <a class="url" href="http://example.com/">
       http://example.com/
    </a>
  </li>
</ul>
```

Figure 21.4 – vCard microformat markup for an address

Figure 21.4 adds subject-domain metadata in the form of the class attributes. For example, it says that the phrase "The Example Company" refers to an organization (`org`). This annotation does not modify or refine the constraint expressed by the `li` tag. It is saying something else entirely.

---

[3] https://en.wikipedia.org/wiki/Microformat

But this is not just about suggesting a different way of interpreting "The Example Company" (as an organization name, as well as as a list item). The vCard markup expresses a complex subject-domain structure. Not only is the list item "The Example Company" annotated as `org`, the list that contains it is annotated as `vcard`. The interpretation of `org` actually depends on that annotation being part of a `vcard` structure.

In other words, the secondary structure created by the annotations in the Figure 21.4 is equivalent to the pure subject-domain markup shown in Figure 21.5.

```
vcard:
    fn: Joe Doe
    org: The Example Company
    tel: 604-555-1234
    url: http://example.com/
```

Figure 21.5 – Subject-domain markup for an address

The microformats overlay a second structure on the list structure. In the world of HTML, this makes sense. HTML needs to be a standardized document-domain language so that browsers can display it for human reading. Humans don't need the vCard annotations to recognize that the content is an address, but algorithms do. So the microformat adds a second, hidden, subject-domain structure to the document for readers that are algorithms rather than people.

Structured writing constrains both the creation and the interpretation of content. Normally, you expect that the creation of content would be just as constrained as the interpretation. After all, it is hard to reliably interpret structure if the creation of that structure is not constrained. However, in this case the interpretation of the data is more constrained than the creation.

Asking writers to add unconstrained subject-domain metadata to document-domain structures would be inefficient and error prone. But this is only a problem if we ask writers to author in this format. If writers author in a format that constrains creation to the same extent that we need to constrain output, it does not actually matter that the resulting output constrains interpretation more than it constrains creation. Our concern as content creators is simply to make sure that any content we produce that promises to abide by a constraint actually does so, whether the format we deliver it in actually imposes that constraint or merely annotates it.

So, you can confidently produce this information using subject-domain markup and then deliver it as HTML with vCard annotation markup using a presentation algorithm like Figure 21.6 (as with all example algorithms in this book, this example is pseudocode):

```
match vcard
    create ul
        attribute class = "vcard"
        continue

match fn
    create li
        attribute class = "fn"
        continue

match org
    create li
        attribute class = "org"
        continue

match tel
    create li
        attribute class = "tel"
        continue

match url
    create li
        create a
            attribute class = "url"
            attribute href = contents
            continue
```

Figure 21.6 – Pseudocode to convert subject-domain address markup to vCard format

Child blocks versus additional annotations

As noted above, sometimes the name of a block is not sufficient to fully describe the constraints it meets. In these cases, you can add additional annotations to a block. But you can also achieve much the same thing by adding child blocks. A child block belongs to the main block, so it is part of it and can constrain the interpretation of the main block just as much as an annotation does. One of the issues in designing data structures for content, therefore, is deciding when to use additional annotations and when to use child blocks.

Different markup languages have different levels of support for additional annotation on blocks, so this issue is affected by the markup language you choose. SAM, for instance, supports only a limited, fixed set of additional annotations on blocks. Therefore, if you want to add constraints on interpretation, you have to use child blocks.

In XML, however, there is broad (though not unlimited) support for additional annotations on blocks in the form of attributes. In the XML world there is considerable choice, and considerable debate, about when and where to use elements versus attributes in your content models.

Consider, for example, this XML element that contains two attributes but no content:

```
<author-name first="Mark" last="Baker">
```

The element is called `author-name`, and it has two attributes, `first` and `last`, which contain my first and last name respectively.

Why is this marked up like this and not like this:

```
<author-name>
    <first>Mark</first>
    <last>Baker</first>
</author-name>
```

Both of these constructs express the same information, and both clearly constrain how you mark up an author's name and how algorithms should interpret that markup.

Is one of these options correct and the other incorrect? When should you use attributes and when should you use elements?

Consider our vcard example. You could write it this way, using just elements:

```
<vcard>
    <fn>Joe Doe</fn>
    <org>The Example Company</org>
    <tel>604-555-1234</tel>
    <url>http://example.com/</url>
</vcard>
```

Or you could write it this way, using attributes:

```
<vcard
    fn="Joe Doe"
    org="The Example Company"
    tel="604-555-1234"
    url="http://example.com/"
/>
```

The first says that `fn`, `org`, `tel`, and `url` are independent structures that belong together as members of a `vcard` structure. The second says that the `vcard` structure has a number of data fields – annotations – that complete its meaning.

Does this distinction matter terribly? Both allow you to get at the information you want. Both constrain the creation and the interpretation of data. There are limits to the version that uses attributes. In XML markup, you can't have more than one attribute with the same name, whereas you could have more than one child block with the same name (multiple `tel` elements for someone with more than one telephone number for instance). Also, XML specifies that attributes are unordered, so you can't restrain either the order in which writers create them or the order in which the parser reports them to a processing application.

Given this, you may be wondering why people bother with attributes, since you can do the same things with elements and have both more flexibility and more capacity to impose constraints. Yet people continue to use attributes extensively when designing markup languages in XML. When people create XML document types for representing data, rather than for writing documents, they almost always use the attribute format, perhaps because it is slightly less verbose and slightly easier to read or, perhaps, because as programmers they are accustomed to representing data as key/value pairs linked with = signs.

But for documents it is more complex. To understand why XML even has attributes and why other languages, such as SAM or reStructuredText, also have similar mechanisms for adding annotations to blocks, you need to go back to the original concept of markup as something written onto a manuscript after the fact. In this view, markup is an addition to the text, not part of it. The content of an element is part of the underlying text. Anything you want to add, therefore, cannot be element content, since that would be adding to the text. Everything else has to be added to element definitions as attributes.

This view is reinforced in the way academics use markup to prepare texts for study. Again, here, the text is preexisting and canonical. The markup is external to the text, and therefore, everything that is external to the original text must be contained in the markup itself (as attributes) and nothing that is part of the original text can be removed or replaced by markup. (This is a form of partitioning in its own right, to serve a particular purpose.) Thus in this fragment (Figure 21.7) of Shakespeare's *All's Well That Ends Well*, marked up by Jon Bosak, you can see that the original text is kept perfectly intact.

```
<ACT>
  <TITLE>ACT I</TITLE>
  <SCENE>
    <TITLE>SCENE I. Rousillon. The COUNT's palace.</TITLE>
    <STAGEDIR>
      Enter BERTRAM, the COUNTESS of Rousillon, HELENA, and LAFEU,
      all in black
    </STAGEDIR>
    <SPEECH>
      <SPEAKER>COUNTESS</SPEAKER>
      <LINE>In delivering my son from me, I bury a second husband.</LINE>
    </SPEECH>
    <SPEECH>
      <SPEAKER>BERTRAM</SPEAKER>
        <LINE>And I in going, madam, weep o'er my father's death</LINE>
        <LINE>anew: but I must attend his majesty's command, to</LINE>
        <LINE>whom I am now in ward, evermore in subjection.</LINE>
    </SPEECH>
    ...
```

Figure 21.7 – Fragment of Shakespeare's *All's Well That Ends Well* marked up in XML

The number and title of scenes can be factored out by refactoring text into markup using the techniques described in Chapter 3 and Chapter 4. If you do that, instead of:

```
<SCENE>
    <TITLE>SCENE I. Rousillon. The COUNT's palace.</TITLE>
```

you can factor out the scene number and the word "SCENE" like this:

```
<SCENE>
    <TITLE>Rousillon. The COUNT's palace.</TITLE>
```

Going further, you might have noted that the introduction of a scene is invariably the name of its location and done this:

```
<SCENE>
    <LOCATION>Rousillon. The COUNT's palace.</LOCATION>
```

or even this:

```
<SCENE location="Rousillon. The COUNT's palace.">
```

And similarly, you could replace:

```
<SPEECH>
    <SPEAKER>BERTRAM</SPEAKER>
```

with the following:

```
<SPEECH SPEAKER="BERTRAM">
```

Making a few changes like this in the markup would leave you with only the words actually spoken by the actors as the text of the play and everything else expressed as elements or attributes.

This actually makes quite a lot of sense, because the stage directions and identification of speakers in a play are metadata that annotates the actor's words, which are the only things the audience is supposed to hear.

So is the right way to markup a play to preserve the original printed text – which includes all of the playwright's metadata – or is it better to separate the playwright's metadata from the speeches that are the 'real' play?

While these questions may be important for the scholarly study of text, they are not nearly as important for structured writing. Our concern is to partition the task of content creation, which means you can simply choose the format that does that in the most functionally lucid way while ensuring that all the information required by the next partition is accurately captured. The fact that such questions exist, however, helps explain why a markup language like XML is structured the way it is and why so many texts are marked up the way they are – and why so many markup languages are designed the way they are.

The way we, as practitioners of structured writing, settle these matters is by asking ourselves which approach best supports the structured writing algorithms that we want to implement, always remembering that the reliability of every other algorithm depends on how well writers conform to constraints, which in turn depends on how well the authoring markup language supports conformance and functional lucidity.

In none of this is there any reason to be in the least concerned about preserving the canonical nature of a preexisting text. There is no preexisting text. Therefore, there is every reason to prefer sub-structures rather than annotations to express things such as our vCard example. In fact, SAM, which is designed specifically for structured authoring, supports only this format:

```
vcard:
    fn: Joe Doe
    org: The Example Company
    tel: 604-555-1234
    url: http://example.com/
```

SAM supports a limited set of annotations on blocks, all of which have predefined meanings. You could, in fact, eliminate annotations on blocks altogether and use child blocks for everything, but I have supported a limited set of common management-domain block annotations in SAM, mostly to improve functional lucidity.[4]

In summary, when defining the mechanical structure of your structured writing, don't get hung up on what is text and what is markup. In each domain, text and markup together form a body of constrained content that can be successfully created by a writer and successfully processed by one or more algorithms. Only when you resolve the content all the way to the media domain do you finally have to sort out exactly which characters appear in which order and which decorations you need to represent that content to a particular audience. When we choose to create content in the other domains, it is precisely because we want to exercise more control over these things and to use algorithms to help us create and manage them. Whether some idea or constraint is expressed by text or markup in those domains should be based solely on what works best in those domains.

---

[4] It is worth noting that XSLT, the language most commonly used to process XML, implicitly assumes that element content in an XML document is the text of the output document and that attributes are not, thus the text of elements passes through to output automatically unless you catch it and suppress it, whereas the text of attributes is ignored unless you catch and output them.

CHAPTER 22

Blocks, Fragments, Paragraphs, and Phrases

If the data structure of a structured document is a set of nested blocks, how do you design a content structure as a set of nested blocks? Let's start by looking at the different types of blocks, their purpose, and how they relate to each other. Most markup systems don't make a high-level distinction between different types of blocks (for example, in XML, they are all *elements*). However, from a language design point of view, it is useful to break down blocks into different types, each of which requires a different design focus.

In this chapter I look at four main types of block. These types are not mechanically different from each other; rather, they play a different design role in the overall structure of a document. The four types are:

- **Structural blocks:** the basic building blocks of structure.
- **Semantic blocks:** blocks that have a meaningful relationship to things writers understand and care about.
- **Information-typing blocks:** semantic blocks used in information typing theory, which is used in some markup systems, particularly DITA.
- **Rhetorical blocks:** blocks that govern the rhetorical structure of a document.

Although the above list is ordered by size, it is actually easier to explain the design roles of these block types if you start in the middle with semantic blocks.

Semantic blocks

If you ask writers to describe the things that make up a document, they will probably name things like paragraphs, tables, and lists. If you ask about a particular type of document, such as a recipe, they will probably name things like introduction, ingredients, and preparation steps. A structured writing language is typically much more fine-grained than this. For instance, a table may be made up of dozens of smaller structures, such as rows, cells, and cell contents. But tables, procedures, and lists are the units that have meaning to writers independent of how those units are constructed internally. Without knowing anything about the mechanics of structured writing, a writer could design the rhetorical structure of a piece of content as a set of such blocks, creating, for instance, a recipe with three main blocks: introduction, ingredients, and preparation.

At the risk of adding further burden to an already overloaded term, I am going to call these recognizable objects *semantic blocks*, because they are blocks that mean something in whichever domain they belong to. (Note that I am not using the term *semantic* to mean subject domain; blocks have semantics in all domains.) High-level markup design is essentially a matter of defining semantic blocks and the ways they go together.

An easy example of a semantic block is a list. A list is a semantic block because *list* is an idea with meaning in the document domain independent of its exact internal structure. A writer can say, "I want a list here," independent of any specifics of markup. If a structure has a name like this in the real world, the block that implements it is a semantic block in the terminology I am coining for this book.

Structural blocks

Semantic blocks generally contain other blocks that you might not talk about independently if they weren't needed to describe the construction of a semantic block. I call these *structural blocks*. Structural blocks are the construction details of a structured document. By analogy, in architectural terms, if a window is a semantic block, the lintel, sash, and jamb are structural blocks.

The distinction between a semantic block and a structural block is not hard and fast. It has more to do with design intent than any concrete characteristic. The main point of making the distinction is to encourage you to think of markup design first in terms of semantic blocks. That is, blocks, with whatever internal structure you require, that capture the structure of something that is real and meaningful to you. Don't get bogged down in the precise internal structure until you figure out which semantic blocks you need.

Different markup languages often construct the same semantic block differently. DITA, DocBook, and HTML all define lists, and each defines them differently. Nonetheless, you can recognize that each implements a list, and for most purposes, you can mark up your content using any one of them without any loss of functionality.

A list is made up of structural blocks that build the shape of a list. Figure 22.1 illustrates this with XML, which makes the blocks easy to see.

```
<ol>
    <li>
        <p>This is the first item.</p>
    </li>
    <li>
        <p>This is the second item.</p>
    </li>
</ol>
```

Figure 22.1 – XML markup for a list

In Figure 22.1, the semantic block is the ordered list bounded by the `ol` tags. The `li` and `p` tags inside are structural blocks that together implement the structure of an ordered list. (Since paragraphs are semantic blocks, albeit relatively simple ones, this illustrates that a semantic block can also be a structural block for a larger semantic structure.)

Other document domain examples of semantic blocks include tables and procedures. For example, DocBook, DITA, and HTML, not to mention S1000D and reStructuredText, all have tables with different internal structures, and both DocBook and DITA have procedures, again internally different. It is possible to disagree greatly about how to structure a semantic block while still recognizing different implementations as examples of the same semantic type.

Semantic blocks in the subject domain

In the subject domain, examples of semantic blocks include an ingredients list from a recipe (Figure 22.2) and the parameter description from an API reference (Figure 22.3).

```
ingredients:: ingredient, quantity, unit
    eggs, 3, each
    salt, 1, tsp
    butter, .5, cup
```

Figure 22.2 – Subject-domain ingredients list

```
parameter: string
    required: yes
    description:
        The string to print.
```

Figure 22.3 – Subject-domain API parameter description

Semantic blocks often repeat as a unit, as Figure 22.4 shows in an API reference entry.

```
function: print
    return-value: none
    parameters:
        parameter: string
            required: yes
            description:
                The string to print.
        parameter: end
            required: no
            default: '\n'
            description:
                The characters to output after the {string}(parameter).
```

Figure 22.4 – Semantic blocks repeated as a unit

Semantic blocks may also be used as a unit in different places in a markup language or in different markup languages. For instance, the ordered-list semantic block may be allowed in more than one place in a document-domain language, such as in a section or in a table cell.

In fact, if you have multiple markup languages in your content system, particularly multiple subject-domain languages, then it makes a lot of sense to define a common set of semantic blocks for use across all of these languages. Thus, if you have five subject-domain languages that all require lists, you can use the same list structure across all five of those languages.

This means that you can define the structure of a list and all of the algorithms that work on lists just once and reuse them for every subject-domain language that shares those structures. It also means that your writers only have to learn one definition for each semantic structure, which makes it much easier for them to learn several different subject-domain languages.

If you do this, then each subject-domain language only has to define, and provide processing algorithms for, those structures that are unique to the subject matter. All the other structures you need are already defined, tested, and ready to use. This makes it much quicker and simpler to design and implement a new subject-domain language when you need one.

Designing in terms of semantic blocks not only helps keep markup design and processing simpler, it also improves functional lucidity. If you present your markup language to writers as a set of familiar objects such as lists and tables or as logical structures such as an ingredients list or parameter description – rather than as a sea of tags – you make their job easier.

Semantic blocks also make things easier for tools developers. An XML editor that implements a WYSIWYG interface to XML authoring can provide tool-bar buttons for inserting semantic blocks such as lists or tables. This allows the writer to enter these blocks as complete structures rather than having to enter every tag separately.

The structure of a semantic block can be strict or loose. A strict semantic block has one basic structure with few options. A loose one allows a much wider variety of structure inside, sometimes to the point that it acts more as a semantic wrapper than a defined semantic block.

DocBook is an example of a language with loose semantic blocks. DocBook has the same high-level semantic blocks as any other generalized document-domain markup language, but it allows so many tags in so many places that none of these objects are simple and easy to understand. Even though you can use DocBook to describe almost any document structure you might want to create, this flexibility reduces functional lucidity and constraint.

How do you balance flexibility with functional lucidity and constraint in creating semantic blocks? Sometimes it is best to have more than one implementation of a particular semantic block. For instance, both DITA and DocBook have two table models, a simple model and a more complex one based on the CALS table model.

Information-typing blocks

We have looked at examples of semantic blocks, which can be in both the document domain (lists and tables) and the subject domain (ingredients list and parameter description). Another way some structured writing systems divide content into blocks is according to the type of information they contain. I call these *information-typing blocks*, since the practice of dividing content into such blocks is commonly called information typing (though this is not the only thing the words information typing could refer to, since all structured writing assigns information to types).

Information-typing blocks are a type of semantic block, but they introduce a degree of abstraction not found with most semantic blocks. Unless they have been trained in an information-typing system such as Information Mapping®[1] or DITA, most writers don't describe their content in terms of information-typing blocks.

---

[1] http://www.informationmapping.com

Information Mapping is a structured writing system based on a theory about how humans receive information. It views all content as being made up of six types of information blocks: Procedure, Process, Principle, Concept, Structure, and Fact. These are information-typing blocks. They don't directly describe a physical or logical element of document structure (except for procedure) nor are they specific to any one subject. They describe the kind of idea that the content conveys.

Which structured writing domain do information-typing blocks belong to? Clearly they are not media-, subject-, or management-domain structures. Are they a kind of document-domain structure or something else again? I believe it is more useful to regard them as document-domain structures than to invent another domain. Information Mapping regards a document as a mapping of information-typing blocks, so Information Mapping's information-typing blocks are components of documents and, therefore, in the document domain.

DITA also adopted this idea of documents being made up of information-typing blocks. In DITA's case, these blocks are named *topics*, which leads to some confusion since the word topic can be used to refer to both an information-typing block and also to a complete document (as in a "help topic" for instance).

DITA has popularized the idea that all content (or all technical content, at least) is made up of just three information-typing blocks: concept, task, and reference, although newer releases of DITA define more topic types than this, and the DITA architecture allows for the creation of even more. This idea is appealing because it is simple, and it is easy to see a correspondence between these three types and the reader activities of learning (concept), doing (task), and looking stuff up (reference). The question is whether this architecture provides adequate or appropriate constraints for the rhetoric of your content and whether it is useful for partitioning the complexity of your content system.

This simple triptych is also appealing because it promises (though it does not necessarily deliver) easy composability for content reuse. Some people also maintain that it makes content easier to access for readers, though others (myself included) criticize it on the grounds that it tends to break content down too finely to be useful and robs content of its narrative thread.

There is evidence that DITA is moving away from this vision of information typing. In DITA 1.3, the technical committee puts the emphasis on topic and map as the core types, rather than concept, task, and reference.

> The DITA Technical Committee wants to emphasize that topic and map are the base document types in the architecture.
>
> Because DITA was originally developed within IBM as a solution for technical documentation, early information about DITA stressed the importance of the concept, task, and reference topics.
>
> Many regarded the topic document type as nothing more than a specialization base for concept, task, and reference.
>
> While this perspective might still be valid for technical content, times have changed. DITA now is used in many other contexts, and people developing content for these other contexts need new specializations. For example, nurses who develop evidence-based care sheets might need a topic specialization that has sections for evidence, impact on current practices, and bibliographic references.
> —http://docs.oasis-open.org/dita/dita-1.3-why-three-editions/v1.0/cn01/dita-1.3-why-three-editions-v1.0-cn01.html#focus-of-dita

Stating that the evidence-based care sheets example would include information from more than one abstract type and proposing this example as a specialization of topic – rather than concept, task, or reference – both suggest a significant shift in thinking on this point and may indicate a shift away from abstract information typing towards a more concrete subject-domain approach.

Since neither Information Mapping nor DITA provide any mechanism for constraining how information-typing blocks go together to form documents, they do not provide much support for creating and maintaining repeatable rhetorical structures. (This does not mean that writers can't create such structures, but the complexity of doing so falls entirely on them, with the added complexity of having to construct the rhetorical structure out of a jigsaw puzzle of abstract information-typing blocks.) DITA, however, does allow you to create much more specific content types. See Chapter 34 for more details.

This abstract information typing is distinct from the subject-based information typing of the subject domain. DITA and Information Mapping's approaches are broad and analytical, concerned with finding commonalities across many different kinds of information. The subject domain is specific and synthetic, concerned with fitting pieces together to successfully describe a particular subject.

However, the two approaches are not as incompatible as this makes them seem. The design role of information types is often to suggest how a particular kind of content should be written. Thus a software procedure, a knitting pattern, and a recipe are all procedural information. At the subject-domain level they are quite distinct, but at the information typing level they are constructed similarly. If you are creating a subject-domain structure, recognizing its information type (according to whichever information-typing scheme you prefer) can help you design its internal structure. The resulting block will be a subject-domain semantic block, rather than a generic information-typing block, but information typing will have informed the construction of that block. DITA attempts to exploit this relationship between the generic information type and the more specific block design of a particular language through its specialization mechanism, which I look at in Chapter 34. However, you can certainly design subject-domain semantic blocks without reference to information-typing theory.

Rhetorical blocks

Neither semantic blocks nor information-typing blocks constitute a piece of content that is rhetorically complete – that persuades, informs, entertains, or enables the reader to act. I use the term *rhetorical block* for a piece of content that is rhetorically complete. A rhetorical block is made of up smaller semantic and structural blocks, but the structure of a rhetorical block controls the completeness and repeatability of a content type. The key design question is, do you want to constrain the structure of your rhetorical blocks? This is an important question because systems based on information typing, such as DITA, provide no mechanism for constraining anything larger than the information-typing block, and document-domain systems, such as DocBook, provide no mechanism for constraining rhetorical structure, even though they allow you to capture complete rhetorical blocks.

As we have seen, there are many advantages to constraining the rhetorical block, but the key advantage is repeatability. This is not simply because repeatability allows you to produce quality content more quickly and reliably. It is also because it makes content quality testable in a repeatable manner. You don't just test the quality of an individual piece, but of a repeatable pattern (as we will see in Chapter 41).

Do all rhetorical blocks have a repeatable pattern? Certainly not. This book is a rhetorical block, but it does not have a repeatable pattern that one could define as a specific data structure. (Hopefully there is a structure to its rhetoric, but it is not one that lends itself to modeling as a set of nested blocks.) Although this book encourages the use of the subject domain for structured writing, it was not written in the subject domain. It was written (in SAM syntax) in a small constrained document-domain language with a number of subject-domain annotations for things such as markup languages, markup concepts, and processing tools.

But not all physical books constitute a single rhetorical block. Many books, such as cookbooks, are collections of related rhetorical blocks, such as recipes. Not only can you define a repeatable rhetorical block for a recipe, doing so provides significant benefits for most of the structured writing algorithms and allows you to effectively partition and direct the content complexity of a cookbook publisher or a magazine publisher that inserts recipes into many different publications.

Where you can reasonably define repeatably structured rhetorical blocks in your content, there are good reasons to do so. One approach is to take the rhetorical blocks that you produce now, identify those that could be repeatably structured, and develop a structure to contain and constrain them. This may involve moving some content around, since the application of a repeatable structure inevitably reveals misplaced, missing, and superfluous content. But overall, this is a reasonably straightforward process.

A second approach is to take material that is currently in long discursive rhetorical blocks (such as text books) and move it to much shorter and more structured rhetorical blocks (such as recipes or encyclopedia articles). Breaking material into smaller rhetorical units is not a new idea. It has been practiced in encyclopedias and periodicals for centuries, and it has greatly accelerated in the age of the web. This has not only improved our access to information, it has radically changed how we seek and use information, creating a phenomenon called information snacking[2] in which readers reach for discrete pieces of information as and when they need them, confident that they will always be able to rapidly find and read what they need when the time arises.[3] In other words, people prefer to consume content in smaller rhetorical blocks (not because they need to know

---

[2] https://www.nngroup.com/articles/information-scent/

[3] I discuss this change of information seeking and consuming habits in my book *Every Page is Page One: Topic-based Writing for Technical Communication and the Web.*

less, but because they need to know less at a time), so it makes sense to refactor your content and its architecture into a collection of smaller rhetorical blocks.

This transformation to shorter rhetorical blocks is often called *topic-based writing*, but that term can be confusing because it is also used to mean a system in which, instead of writing in long rhetorical blocks, individual writers write independent information-typing blocks that are then compiled into long rhetorical blocks, often by somebody else. The confusion caused by giving these two different approaches the same name is that some people have started to think that topic-based writing means writing information-typing blocks and then publishing them separately as if they were rhetorical blocks. This confusion is compounded because DITA calls its information-typing blocks *topics* (Information Mapping calls them *blocks*) and because publishing each inform-ation-typing block separately is the default behavior when you publish a DITA map to the web.

But while a single information-typing block may sometimes be a rhetorical block all by itself, this is often not the case. Whether one believes in the usefulness of information-typing theory or not, it is clear that most useful rhetorical blocks contain more than one type of information.

I have attempted to distinguish topics as complete rhetorical blocks from topics as information-typing blocks by coining the term "Every Page is Page One topic." Why "Every Page is Page One"? Because a complete rhetorical block, regardless of its length, is a block that an information-snacking reader will consume independently of any larger work or collection in which it is em-bedded. It is page one for that reader. And since information-snacking readers use search to dive directly down to the individual page they want, every rhetorical block in your collection is going to be page one for some reader. I explore the design of Every Page is Page One topics in my book *Every Page is Page One: Topic-based Writing for Technical Communication and the Web*.

But does an Every Page is Page One topic have any more reason to follow a constrained rhetorical block structure than a book? Yes, topics tend to demonstrate a consistent rhetorical pattern once they become smaller and are accessed in an Every Page is Page One fashion. This makes the in-formation in these topics more accessible to the information-snacking reader, both because it makes them easier to find and recognize and because it ensures that they do the job they are supposed to do more completely and consistently.

Additionally, as I explored in Chapter 8, Every Page is Page One topics are part of the way you create a bottom-up information architecture – one in which readers enter by search or by following a link and navigate the information set from the point they arrived. Building and maintaining a bottom-up information architecture is much easier if you do it algorithmically using subject-domain linking and information architecture algorithms. Thus, creating subject-domain rhetorical block types for as much of your content as will fit a repeating pattern is key to creating and managing a bottom-up information architecture.

Deciding if you want to model and constrain your rhetorical blocks is therefore one of the most important decisions you make in designing your content system and determining how complexity will be partitioned and directed in that system.

Granularity

There can be a conflict between ease of writing and ease of content management. The designers of content management systems typically want to manage content down to a fine level of granularity, especially for purposes of content reuse. Content management algorithms may be best served by managing fairly small chunks of content – semantic blocks rather than rhetorical blocks. But for writers, writing something less than a rhetorical block can be difficult. Writers can find it difficult to get a sense of how well a particular semantic block will meet a reader's needs when they don't see the rhetorical block that block will fit into.

It is hard for writers to create parts of a larger structure in isolation, unless the parts are really well defined. Writers are trained to construct entire essays and can structure them well as a whole. But it is hard to correctly structure a part without clear and explicit guidance – the kind of guidance you do not get from a generic information-typing block such as concept, task, or reference.

Fragments

Another division of content, found mostly in relationship to the management domain, is the fragment. By fragment I mean a chunk of text that is not a semantic block, an information-typing block, or a rhetorical block. A fragment is an arbitrary piece of text that you want to manage independently of the surrounding text. In other words, it is a block that has no rhetorical role and exists purely for management purposes.

An existing structural block may also be treated as a fragment. For example, in a content-reuse scenario, you might want to make an item in a list a fragment that can be conditionally included based on which version of a product it applies to.

Individual list items are not semantic blocks. They are just structural blocks of a list. When you make list items conditional, you are actually creating multiple separate lists with some items in common and recording them as a single list. You might be able to attach reasonably informative metadata to any one of those lists as a whole, but there is usually not a lot you can say about individual list items. They are fragments of a list. When you apply conditions to them, you are applying those conditions to fragments.

Sometimes a fragment is an arbitrary chunk of text within a paragraph – three words in a sentence for instance. The block that contains those words is a fragment. Some reuse systems allow you to reuse arbitrary bits of text from other parts of the content set, simply because the text is the same in each case. Those bits of text are fragments.

Where a fragment coincides with an existing structural block, you can sometimes just attach the fragment metadata to the block. Where the fragment doesn't coincide with an existing block, you have to introduce additional markup into the document to delineate the fragment.

Fragments definitely solve some problems, but they are also inherently unstructured and unconstrained. It is easy to get into trouble with fragments. It is easy to create relationships and dependencies that are hard to manage because they don't follow any structural logic. You should use them with great caution and restraint.

Paragraphs and phrases

Paragraphs are the one thing that makes structured content different from other computable data sets. This is not because of the paragraph structure per se, but because of the way we annotate phrases within the paragraphs. It is rare in any other data set to see a structure floating within the value of another structure. But that is exactly what happens when you annotate phrases in a paragraph.

```
In {Rio Bravo}(movie), {the Duke}(actor "John Wayne")
plays an ex-Union colonel.
```

In this example, the annotation on the phrases "Rio Bravo" and "the Duke" float in the middle of the paragraph block. Here is the same thing in XML:

```
<p>In <movie>Rio Bravo</movie>,
<actor name="John Wayne">the Duke</actor>
plays an ex-Union colonel.</p>
```

Here the movie and actor elements float in the content of the p element. In XML parlance, this is called *mixed content*.

XML breaks the structure of elements down into three types:

- **element content:** Elements that contain only other elements.
- **data content:** Elements that contain only text data.
- **mixed content:** Elements that contain both text data and elements.

Mixed content is the reason that most traditional data formats are not a good fit for content. They may be able to model element content and data content, but they lack an elegant way to model mixed content.

Even conventional programming languages have trouble with mixed content. Most libraries for XML processing invent an additional wrapper around each string of characters in a mixed content element, effectively representing it as if it were written like this (without mixed content):

```
<p><text>In </text><movie>Rio Bravo</movie><text>,
</text><actor name="John Wayne">the Duke</actor><text>
plays an ex-Union colonel.</text></p>
```

But while this makes the content easier to process for programming languages, it is clearly false to the actual structure of the document. Structured writing is essentially about reflecting the structure of thought or presentation in a narrative, and narratives have a structure that is not shared with other data. Indeed, you might say that all other data formats exist as an attempt to extract information from the narrative format to make it easier to process.

Thus, if you are presented with a problem in this format:

> John had 4 apples and Mary had 5 apples. They place their apples in a basket. Bill eats 2 apples. How many apples are left in the basket?

You solve it by first extracting the data from the narrative:

```
4 + 5 - 2 =
```

But with content, while you can sometimes extract the data from the content, you also need to retain the content or the means to recreate the content, since the content (not the answer) is the product you are trying to produce.

When you move content to the subject domain, you will, in some cases, break down paragraphs and isolate data with the intention of recreating paragraphs algorithmically on output or switching from a narrative to a data-oriented reporting of the subject matter. Either way, this process makes the data easier for algorithms to handle and, thus, makes most of the structured writing algorithms work better. (You may have noticed that the subject domain provides the most constrained and elegant solution to many structured writing algorithms.)

Even so, it is not always possible to do a complete breakdown of all paragraphs when refactoring content to the subject domain. Therefore, most subject-domain markup languages use paragraphs and other basic text structures and allow writers to annotate phrases within paragraphs. Only narrative is capable of expressing the full variety and subtlety of the real-world relationships between things, and only narrative is capable of conveying these things effectively to most human readers. Even things that can be fully described to algorithms with fielded data must be described to human audiences with narrative.

Subject-domain structured writing extends the reach of more conventional algorithms into the world of narrative to enable specific structured writing algorithms and to provide rhetorical constraints to improve the quality of the writing. Subject-domain structured writing does not attempt to capture the whole semantics of a narrative, just to discipline and structure narrative to achieve specific content creation objectives – a particular partitioning of content complexity.

Phrases

Every structured writing domain needs to annotate phrases. In structured writing, a phrase is any string of text below the level of a paragraph that expresses some meaning that you need to annotate. It may be a single word, a sentence, or even multiple sentences. Media-domain structured writing needs to annotate phrases to describe formatting; the document domain to describe their role in the document; the management domain to assign conditions or extract content for reuse; and the subject domain to describe the subject the phrase refers to. Figure 22.5 contains phrases annotated with subject domain, document domain, and media domain annotations.

```
{Moby Dick}(novel), {Gone With the Wind)(citetitle), and {War and
Peace}(italic) are classic novels.
```
Figure 22.5 – Phrases annotation in three different domains

Some of the most important subject matter that you need to model and make available to algorithms cannot effectively be factored out of paragraphs, particularly while maintaining functional lucidity. Be prepared, therefore, to think seriously about the types of phrases you need to annotate and exactly which domain those annotations should be in.

CHAPTER 23

Wide Structures

The notion of separating content from formatting works quite well when the content is a string of words. A string of words has only one dimension: length. A printed string, of course, has two dimensions: length and height, since each letter has a height and a width. But the height and the width of letters is a pure media-domain concern.

Fitting a one-dimensional string of characters into a two-dimensional font on a two-dimensional page is one of the first things that gets factored out as you structure content. When you separate content from formatting, you separate the font from the character and are left with a string of characters whose length is measured not in inches or centimeters but in character count.

Once these formatting dimensions are factored out, it becomes easy to create and manage text in the document and subject domains without thinking about how it will eventually flow onto a page or screen. But when it comes to content that has dimensions that cannot be factored out, things get more difficult.

The main problem cases are:

- tables
- pre-formatted text, such as a program listing, that has meaningful line breaks
- graphics and other media

Tables

Tables are one of the more complex problems in structured writing, particularly in the document domain. A table laid out for presentation in one publication can easily get messed up when an algorithm tries to fit it into another, as in Figure 23.1, which comes from a commercially published book on my Kindle.

Figure 23.1 – Broken table formatting

This table is difficult because one wide thing (a table) contains another wide thing (pre-formatted program code). I don't know how this table was marked up, which domain the content was written in, or how the formatting algorithm failed, resulting in the mess in Figure 23.1. However, including pre-formatted text in a table cell creates a no-win situation for a rendering algorithm. Does it:

- violate the formatting of the program code by introducing extra line breaks.
- give the code the space it needs by squeezing all the other columns impossibly narrow.
- resize the columns proportionally and let the pre-formatted text overlap the next column, but truncate it at the edge of the table.
- resize the columns proportionally and truncate the pre-formatted text at the column boundary.
- shrink the entire table so everything formats correctly, even if it is shown in three point type.
- let the table expand outside the viewport so that it is either cut off or the reader has to scroll horizontally. (Web browsers tend to take this approach, but will it work on an e-reader? It certainly won't work on paper.)
- make the table into a graphic so readers can pan and zoom like they do with a large picture. (Some ebooks take this approach.)

If you are thinking that there is no good option in the bunch, you appreciate the extent of the problem. For existing books being transferred to e-readers, you can't do much to salvage the situation, short of completely restructuring the file. Those books were probably prepared in a word processor on the more abstract edge of the media domain, and the tables were prepared for a known page width in the printed book.

Because many people read on small devices such as tablets, e-readers, and phones, wide tables cause problems. On a phone, the amount of a table that is visible on screen at any one time may be so small as to make the table essentially unnavigable and useless for such common tasks as looking up values or presenting an overview of a subject at a glance.

Tables can cause problems with height as well as width. While most writers would never import a graphic that was six feet tall, we sometimes create tables that are that long or longer. On a web browser, readers can simply scroll the table. But as soon as you start scrolling, you lose sight of the column headers and it becomes harder to read data across the table. On paper, you can repeat the headings at the top of each page when a table flows over several pages. This works, and you can imitate the effect in a web browser by placing the body of the table in a scrollable frame under a fixed heading. But what happens if the height of a table row is larger than the height of the page? How do you break the text in each cell in that row? In traditional typesetting, you can make adjustments by hand on a case-by-case basis, but getting a rendering algorithm to do this gracefully in every case is very challenging.

Creating tables in the document domain creates problems even when the intended output is paper, and you assume a sufficiently wide viewport. Since a table divides content into multiple columns, there is always a question of how wide each column should be relative to the others and whether or not the table should occupy the full width of the viewport. For example, you probably don't want a table with just a few numeric values to span a full page width because that would spread the numbers out too far and make comparisons difficult. One the other hand, a table with a lot of text in each cell needs to be full width and needs to have column widths roughly proportional to the amount of text in the each column. However, suppose the first column is a side head with far fewer words than the other columns. You probably don't want to compress that column proportional to its word count because then the side headings will be unreadable.

In a media-domain editor, which shows the formatting of the content as it will appear on paper, a writer can create a table at a fixed width and drag the column widths around to get the aesthetics of column boundaries right by eye. But tables created like this are not likely to format correctly on other devices, as Figure 23.1 shows. And if you move content creation out of the media domain and into the document domain, you can no longer present writers with a WYSIWYG page width that would allow them to adjust column widths by eye. (You can fake it, but adjustments made on screen bear no relationship to how the table displays on any output page.) At this point you have to leave column-width calculation to the rendering algorithm. The best you can do it to give it some hints.

```
<table>
  <title>Table title</title>

  <tgroup cols="3">
    <colspec colname="_1" colwidth="1*"/>
    <colspec colname="_2" colwidth="3*"/>
    <colspec colname="_3" colwidth="2*"/>

    <thead>
      <row>
        <entry>1st cell in table heading</entry>
        <entry>2nd cell in table heading</entry>
        <entry>3rd cell in table heading</entry>
      </row>
      <row>
        <entry>1st cell in table heading</entry>
        <entry>2nd cell in table heading</entry>
        <entry>3rd cell in table heading</entry>
      </row>
    </thead>

    <tbody>
      <row>
        <entry>1st cell in row 1 of table body</entry>
        <entry>2nd cell in row 1 of table body</entry>
        <entry>3rd cell in row 1 of table body</entry>
      </row>

      <row>
        <entry nameend="_2" namest="_1">cell spanning two columns</entry>
        <entry morerows="1">cell spanning two rows</entry>
      </row>

      <row>
        <entry>1st cell in row 3 of table body</entry>
        <entry>2nd cell in row 3 of table body</entry>
      </row>
    </tbody>
  </tgroup>
</table>
```

Figure 23.2 – CALS table markup language column and row span example (from Wikipedia)

The need to give the rendering algorithm hints about how to fit tables to pages has resulted in the creation of some very complicated table markup languages. Figure 23.2 contains an example using the CALS table model. It shows a table that has one cell spanning two columns and one cell spanning two rows. As you can tell, this is not exactly obvious from the markup. In practice, few writers create CALS tables by writing the markup by hand. Instead, they use the table drawing tools in a graphical XML editor.

However, although graphical XML editors can display tables in a manner that looks the same as a word processor, they cannot allow writers to make the kinds of media-domain adjustments that they can make with a word processor. A word processor's graphical display is based on the actual page currently set up in printer settings and on the font the document will be printed in. Therefore, a writer can see how a table fits on an actual page and make media-domain adjustments. An XML editor cannot know the page size or the font. So while the display looks like it allows writers to make the same media-domain adjustments, this is an illusion, and the table will not print as shown on screen. This gives a false impression of the real complexity of the table problem, hiding complexity from the person who is supposed to deal with it.

Other markup languages take a different approach to tables. For instance, reStructuredText allows you to create a table using markup that looks like Figure 23.3.

```
+------------+------------+-----------+
| Header 1   | Header 2   | Header 3  |
+============+============+===========+
| body row 1 | column 2   | column 3  |
+------------+------------+-----------+
| body row 2 | Cells may span columns.|
+------------+------------+-----------+
| body row 3 | Cells may  | - Cells   |
+------------+ span rows. | - contain |
| body row 4 |            | - blocks. |
+------------+------------+-----------+
```

Figure 23.3 – reStructuredText table markup language column and row span example

Like the CALS example, reStructuredText allows you to span rows and columns, and in this example, the effect is obvious. Equally obvious is that editing or creating a table with any significant amount of text in the cells is going to be very difficult. Nor does reStructuredText solve any of the table rendering challenges described above.

Alternatives to tables

Structured writing is about partitioning and redirecting the complexity of content so that it is always handled by someone with the attention, knowledge, and resources to handle it. As the discussion above demonstrates, table markup dumps a lot of complexity on writers, complexity they are not fully able to handle because they don't control the final formatting of content in all the media and devices it will be presented on. The best way to partition the problem, therefore, is to factor out the decision to present information as a table. Let writers capture information as data, and move the decision on whether to present that information as a table or in some other form down the road.

What can you factor the content into? There are a number of alternatives, depending on what you are using the table for.

Alternate presentations

In many cases, tables simply aren't necessary. There are other ways to present the content with no loss of comprehensibility or quality. Sometimes, tables are just used to format lists, particularly lists with two levels of nesting. If a list is an equally effective way of presenting content, choose a list rather than a table when writing in the document domain. Sometimes only part of the content needs to be in tabular format and can be expressed just as well using other, simpler structures that contain a table or tables within them.

Subject-domain structure

One way to present the list of ingredients in a recipe is to create a table with the ingredient name aligned left and the quantity aligned right. But as we have seen in our recipe examples, you can create a subject-specific ingredient list structure to capture your ingredient information, which you can then format any way you like for output.

```
ingredients:: ingredient, quantity, unit
    eggs, 3, each
    salt, 1, tsp
    butter, .5, cup
```

Figure 23.4 – Subject-domain ingredient listing

A structure such as the one shown in Figure 23.4 is a table in a different sense of the word: it is a database table and the `ingredients` structure creates a mini database table inside the body of the content. The difference between this table and a media-domain table is that we know exactly what type of information each column contains. This allows the presentation algorithm to make intelligent choices about column widths and all the other rendering issues that arise with tables and then pass on appropriate hints to the rendering algorithm. Of course, this structure also supports presentation using a list rather than a table.

Tables are also sometimes used in procedures. With a table, you can create side heads for step numbers in one column and a description for each step in another column. However, instead of this, consider using explicit procedure markup, which can be formatted differently depending on the output medium. If you choose a table as the output format, knowing that the contents are a procedure allows the formatting algorithm to provide appropriate layout hints to the rendering algorithm.

Record data as data

Many reference works have traditionally been presented as tables on paper. But most such works are really databases. They are designed to be queried, not read. That is, they are used to look up individual pieces of data in a large set. For a database of this sort, differential single sourcing requires that you provide the best method of querying the data for each output medium; that is, the method whose interface fits best in the available viewport. To support differential single sourcing, you should not record data in tables, at least, not in media-domain tables. You should record it in whatever database format best suits the data and the kinds of queries your readers want to make.

If you want to support printed tables for print media, then you should extract content from the database to create the printed table. The additional semantic information available from the database structure allows the formatting algorithm to supply appropriate rendering hints to the rendering algorithm.

When you have done all of that, you will probably be left with two kinds of tables to deal with: small ad-hoc grid layouts and one-of-a-kind database tables. These are tables that occur just once and don't justify the overhead of creating a subject-domain data structure. For these, you need some form of document-domain table markup. Which markup you choose will come down to how much fancy formatting of tables you require and how willing you are to let the rendering algorithm format your tables without extensive hinting.

Computer code

There are some texts, particularly computer code and data, in which line breaks are meaningful. (Poetry is another example, but its issues are simpler than those of code, so I will stick to talking about code.) Code is a form of structured writing and, in many languages, whitespace – meaning line breaks, spaces, and indentation – is part of the markup that defines the structure of the program. Therefore, when you present code in a document, you have to respect whitespace.

Furthermore, programmers usually work in a fixed-width font, meaning that all the letters are the same width when displayed. Programmers align similar structures with whitespace to make them easier to read; using a proportional-width font for code in documentation messes up formatting and looks weird to programmers. It also makes the code less recognizable as code, which could reduce information scent.

Therefore, computer code, data, and other similar formats where whitespace is meaningful must be presented in a fixed-width font with line breaks placed where they are supposed to be. That makes code samples wide objects, just like tables, with many of the same issues when it comes to rendering them on small devices. One saving grace is that programming examples are less likely to have issues with height.

There is not much you can do to help the rendering algorithm when it comes to code. The options for fitting wide code on a narrow display are: shrink to fit, scroll to view, truncate, or introduce extra line breaks, which may or may not follow coding rules. You probably do not want your rendering algorithm to make a different choice for different kinds of code. Needless to say, putting a code block inside another wide structure, such as a table, is a recipe for disaster, as Figure 23.1 shows. It would be wisest not to allow this in your markup language design.

What is essential is that your document-domain or subject-domain markup clearly indicate when a piece of text is code. Preferably it should also indicate what kind of code it is, since knowing this can allow the formatting algorithm to do syntax highlighting for code in a known language and can allow the linking algorithm to detect and link API calls to the API reference. In some cases it might even allow an algorithm to validate the code to make sure it runs or uses the current version of the API.

Pictures and graphics

Pictures and graphics are naturally wide objects. There are two basic formats: raster and vector. Raster graphics are made up of pixels, like a photograph, and have a fixed resolution. Vector graphics are stored as a set of lines and curves and can be scaled to meet any output requirement.

The publishing algorithm needs to know how big the graphic is and how large it is supposed to be on the page. With raster files, the resolution – that is, the number of pixels – is set. However, its size on the page can vary. Is a graphic that is 600 pixels by 600 pixels a 1x1 inch picture at 600 dpi, a 2x2 inch picture at 300 dpi, or a 6x6 inch picture at 100 dpi? This is important if you are inserting a headshot into a document that will be published on both paper and the web. Rendering that photo on paper as a 1x1 inch image makes sense, but you don't want that photo to blow up to a 6x6 image when you add it to a web page, where it will be displayed it at a typical 96dpi unless something intervenes to scale it appropriately.

Then there is the question of the intended size of the image, which is a design consideration independent of the resolution of the raster file. The intention of the graphic artist who created the picture and the intention of a writer using it to create a deliverable both play a role here. Diagrams showing complex relationships should not be shrunk down to where the relationships are unreadable. Simple diagrams should not be blown up to the size of a full page. Diagrams containing text should not be reduced or expanded so that the text becomes invisible or out of proportion with the text on the page.

The writer may have some discretion, based on the role the graphic has to play, but the result should stay within the range prescribed by the artists's intention. In other words, you have to correctly partition the concerns of both the creator and the user of a graphic, and you have to make sure both parties can communicate effectively so that none of the complexity of the relationship gets dropped.

If the rendering algorithm does not know how big a graphic should be, it has limited choices, depending on whether the graphic is too big or too small for the viewport:

- Show a raster graphic at 100% of its resolution, regardless of whether it fits in the viewport (which may require cropping it or forcing readers to scroll it).
- Scale the graphic to the viewport (which may require stretching or shrinking it).

Since neither of these options produces consistently good results, you generally need to give the rendering engine information to help it render the graphic appropriately.[1]

The simplest way to supply this information is to include it in the markup that inserts the graphic. Thus, HTML lets you specify the height and width of a graphic (see Figure 23.5).

```
<img
    src="http://www.example.com/images/example.png"
    height="150"
    width="140" />
```

Figure 23.5 – HTML image markup

But do these values represent the size of the graphic or the size at which is it to be displayed in a particular medium? In other words, do they tell a processor the size of the image itself or do they describe the size of the box (viewport) the image should fit in?[2]

DocBook allows you to make this distinction. Its `imagedata` tag supports attributes for specifying the size of the viewport (`height` and `width`) and the size of the image (`contentheight` and `contentwidth`). The specification also contains additional attributes related to scaling and alignment and complex rules about how the rendering algorithm is supposed to behave based on which combination of these attributes you specify.[3] In other words, DocBook contains a sophist-icated language to describe the sizing and scaling of graphics. It not only deals with media-domain properties, it actually gives media-domain instructions.

Of course, working in the media domain can cause problems. It interferes with functional lucidity and makes differential single sourcing more difficult. But there is another issue to consider. In addition to needing different scaling values for different output media, sometimes the best approach to differential single sourcing is to use different versions of the same graphic for different media. For instance, you may want to use a vector format for print and a raster format for online media.

For all these reasons, we need to distinguish the source of an image from the rendering of that image. For raster images, the source is the original high-resolution file recorded by a camera, a

---

[1] Many web designers take an opposite approach, preparing a graphic to the exact size required for a specific web page layout. This is strictly a media-domain approach, of course. In structured writing, we need a more flexible solution to avoid what we have all seen happen when a meticulously designed desktop website gets displayed on a phone screen.

[2] This is actually quite a complicated question, and the meaning has changed between various version of HTML. For some hints of the complexities involved, see http://www.w3.org/TR/html5/single-page.html#attr-dim-width.

[3] http://www.docbook.org/tdg/en/html/imagedata.html

screen shot, or a raster file produced by an image editing program. For vector graphics, it is the original vector drawing file. From these source images, you can render a wide range of images.

What if you need a vector version for some media and a raster version for others? One approach is to generate a raster version of the appropriate size from the vector version. This can be done at build time, but most of the time you store multiple versions and select the right one to publish. To create the best image in each format, the artist may even create several renderings of the same image idea, optimizing each for different uses. For instance, you may need a separate gray-scale version of your company logo, because the automatic gray-scale rendering of a color logo may not look good. You may also want to use different resolutions of the same raster graphic for different media or for different purposes. And you may want to manually redraw a graphic at different resolutions to add or remove detail, rather than simply scaling it mechanically.

In these cases, how do you include the image in your source content? You can't simply include the source file and scale it, since there are now several source files. You have to go back to the idea of the image – the image that was in the artist's head – rather than any of the individual renderings of that image. How do you do that?

In DocBook you can use conditional processing, as shown in see Figure 23.6. In this example, the `condition` attribute on the `imageobject` element specifies a different file to be used for two versions of a book (this book, actually). The `epub` version is for e-readers, most of which cannot render SVG drawings, and so require a raster format (PNG in this case), while the `fo` version is for print publication using the XSL-FO page description language and uses the vector format SVG for high-resolution rendering in print.

```
<mediaobject>
  <imageobject condition="epub">
    <imagedata
        fileref="../graphics/assemble.png"/>
  </imageobject>
  <imageobject condition="fo">
    <imagedata
        fileref="../graphics/assemble.svg"
        contentwidth="4in"
        align="left"/>
  </imageobject>
</mediaobject>
```

Figure 23.6 – DocBook markup for conditional images

However, this approach not only uses media-domain markup, it combines it with management-domain markup. Is it possible to factor all of this out of the authored format?

Yes, if the writer includes the idea of the image rather than a rendering of the image. There are several ways to do this. In fact, this is the same method described for the reuse algorithm where we factored out the filename of the content to be included and replaced it with a semantic representation of the reason for the content.

In Chapter 5, we factored out an explicit filename from this example:

```
procedure: Blow stuff up
    >>>(files/shared/admonitions/danger)
    step: Plant dynamite.
    step: Insert detonator.
    step: Run away.
    step: Press the big red button.
```

and replaced it with a management-domain key in this example:

```
procedure: Blow stuff up
    >>>(%warn_danger)
    step: Plant dynamite.
    step: Insert detonator.
    step: Run away.
    step: Press the big red button.
```

Then we refactored that into a subject-domain assertion of fact in this example:

```
procedure: Blow stuff up
    is-it-dangerous: yes
    step: Plant dynamite.
    step: Insert detonator.
    step: Run away.
    step: Press the big red button.
```

You can apply these techniques to insert graphics as well as text. Suppose you have a constraint that whenever a procedure mentions a dialog box, you must include a picture of that dialog box, and you need to support three different platforms, each of which has different dialog boxes.

You can factor out the platform-specific version of the image by using a key to insert the image:

```
procedure: Save a file
    step:
        From the File menu, choose
        Save. The *Save As*
        dialog box appears.

        >>>(%dialog.save-as)
```

Now the presentation algorithm can use a key lookup table to select the right version of the **Save As** dialog box for the version of the documentation you are building. If you port the product to a new platform, all you need is a new set of screen shots and a new key lookup table. You don't have to change the content at all.

There is a simple rhetorical pattern at work in this passage. When a step mentions a dialog box, you show a picture of that dialog box. You can exploit this pattern to factor the insert command out of the content altogether by annotating the mentions of UI components in the text:

```
procedure: Save a file
    step:
        From the {File}(menu) menu, choose
        {Save}(menu-item). The {Save As}(dialog-box)
        dialog box appears.
```

Now you can insert the correct screen-shot graphic for the current platform with an algorithm:

```
match procedure/step/dialog-box
    $dialog-box-name = contents
    $graphic = find graphic where type = dialog box
        and name = $dialog-box-name
        and platform = $current-build-platform
    insert $graphic
```

There are several benefits to this partitioning of the image problem:

1. Writers do not have to worry about finding or inserting graphics or remembering the rules about when they are supposed to use screenshots. They just have to remember to mark up the names of dialog boxes when they mention them. If you only want to show screen shots for certain screens, you supply the screen shots for only those screens and have your algorithm pass silently over any `dialog-box` entry that does not have a screen shot in the collection. This allows you to adjust to reader feedback about how many screen shots are needed by

simply adding them to or removing them from the collection; you don't need to change the content at all – partitioning at work.

2. Updating graphics for UI changes is simpler because you just update the catalog of images. You don't have to search through docs to find images that are affected by the change.

3. You don't need any conditional logic in the text to include the right graphic for the platform.

4. In media where the screen shot would not fit in the viewport, you can suppress the image or handle it a different way.

5. If you decide that most readers won't need to see the screenshot, you can use the markup to create a link to a topic describing the dialog box instead of putting a screenshot inline. In a reuse scenario, you might make different choices for content aimed at different levels of users (for instance, including screenshots only in material intended for novices).

In principle, there isn't any difference between factoring out text and factoring out graphics. Indeed, beyond factoring out text or graphics, this partitioning also factors out the decision about whether to express a particular idea with text or a graphics. This means you can choose whether to use text or graphics depending on the audience or medium and implement that choice without changing your content.

However, when factoring out graphics, you still need to supply the metadata needed to render those graphics. One way to handle this is to create a metadata file for each image that provides the data for multiple renderings of the image and a path to each of the renderings.

The simplest way to implement this is to use an include instruction that points to the metadata file instead of the image file. This is what I did in writing this book. This strategy uses the DocBook technique shown in Figure 23.6, which conditionally includes two different versions of a graphic, one for epub and one for print. But this book is not written in DocBook, it is written in SAM. In the SAM source file, the image insertion looks like this:

```
>>>(image ../graphics/assemble.xml)
```

This is not the full factoring out of the graphics as described in the previous example, because the inclusion still refers to an image, and I haven't factored out the inclusion code. However, the file being included is not a graphics file; it is an XML file (see Figure 23.7).

The XML file in Figure 23.7 describes the idea of the image, listing not only its source file and both of its renderings, but even a text description for use when the graphic cannot be displayed. By including this file instead of an image file, I was able to include the idea of the graphic in my content. When the content is processed, the presentation algorithm loads and reads the `assemble.xml` file and uses the information in it to generate the conditionalized DocBook file that becomes the source file for the formatting algorithm, which is implemented by the publisher's existing DocBook tool chain.

```
<?xml version="1.0" encoding="UTF-8"?>
<image>
    <source>assemble.svg</source>
    <fo>
        <href>assemble.svg</href>
        <contentwidth>4in</contentwidth>
        <align>left</align>
    </fo>
    <epub>
        <href>assemble.png</href>
    </epub>
    <alt>
        <p>A diagram showing multiple pieces being
        combined in different ways to produce different
        outputs.</p>
    </alt>
</image>
```
Figure 23.7 – Graphics metadata file example

Could I have factored out the filename `assemble.xml` as well? Certainly. There are a number of other ways that I could have chosen to represent the idea of the graphic in the content. There are times when it makes a lot of sense to do that. If you are including screen shots in a procedure, for instance, the name of a dialog box is a good way of representing the idea of a graphic that is semantically relevant to the procedure itself.

But in the case of the images in this book, their relationship to the text is a little more arbitrary than the relationship of a screen shot to a step in a procedure, so factoring out the filename would have created an abstraction that was actually more difficult to remember as a writer. The point is not to be as abstract as possible, but to combine the highest degree of functional lucidity with the constraints that improve content quality, and that will be different for different kinds of material and for different circumstances.

Inline graphics

One further wrinkle with graphics is that writers sometimes need to place small graphics in the flow of a sentence, rather that as a separate block object. For instance, when describing instructions that involve a keypad or keyboard, some writers use graphics of the keys rather than simply printing the character names. Under certain circumstances, this can make the content easier for a reader to follow.

Inline graphics can cause rendering problems. For instance, they may cause line spacing to be thrown off if the height of the graphics is greater than that of the font used. Writers can control and make judgments about inline graphics when writing in the media domain, but using them may lead to unexpected and unwelcome consequences in document- or subject-domain content.

There are two techniques you can use to minimize problems with inline graphics. The first is to avoid them altogether, wherever practical. If you can present the same material just as effectively in some other way, choose that option.

The other technique is to factor out the graphic by using a structure to record its semantics. For instance, instead of including an **Enter** key graphic like this:

```
3. Press >(image enter_key.png) to confirm the selection.
```

Do this:

```
3. Press {Enter}(key) to confirm the selection.
```

This leaves open the choice of how to represent the key in the output and allows for differential single sourcing. For example, on a display that did not support graphics, or where graphics would be too fussy, the presentation algorithm could render this as:

> 3. Press [Enter] to confirm the selection.

But for media where the use of a graphic is appropriate, the presentation routine could use a lookup table of key names and graphics to select the graphic file to represent the **Enter** key.

> 3. Press to confirm the selection.

This approach allows the document designer to select graphics that work best on different displays or at different scales. This partitions the graphic maintenance problem to the most qualified person. And writers don't have to stop to think about which graphic to use.

The same approach could be used in another common case, which is describing tool bar icons in a GUI application.

```
4. Press {Save}(button) to save your changes.
```

This has the same advantages, with the additional benefit that if the interface designer decides to change an icon or to redefine the whole set of icons, you only have to update the lookup table used by the presentation algorithm. This could also be used to substitute different icons for different platforms if your application is run on more than one operating system. This is much more efficient than using conditional text to import different graphics for different configurations.

This is a good example of using the idea of the graphic rather than the graphic. The idea of the graphic is to represent the **Enter** key or **Save** button. You can do this in a number of ways, including using a photograph of the key, using a special font that creates the look of a key, representing the key using text such as [Enter]. The idea of a graphic is to represent a subject. So while you can insert the idea of a graphic in the form of a key or a reference to a file that records the idea of the graphic and its implementations, you can also simply identify the subject itself.

As always, a common principle is at work here: it's better to capture the subject than a resource that represents the subject. Resources change more often than subjects, and you may want different resources to represent a subject under different circumstance. But as long as the content remains current with it subject matter, the identification of the subject will not change.

CHAPTER 24

Subject-Domain Structures

Not all content can be meaningfully moved into the subject domain. The subject domain requires a repeatable rhetorical structure. Sometimes the nature of your content, your subject matter, or your argument does not lend itself to a repeatable structure. In those cases, the best you can do is work in the document domain. However, you can add subject-domain annotations (as described in Chapter 18) – marking up significant subjects in your content by their type and value – across all your subject-domain and document-domain content, allowing you to apply the conformance, change management, linking, and information architecture algorithms across your entire body of content.

But, as I noted in Chapter 22, you can move a lot of your content into the subject domain, even if it does not currently obey a closely defined rhetorical type, and you can move your content into smaller rhetorical blocks that are easier to apply a repeatable rhetorical structure to. In this chapter I walk through the steps of defining a repeatable rhetorical type using the recipe example we have been looking at throughout the book.

To begin, let's take a step back and look at Figure 24.1, which shows what a recipe might look like if it were not presented in its familiar rhetorical pattern.

```
Hard-Boiled Eggs
================
A hard-boiled egg is simple and nutritious.
Place 12 eggs in a pan and cover with
water. Bring water to a boil. Remove
from heat and cover for 12 minutes.Place eggs
in cold water to stop cooking. Peel and
serve. Prep time, 15 minutes. Serves 6.
```

Figure 24.1 – Unstructured recipe for hard-boiled eggs

If you were discerning the rhetorical structure of a recipe for the first time, you might look at several examples, try making the dishes or observe others doing so, and conclude that it is easier to find and follow a recipe if the ingredients are listed separately and the preparation steps are presented one at a time. You might then come up with a structure that looks like Figure 24.2.

```
Hard-Boiled Eggs
================
A hard-boiled egg is simple and nutritious.
Prep time, 15 minutes. Serves 6.

Ingredients
-----------
======  ========
Item    Quantity
======  ========
eggs    12
water   2qt
======  ========

Preparation
-----------
1. Place eggs in pan and cover with water.
2. Bring water to a boil.
3. Remove from heat and cover for 12 minutes.
4. Place eggs in cold water to stop cooking.
5. Peel and serve.
```

Figure 24.2 – Basic structured recipe for hard-boiled eggs

At this point you have the basic rhetorical pattern of a recipe, but you don't have any formal constraints for writers to make the format repeatable. To fix this, you could create a rhetorical block named `recipe` that contains three semantic blocks: `introduction`, `ingredients`, and `preparation`. This step factors out the titles of those sections. You might also decide to present the ingredients as a bulleted list and the preparation as a numbered list (these lists are document-domain semantic blocks, composed of structural blocks that we need not delve into here). Figure 24.3 shows the result of these steps.

This defines the basic rhetoric of a recipe – it enforces the rhetorical constraint. This structure stands at the intersection between the document and subject domains. It is document domain in the sense that it describes the structure of a document, but also subject domain in that it is specific to a single subject and defines the document structure in terms of the rhetoric of that particular subject. This structure provides guidance for writers, which helps with repeatability, but when you look closer, there are some other things that can be made repeatable in the rhetorical model to make sure they are included every time.

```
recipe: Hard-Boiled Egg
    introduction:
        A hard-boiled egg is simple and nutritious.
        Prep time, 15 minutes. Serves 6.
    ingredients:
        * 12 eggs
        * 2qt water
    preparation:
        1. Place eggs in pan and cover with water.
        2. Bring water to a boil.
        3. Remove from heat and cover for 12 minutes.
        4. Place eggs in cold water to stop cooking.
        5. Peel and serve.
```

Figure 24.3 – Hard-boiled egg recipe as a rhetorical block

Two obvious items are prep time and the number of servings. These are important items that should be in every recipe, but currently they are only in the introduction, where writers could forget them. Also, algorithms can't access them, so if you wanted to make a collection of recipes that took less than 30 minutes to make, you would not be able to find them algorithmically. So let's pull those items out into separate fields. Now you can verify that they are included and that algorithms can find the information (see Figure 24.4).

```
recipe: Hard-Boiled Egg
    introduction:
        A hard-boiled egg is simple and nutritious.
    ingredients:
        * 12 eggs
        * 2qt water
    preparation:
        1. Place eggs in pan and cover with water.
        2. Bring water to a boil.
        3. Remove from heat and cover for 12 minutes.
        4. Place eggs in cold water to stop cooking.
        5. Peel and serve.
    prep-time: 15 minutes
    serves: 6
```

Figure 24.4 – Hard-boiled egg recipe with prep time and servings added

But if you look a little closer, you will see more information that should be provided consistently and repeatably – the list of ingredients. You probably want to make this information accessible to algorithms, so you can handle it differently for different media or audiences.

There are three elements to an ingredients list: the name of the ingredient, the quantity required, and the unit of measurement for that quantity. Different units of measure are used for different types of ingredients and for large and small quantities in cooking, so you have to specify the unit of measure each time. The ingredients section of Figure 24.5 uses pure subject-domain markup – the rhetoric has been factored out leaving pure data.

```
recipe: Hard-Boiled Egg
    introduction:
        A hard-boiled egg is simple and nutritious.
    ingredients:: ingredient, quantity, unit
        eggs, 12, each
        water, 2, qt
    preparation:
        1. Place eggs in pan and cover with water.
        2. Bring water to a boil.
        3. Remove from heat and cover for 12 minutes.
        4. Place eggs in cold water to stop cooking.
        5. Peel and serve.
    prep-time: 15 minutes
    serves: 6
```

Figure 24.5 – Hard-boiled egg recipe with ingredients list added

Once you have put all of the current content into repeatable, accessible structures, it is time to think about whether there is any other information that should be included.

Usually during this process, people start to say things like "Wait, shouldn't we tell them X?" Sometimes X is included in one or two of the examples that you are looking at, and you realize it would be useful in all of them. Sometimes it occurs to people for the first time that X might be valuable. The process of formalizing the information you are proposing to offer often produces the realization that there is valuable information that you have not been providing or not providing consistently in your current content.

Also, it is at this stage that you should think about differential single sourcing, where you might want to capture content that should be presented differently in different domains and reuse scenarios or where you have information presented in one publication but not another.

Therefore, you might decide to add wine and beverage matches and nutritional information to the model (see Figure 24.6).

```
recipe: Hard-Boiled Egg
    introduction:
        A hard-boiled egg is simple and nutritious.
    ingredients:: ingredient, quantity, unit
        eggs, 12, each
        water, 2, qt
    preparation:
        1. Place eggs in pan and cover with water.
        2. Bring water to a boil.
        3. Remove from heat and cover for 12 minutes.
        4. Place eggs in cold water to stop cooking.
        5. Peel and serve.
    prep-time: 15 minutes
    serves: 6
    wine-match: champagne and orange juice
    beverage-match: orange juice
    nutrition:
        serving: 1 large (50 g)
        calories: 78
        total-fat: 5 g
        saturated-fat: 0.7 g
        polyunsaturated-fat: 0.7 g
        monounsaturated-fat: 2 g
        cholesterol: 186.5 mg
        sodium: 62 mg
        potassium: 63 mg
        total-carbohydrate: 0.6 g
        dietary-fiber: 0 g
        sugar: 0.6 g
        protein: 6 g
```

Figure 24.6 – Hard-boiled egg recipe with beverage and nutrition information

Once you are satisfied that the overall structure of the rhetorical block is correct, it is time to think about subject annotation. Annotating subjects that are significant to the subject matter of the information set makes that information available for many purposes, from linking and formatting to validation and auditing. To do this, go through the text looking for mentions of subjects related to cooking. As you find mentions of food, utensils, and common cooking tasks, add annotation markup for them to your model. Figure 24.7 contains annotations for items that can be classified as food, utensils, and tasks.

```
recipe: Hard-Boiled Egg
    introduction:
        A hard-boiled {egg}(food) is simple and nutritious.
    ingredients:: ingredient, quantity, unit
        eggs, 12, each
        water, 2, qt
    preparation:
        1. Place eggs in (pan){utensil} and cover with water.
        2. {Bring water to a boil}(task).
        3. Remove from heat and cover for 12 minutes.
        4. Place eggs in cold water to stop cooking.
        5. Peel and serve.
    prep-time: 15 minutes
    serves: 6
    wine-match: champagne and orange juice
    beverage-match: orange juice
    nutrition:
        serving: 1 large (50 g)
        calories: 78
        total-fat: 5 g
        saturated-fat: 0.7 g
        polyunsaturated-fat: 0.7 g
        monounsaturated-fat: 2 g
        cholesterol: 186.5 mg
        sodium: 62 mg
        potassium: 63 mg
        total-carbohydrate: 0.6 g
        dietary-fiber: 0 g
        sugar: 0.6 g
        protein: 6 g
```

Figure 24.7 – Hard-boiled recipe with annotations

This process can raise some interesting questions, as noted in Chapter 23. You might notice that every recipe uses utensils. If that's the case, why not make a list of utensils? Wouldn't such a list help readers determine if they have the equipment needed to make a recipe? And you might notice that a close cousin of the recipe, the knitting pattern, does list equipment: the precise knitting needles used.

So, should you change the structure of the rhetorical block to include a list of utensils? Specifying subject affinities formally forces you to think systematically about the correct rhetoric for your subject matter and audience. Sometimes the answer to questions like this is no. Over the years, cooks around the world have determined that while a list of ingredients is an essential part of the rhetorical structure of a recipe, a list of utensils is not. Don't create additional structures just because you can. Create them because they enhance your rhetoric or process.

Finally, you may want to think about whether there is any management-domain metadata that you need to add to the model. In Chapter 25 and Chapter 36, I consider how much management metadata belongs in the content model itself, as opposed to an external system. Here it is sufficient to note that you may need additional management metadata to support your processes. Figure 24.8 adds some basic tracking metadata, such as the author's name and copyright information.

```
recipe: Hard-Boiled Egg
    author: bcrocker
    rights: full
    season: winter, spring, summer, fall
    introduction:
        A hard-boiled {egg}(food) is simple and nutritious.
    ingredients:: ingredient, quantity, unit
        eggs, 12, each
        water, 2, qt
    preparation:
        1. Place eggs in (pan){utensil} and cover with water.
        2. Bring water to a boil.
        3. Remove from heat and cover for 12 minutes.
        4. Place eggs in cold water to stop cooking.
        5. Peel and serve.
    prep-time: 15 minutes
    serves: 6
    wine-match: champagne and orange juice
    beverage-match: orange juice
    nutrition:
        serving: 1 large (50 g)
        calories: 78
        total-fat: 5 g
        saturated-fat: 0.7 g
        polyunsaturated-fat: 0.7 g
        monounsaturated-fat: 2 g
        cholesterol: 186.5 mg
        sodium: 62 mg
        potassium: 63 mg
        total-carbohydrate: 0.6 g
        dietary-fiber: 0 g
        sugar: 0.6 g
        protein: 6 g
```

Figure 24.8 – Hard-boiled recipe with management-domain information

Management-domain metadata may not be part of your current content examples. This is alright, because adding management metadata should be the last step in this process. Once you have formalized the rest of the model, you may find that the subject-domain data you have delineated is usable for management and tracking purposes as well, and you don't need to add fields just for management. For instance, you can use subject-domain information to manage which recipes you include in particular publications. For example, with the information in these examples, you could assemble quick and easy meals (less than 20 minutes prep time, fewer than 6 ingredients, fewer than 5 steps) or a low-cal vegetarian cookbook (fewer than 160 calories, no meat ingredients).

Keep it simple and lucid

Most subject-domain languages are small, simple, and fairly strict in their constraints. This is as it should be. Since you have to design them, and the algorithms that translate them into the document domain for publishing, you don't want them to be elaborate or full of different structural permutations. The point of a subject-domain language is to partition the gathering of information about a subject from all the processes that you might want to perform on that content, from differential single sourcing to linking to information architecture to content reuse. Thus, it removes all of the complexity associated with those functions and simply captures the metadata that describes the subject itself. If your rhetoric is well defined, this should result in simple, straightforward structures.

If you find yourself needing a similar language for a related subject, it is usually better to create a new, equally small and strict language for that subject rather than trying to make one language cover both. Subject-domain languages get both their power and ease of use from the simple and direct way that the language relates to its subject matter. Trying to make one language cover more than one subject takes away from these properties. If you have ever tried to fill out a government form in which different people or entities are supposed to fill out different fields in different ways, you know how difficult it can be to be sure you have filled out all the sections, and just the sections, that apply to you. It is far better to have one form for each case (though obviously you need to make it clear which case each form applies to).

A subject-domain language should communicate with writers in terms they understand. This means that the names of structures should make sense to them, but it also means that the way that the formal structures break things up should make intuitive sense as well. For writers with experience in the field, a subject-domain language should be such a good fit that they don't really feel like they have to learn anything to use it. This vastly increases the functional lucidity of the language, leaving more of the writer's attention free to focus on the subject matter and the aspects of rhetoric that cannot be modeled, while at the same time providing constraints and guidance to make sure that the resulting content is complete and consistent.

Be careful not to take things too far. Once you get started, it is easy to get carried away and formally describe the subject matter in finer and finer detail. Remember that all of this effort is wasted unless it helps improve your rhetoric or process. If you get to the point where more precisely modeling the subject matter makes your markup mysterious, tedious, or difficult to create, you can do more harm than good, reducing functional lucidity without any compensating increase in quality or efficiency.

CHAPTER 25
Metadata

We live in the age of metadata, so much so that the word metadata has almost come to replace the word data itself and has come to be applied to almost any form of data that describes a resource. For example, we hear a lot about law enforcement getting access to metadata related to phone calls, which simply means the data about which number called which number and for how long.

The standard definition of metadata is "data that describes data," but that definition misses the central point. Metadata does not merely describe data; metadata creates data. Metadata turns an undifferentiated set of values into useful data.

In the document domain, the ingredients of a recipe are just list items, strings of characters.

```
section: Ingredients
    * 12 eggs
    * 2 qt water
```

Adding subject-domain markup lets you tell algorithms exactly what the strings mean.

```
ingredients:: ingredient, quantity, unit
    eggs, 12, each
    water, 2, qt
```

This subject-domain markup is metadata, and it turns unidentified list items into ingredient data. It is not that the data existed and the metadata came along afterward to describe it. The data exists as data only because metadata describes it. While a human reader can recognize that the list items are ingredients, an algorithm sees only strings of characters until you apply metadata that tells the algorithm that the characters are ingredients.[1]

Structured writing obeys constraints and records the constraints it obeys, creating an interface between partitions in your content system. It applies metadata to constrain the interpretation of values in content, making those values accessible as data.

---

[1] I am talking here about ordinary algorithms: simple rules that govern the processing of well-defined data. Advances in artificial intelligence (AI) are making progress towards developing algorithms that read the way humans do. A sufficiently advanced AI algorithm might not need this level of explicit metadata to recognized list items as ingredients. However, such an AI would still depend on metadata. It would just be using the same metadata that humans use, which we might fairly, if briefly, characterize as a combination of grammar and memory. Until such AIs are available to us, though, structured writing enables us to add more explicit metadata to content to make it accessible to simpler algorithms.

Partitioning the content system means moving complexity – decisions – from one partition to another. Metadata enables you to record and transfer the information required to do that. If you want to transfer a decision from the writer to another person or process, you have to ask what information the receiver needs in order to make and execute that decision. That information is the metadata that needs to be added to the content.

This is the basis for all partitioning of the complexity of content creation. When you attach metadata to content, you can pass that content to people and processes without dropping any of the complexity. You cannot partition complexity safely if you drop any of the complexity in the process. Metadata ensures that all the complexity transfers successfully from one partition to another.

Metadata preserves the information needed to make decisions about a piece of content. Preserving the fact that a piece of text records an ingredient lets you make decisions later about how to format ingredient listings or about which recipes to include in a collection. Explicitly recording which parts of that string are the quantity and unit of measure lets you defer a decision on which set of weights and measures to use when presenting the recipe to a reader, allowing you to publish the content in other markets.

The subject-domain `ingredient`, `quantity`, `unit` markup is metadata that turns the ingredients list into ingredients data. But this is is not to say that a `list` structure is not metadata also. `list-item` is document-domain metadata. It allows you to partition the job of formatting lists from the job of the writing recipes and other kinds of content. This makes it easier to write algorithms that recognize lists and format them for whatever medium you choose. The operations you can perform on list-item data are far less sophisticated than those you can perform on ingredient data, but list-item data is still data and still created by metadata.

There is an important point here: The same set of values – the same string of letters, words, and numbers – can be turned into different kinds of data by applying different kinds of metadata. This means that you can choose what kind of data you turn your content into by choosing what type of metadata you apply to it. Moving content from one structured writing domain to another means turning it into different kinds of data by applying different metadata to it. By turning it into different kinds of data, you make it accessible to different algorithms.

The recursive nature of metadata

Metadata is a confusing concept because metadata is recursive. If metadata is the data that creates data, then as data it too must be created by metadata. In other words, if the line `ingredients:: ingredient, quantity, unit` turns the ingredient lines into metadata in this markup:

```
ingredients:: ingredient, quantity, unit
    eggs, 12, each
    water, 2, qt
```

Then what makes `ingredients:: ingredient, quantity, unit` a piece of metadata and not just another string of characters? There is another piece of metadata that says that that string is a record set definition, and, in this example, that metadata is the SAM syntax specification. This metadata is not part of the same file as the data it defines, but metadata can be both inline and separate. Thus one piece of data can have a whole cascade of metadata defining it, metadata defining that metadata, and so forth.

In structured writing, you add structure to content to replace the things you have factored out. That structure is metadata to the data that is the text of the file. But if you store that file in a repository, the information that identifies the file in that repository is metadata to the file as a whole. If the structure of the file is described by a schema, the schema is also metadata for the file.

But we're not done yet because the specification of the schema language is the metadata that tells you what the schema means. And then of course, there is the specification of the markup to consider. The XML specification is part of the metadata tree for every XML document in existence. And because the XML specification uses a formal grammar description language called EBNF, the EBNF specification is metadata for the schema language description.

How do we break out of this seemingly infinite pattern? Data is information that has been formalized for interpretation by algorithms. Fortunately, human beings can understand natural language without that degree of formalization. Eventually, then, we reach a point where the last piece of metadata is not described by metadata but by narrative. That narrative document essentially bootstraps the whole metadata cascade that eventually yields pieces of data that can be unambiguously interpreted by algorithms.

So, every piece of data has a spreading tree of metadata supporting it, which, if traced to its roots, eventually leads to narrative documents that explain things in human terms. Thus the XML specification combines narrative definitions with EBNF, and if we go to the EBNF specification we will find the plain narrative that describes EBNF. Data is a formalization of narrative, and it is this relationship that allows structured writing to move information back and forth between data and narrative forms.

Where should metadata live?

One of the great questions about metadata is where it should live: with the data it describes or separate from it? As we noted above, much of the metadata in the metadata cascade is stored separately. Either the data points to the metadata (as when an XML document declares which schema it uses), the metadata points to the data (as when a content management system stores metadata), or the metadata is embedded in the content itself (as in structured writing).

The issue of where metadata should live is closely related to the issue of how you partition responsibilities in your content system. Since metadata transfers complexity from one partition to another, the responsibility for creating the metadata lies with the person or process in the originating partition, and the metadata requirements are dictated by the needs of the receiving partition. By adjusting how your system is partitioned, you can adjust how onerous the metadata requirements are on any one actor in the system. Therefore, the partitioning dictates the location of the metadata.

For example, most early graphic file formats stored only the image. Most modern formats also store extensive metadata about the image. The pictures you take with your digital camera include information about the camera and the settings that were used to take the shot, all of which helps rendering algorithms and graphic editing applications handle the raw image data better. Having that metadata embedded in the file ensures that the picture and its metadata stay together. Separating them would greatly complicate the system.

Unfortunately, content tools are often designed with other priorities in mind. For one thing, many tool developers think almost exclusively in relational database terms. The idea that you could store metadata anywhere other than in relational tables is foreign to them. For another, system vendors have a vested interest in a partitioning that requires every user to interact with their system all day long, because this forces companies to buy a software license for every contributor. Both factors encourage system vendors to implement models that separate metadata from content and store the metadata in their repository, ensuring that you must use their system to have access to the metadata.

For example, should the history of a file be stored in the file or in the repository? Storing it in the file lessens the file's dependence on the repository and makes it more portable. But a repository vendor may prefer to store file history in their workflow system, making it hard for you to move away from that system. If metadata is stored in the file, it is easier to edit content when you aren't connected to the system, which can save you on licenses.

In the case of photos, the metadata is in the file because the camera is the best, and possibly only, instrument for recording it. This partitioning of complexity is best for creating, accessing, and managing the information. The location of the metadata should be determined by the best partitioning of the content system, not the convenience of a tool vendor.

Writing your content in the subject domain means that more of your metadata is stored in the same file as the content, increasing its independence and portability. As we have seen, using subject-domain structures can lessen the need for management-domain structures to support algorithms such as single sourcing and content reuse, which reduces the need for external management-domain metadata. All of this contributes to improved functional lucidity, referential integrity, and change management.

But this does not mean that all metadata belongs in the content file. For example, when you import a graphic into a document, you often give it a caption and specify the display size. It doesn't make sense to include this information in the graphic file, because this information doesn't describe the graphic itself; it describes the graphic's relationship to the current document. In a different document, the same graphic might be displayed at a different size with a different caption. Therefore, it is better to include this metadata in the file that imports the graphic, not the graphic file itself. In Chapter 23 I looked at another way of partitioning this metadata, which is based on the nature of the relationship between the files and on the way that the complexity of the content system is partitioned.

This example shows that there are definitely types of metadata that belong in a repository or content management system. If you store your content in a version control system (VCS) such as Git (something that is increasingly popular in the "treat docs like code" movement), the VCS records the differences between each version of the file as well as who committed each version. This allows the VCS to handle management tasks such as telling you who changed a file and when it changed. Storing such metadata in the file would make it much more complex to write the algorithm for this task. Chapter 36 has more information about deciding where to store metadata.

Ontology

Finally it is worth saying a word about ontology. Ontology (in the information processing sense) attempts to create a formal mapping of the relationships between entities in the real world such that algorithms can draw inferences and reach conclusions about them.

In many ways, an ontology does for algorithms what narrative does for humans. After all, we read so that we can understand the world better. By understanding what various objects and institutions are and how they relate to each other, we can decide what to do.

In some sense, therefore, ontology is the ultimate in subject-domain markup. Indeed, one should be able to generate human-readable narrative from an ontology, given a sufficiently sophisticated algorithm and a sufficiently sophisticated ontology.

All of this is outside the scope of this book. Subject-domain markup attempts to capture certain aspects of the subject matter of a work, but it does not attempt to model the argument of a work. Consider this passage:

```
In {Rio Bravo}(movie), {the Duke}(actor "John Wayne")
plays an ex-Union colonel.
```

Here the subject-domain markup formalizes the fact that Rio Bravo is a movie and that "the Duke" refers to the actor John Wayne. It does not model the relationship between them. With an ontology, you would model the "starred in" relationship between John Wayne and Rio Bravo, whereas with subject-domain structured writing you would normally leave this to the text.

Similarly, this subject-domain markup does not bother to identify Union as a reference to both a country and its armed forces or that colonel is a rank in those armed forces. It does not identify these relationships because this particular markup language is concerned with movies, and these facts are incidental to the movie business. Actors, directors, and movies are significant subjects in the movie review domain. The names of nations and armies that figure in the plot of individual movies are incidental in that domain. A full ontological treatment of the passage above, however, would need to model those relationships.

Structured writing does make certain aspects of content clear to algorithms, but not with the intention of making it possible for the algorithms to make real-world inferences and decisions based on the information in that content. It only does what is necessary to partition and redirect content complexity in a content system in which human writers use algorithms as tools to improve the quality of the content they prepare for human readers.

CHAPTER 26
Terminology

Whether you store metadata internally in the content or externally in a content management system (CMS), it is important to keep terminology consistent. Similarly, it is important to name subjects in the content consistently to avoid confusing readers. And if you provide a top-down navigation scheme for your content, you need to choose the right terms to name the subjects so that readers can find them.[1]

For all of these reasons, large-scale content projects need to control terminology. Managing terminology is complex, and taking a simplistic approach can result in dropped complexity and compromised rhetoric in the form of incomprehensible language, incorrect classification, or poor connections between units of content.

The biggest danger is thinking of terminology as a simple data-management problem. Separating words from their role in sentences and paragraphs makes them easier to fit into the traditional rows and columns of data management, which makes the problem look simpler than it is. If you approach terminology management in this way, it might seem straightforward to get a bunch of people in a room, spin up a list of words and their definitions, and declare the result as your corporate taxonomy, but this is a false partitioning of the problem.

The principal difficulties in establishing terminology include the following:

- Human beings have a fairly small active vocabulary, and we reuse words all the time. In safety critical functions like air traffic control or the operating room, we train everyone to use a special unambiguous vocabulary that is generally undecipherable to the layperson.[2] But usually, the words you want to control are used in multiple ways that are not easy to disambiguate formally.
- People in different fields (even within the same organization) often use different terms for the same concept. What the chef calls *pork*, the farmer calls *pig*. What the English call *boot*,

---

[1] Control of terminology is important for translation as well, but that is outside the scope of this book.

[2] In fact, we not only train them, we also test and certify their knowledge before we let them in the room. Getting people to use a specialized vocabulary consistently and in a precise way is so difficult that it becomes a major factor in the development and licensing of professionals in safety critical functions. Merely issuing a glossary or taxonomy to an organization is not going to do the trick.

North Americans call *trunk*. Forcing everyone to use the same term means forcing them to say things that don't make sense in their own field.

- People in different fields (even within the same organization) often use the same words to mean different things. What the conference organizer calls a *function* is not what the programmer calls a function. What a programmer calls a function is (in more subtle ways) not what a mathematician calls a function.

- People in one field may have ten terms that make fine-grained distinctions among things that people in another field lump together under a single term. For instance, programmers make a distinction between subroutines, functions, methods, and procedures. I know of one documentation project that mandated that all of these should be called routines. This worked most of the time, but became a problem when function pointers were introduced into the product, because you can't use the modifier *pointer* with any of the other terms. Enforcing a single term obscured an important distinction.

In short, very little of our terminology is truly universal. The meaning of words changes depending on the context and the audience. This is at the core of how language works and why language is different from other forms of data. Content tells stories, and what words mean depends on the context in which they are used in the story. Stories are not as precise as formal data, but the only way to define formal data is with stories. This is why the spreading tree of metadata that supports and explains any point of data always ends with narrative. Structured writing attempts to apply some of the orderliness and manageability of data to stories in order to improve their quality and consistency, allowing us to communicate more effectively.

Terminology control is key to making your structures and metadata work. But, at the same time, you must use care and sensitivity and understand the limits within which you can control the terms you use to tell stories. Good taxonomies define their terminology within the confines of a specific domain, but in all but the most strictly controlled domains (such as the operating theater or the control tower), individual words have shades of meaning that can be fully disambiguated only in the context of a particular story.

In Chapter 10, we saw how you can use context to determine the meaning of block names. Terms don't have to be universally unique as long as you can identify them unambiguously in context. Terms that cannot easily be controlled universally can often be controlled in context. Appropriate partitioning makes the terminology problem much more tractable. Structured writing helps you partition the terminology problem by providing the context in which terms are understood.

In Chapter 18, we saw that subject annotations can be made more precise by identifying the type of subject being named. Here again you are adding context to terminology to control it better in its local domain. One of the virtues of the subject domain is that every subject-domain content type and indeed every subject-domain block type provides context for controlling the interpretation of the terminology within it. This is another instance of using structured writing to constrain the interpretation of content.

Top-down terminology control

There are two ways to partition the terminology control problem: top down or bottom up. The principal tools of top-down management are controlled vocabularies and taxonomies. A controlled vocabulary is essentially a list of terms and their proper usage within a specific domain.

A taxonomy is a more elaborate scheme for controlling and categorizing the names of things. Taxonomies are frequently hierarchical in nature, defining not only the terms for individual things but the names for the classes of things. Thus, a taxonomy does not just list sparrows and blue jays and robins, it also classifies them as birds, birds as animals, and animals as living things. A good taxonomy should be specific to the domain for which it is intended. Blue Jays and Cardinals occupy a very different place in a baseball taxonomy than in a ornithological taxonomy.[3] As a classification scheme, you can use a taxonomy not only as the basis for controlling vocabulary, but also as a basis for top-down navigation of a content set.

Alternatively, you can control terminology from the bottom up. To do this, you use subject-domain annotations in your content to highlight key terms and place the usage in the appropriate domain.

```
In {Rio Bravo}(movie), {the Duke}(actor "John Wayne") plays
an ex-Union colonel.
```

In the passage above, the annotations call out the fact that "Rio Bravo" is the name of a movie and "the Duke" is the name of an actor called John Wayne.

This is taxonomic information. It places Rio Bravo in the class `movie` and John Wayne in the class `actor`, with the added information that the Duke is, in this context, an alternate term for

---

[3] This reference to Blue Jays and Cardinals is culturally dependent. Readers outside North America may not recognize these as the names of baseball teams. This is a concern for terminology management and could call for differences in rhetoric in different markets, such as a footnote to explain the reference or a totally different example that better explains the concept to another culture.

the actor John Wayne. Placing these terms in these classes makes it clear that Rio Bravo, in this context, is not the Mexican name for what Americans call the Rio Grande or any of the several towns named Rio Bravo and that the Duke refers to John Wayne and not to the Duke of Wellington (who was often called by that nickname) or any of the other possible meanings of Duke. In other words, the taxonomy is embedded in the content. The subject domain internalizes taxonomic metadata.

To appreciate why this might be useful, let's look at some of the challenges of maintaining and enforcing vocabulary constraints.

As mentioned above, constraining vocabulary is hard because the same term can mean different things in different contexts, and different terms can mean the same thing in different contexts. If you attempt to build a taxonomy from the top down it can be difficult to anticipate the various meanings a term may have in different contexts. Even if you study existing content – a tedious and time consuming activity – there is no guarantee you will exhaust all the possibilities.

Secondly, after you define your top-down taxonomy, how do you enforce it? You can require writers to use terms from the taxonomy, but how is that going to work? Do you expect them to carry the entire taxonomy around in their heads? Do you expect them to recognize when they attempt to use a word that has a synonym in the taxonomy but isn't in the taxonomy itself? Such requirements create mental overhead for writers, and, as I noted in Chapter 9, dividing a writer's attention has a negative impact on content quality. A top-down taxonomy dumps a complex task, which depends on a huge amount of data, onto writers and requires them to pay attention to it continuously in every word they write. This is the antithesis of good partitioning of complexity.

There are mechanical solutions that attempt to catch terminology problems, but the fact that words can mean so many things in different contexts means no such process can get it right all the time.

Bottom-up terminology control

An alternative to defining and enforcing a taxonomy from the top down it to let it emerge, in a disciplined way, from the content itself. The key to this, particularly in the subject domain, is for writers to annotate significant subjects as they write. If the terminology you are trying to enforce is not significant to your content, you are wasting your time.

If writers annotate the significant subjects in their content as they write, they will be annotating the terms you want to control. So, when they mention the name of a bird like blue jay or robin, they should annotate it as {blue jay}(bird) or {robin}(bird). By specifying the type of the subject, they establish the taxonomic category of the subject in context.

Of course, this does not ensure that writers use the right terms for birds. It only highlights the terms they actually use. To achieve consistency, you need to audit the list of birds mentioned in your content. To get the current list of bird names from the content, you have an algorithm scan your content for terms annotated as bird and compile them into a sorted list.[4] You can then quickly see if any incorrect or unexpected terms are being used, and you can either edit them or call them to the writer's attention. This partitions the vocabulary control problem in a way that makes it easier for both the writer and the person managing terminology – the *terminologist* – to do their jobs.

Notice that you are not introducing a new task here. Even if you demand that writers follow the taxonomy up front, they will make mistakes, and you will need to audit anyway, unless you are willing to live with the mistakes that result from that dropped complexity. This approach simply makes the audit easier to perform by highlighting the use of terms, and assigning them to the correct domains, in the text.

Creating and maintaining a taxonomy

This approach not only works to audit conformance to an existing taxonomy, it can also be used to create and maintain a taxonomy. If a writer mentions a new bird that you didn't include in your taxonomy, it will show up in the list for the next audit. You can then decide if that bird should be added to the taxonomy or not.

Adding terms to the taxonomy does not necessarily imply that you maintain a top-down taxonomy that is separate from the content itself. You can regard the current annotated content set itself as the taxonomy. After all, it contains all the approved terms and their types. Any list of terms you generate from the content are just reports on the taxonomy, not the taxonomy itself. By storing a report from the previous audit, you can compare to see if any terms have been added to the taxonomy since the last audit.

---

[4] This is one of those query-oriented algorithms that I mention at the end of Chapter 11 and an application of the content-generation algorithm discussed in Chapter 14.

This approach may not always give you sufficiently firm control over your taxonomy, so you may, instead, use annotations in your content as an indication that writers are using new words, and you can then add them (or not) to the official taxonomy list.

If you maintain an official taxonomy separate from the content, it is trivial to have an algorithm compare the audit list to the official taxonomy and alert you every time a new bird is mentioned. Effectively, now, your taxonomy is bubbling up from your content. Your writers do not need to worry about whether the terms they use are in the taxonomy or not, as long as they mark up what type of thing they are naming. You can run the audit-and-compare algorithm regularly (nightly perhaps) and have it alert your terminologist every time a new term is added to the content. The terminologist can evaluate whether the term is being used correctly and whether or not it should be added to the taxonomy. This way, the terminology audit is almost entirely automated, with action required from the terminologist only when a writer uses an unfamiliar term.

Writers may forget to annotate some birds or may annotate them incorrectly (as something other than a bird). But you can easily catch most of these mistakes as well. Incorrect annotations tend to show up as anomalous entries in other annotation categories. If "sparrow" shows up in the `bard` category next to Shakespeare, the terminologist will get an alert that a new bard has been mentioned and can easily see and fix the mistake.

When there are genuine name conflicts between two different domains, you can list the conflicting names, use an algorithm to compile a list of all tagged instances of those names, and review them for incorrect tagging. To catch omitted tags, you can have an algorithm scan the entire content set for unannotated instances of annotated terms. If you get a lot of false hits (the same words used in a different context), you can annotate them to be ignored:

```
It reminded me of {Robin}(ignore) Hood.
```

Centralized versus distributed allocation of complexity

The point of partitioning and redistributing the complexity of a problem is to direct the complexity to those best placed to handle it. This sometimes means directing complexity to a central person or process with specialized knowledge, such as knowledge of the organization's formatting standards. But just as often it means distributing the complexity outward or down to people in the field who have the contextual knowledge needed to make the correct decision.

Taxonomy is a third case that combines distribution of complexity and centralization. Only the writers in the field can tell you what ideas need to be expressed, and only a central process can coordinate and disseminate information about what terms have been added to the taxonomy. Any process that does not allow communication in both directions, and does not allow writers to make good decisions in context, will damage rhetoric and make content less accessible to readers. Equally important is how complex you allow each person's task to become and how many things you ask them to do at the same time. Even if writer and terminologist are the same person, the approach to terminology control outlined above separates the tasks in time, so that writers do not have to keep the taxonomy in their head as they write. In fact, it means that no one has to keep the taxonomy in their head at all.

Stop lists

Another useful conformance tool that can come out of the bottom-up approach is a stop list. A stop list is a list of terms that should not be used. You can have an algorithm scan content against a stop list to identify inappropriate vocabulary. Stop lists can only really be created bottom up. You can't anticipate or ban every term anyone might ever come up with. You should only ban terms that are both problematic and occur frequently. The chance of false hits – of banning terms that are perfectly legitimate in other contexts – rises with every word you add to the stop list. With a bottom up approach to terminology control, you get an accurate measure of which terms are being misused and the frequency and nature of the misuse. This is an excellent basis for compiling a useful stop list.

Also, because subject annotation can specify the type of a term (that is, distinguish between {Blue Jays}(baseball-team) and {Blue Jays}(bird) you can make a type-specific stop list, banning terms used in one sense but not another. This can greatly reduce false hits, which is important because people won't use conformance tools if they produce too many false hits, as you can see from how infrequently writers use the grammar checkers in word processors.

The bottom-up approach does not make all vocabulary constraint problems go away, but it does have a number of advantages.

- It turns an up-front taxonomy development effort into an ongoing, permanent part of the content development process. This not only reduces the up-front spike in effort, it also helps ensure that your taxonomy is based on real experience writing real content and that it is continually maintained as subjects and business objectives change. Indeed, if you already annotate your content to support any of the other structured writing algorithms, you essentially get taxonomy development, maintenance, and control almost for free. Of course, someone does have to review the reports, make the edits, and add to the canonical taxonomy.

- It improves functional lucidity by not forcing writers to refer to the taxonomy while writing. If you are already annotating subjects for other reasons, you are imposing no additional burden on writers at all.
- By improving the consistency of the annotations, it makes the algorithms that rely on annotations more reliable. This is one of the greatest virtues of the subject domain. Subject-domain markup can serve multiple algorithms, meaning you get the benefits of multiple algorithms with less cost.

CHAPTER 27

Relevance

Establishing the relevance of content is essential for readers and for many of the structured writing algorithms.

- Readers looking for content have to decide if a given document meets their needs.
- Search engines need to determine if a document is relevant to a reader's search.
- The reuse algorithm needs to determine if content relevant to a certain point exists.
- The content-generation algorithm needs to find all of the content relevant to a purpose.
- The linking algorithm needs to find any documents relevant to a subject mentioned in the current document, so it can create links to them.
- The conformance algorithms need to determine if there is information in one document that should conform to information in another document and ensure that a document contains all the information relevant to the reader's needs.
- The algorithms that implement your information architecture need to discover all of the connections between pieces of content, the types of those connections, and the constraints they obey in order to build a navigable information set.

Managing relevance has a twofold role: first, alleviate as much as possible the complexity the reader faces in establishing relevance, and second, make sure that algorithms that need to establish the relevance of content do so correctly and unambiguously.

Structure supports relevance in two ways: being relevant, and demonstrating relevance.

Being relevant

Relevance is too often neglected. Content management systems (CMS) attach metadata to unstructured documents, asserting their relevance to certain topics or requirements without verifying that those documents actually meet those requirements. In many cases, the CMS has a cataloging system that requires writers to complete certain metadata fields, often from a predetermined list of values. Developing a good set of relevance criteria and values is complex, and development efforts often do not yield sound relevance criteria for readers. Defining the taxonomy by committee often imposes an abstract logic that does not match the concerns or language of readers.

If developed correctly, such metadata may have the potential to define the relevance of content, but these requirements are not always communicated to writers in advance. Instead, writers are left to compose content without any guidance as to what would make it relevant, and then they are asked to tag content with CMS metadata after the fact. In many cases, this tagging is done hastily and lazily, simply picking whatever terms seem appropriate at first glance without really thinking about whether the piece fulfills the promise the metadata is making.

As a result, content returned by the CMS is often not relevant to the reader's needs because it simply does not contain what the CMS metadata claims it does. In many cases, bypassing the CMS's navigation mechanisms and doing a simple site search produces better results, since the search engine looks at the content itself to determine what it is relevant to. But while this can rescue the reader's quest for relevant content, where such content exists, it does nothing to ensure that relevant content actually gets created. Without this assurance, the CMS may be telling lies to its owners as well as to it readers.

If you are going to implement strict standards of relevance and a strict vocabulary for proclaiming relevance, you must communicate this to writers long before they try to submit their content to the CMS. One way to do that is through structured writing.

Showing relevance

Having relevant content is important, but it is equally important that your content show that it's relevant. Readers and algorithms can't find your content if it doesn't show its relevance.

Actually, that is not entirely true. Search engines can determine the relevance of content with an amazing degree of accuracy simply by reading the text and looking at how the content is used and linked to. This attests to the first importance of being relevant.

However, if the search engine decides a topic may be relevant and adds that topic to search results, but the reader looks at it and can't see the relevance, you still haven't reached that reader. In addition, the search engine may take note as well and downgrade that topic's relevance ranking.

Showing relevance is all about establishing your subject matter and context and doing so clearly, unambiguously, and quickly. Otherwise, the reader will dismiss the results and continue searching. When developing a markup language (or other structured content container) for your content, one of the most important things to consider is how the reader will recognize what your documents are relevant to. This is determined by the rhetorical structure of each piece of content and whether that structure helps the reader see the relevance of that content. I deal with this question at length in my book, *Every Page is Page One: Topic-based Writing for Technical Communication and the Web.*

Relevance is not established the same way for all readers or for all subjects. For a recipe, for instance, a picture may do a lot to establish relevance. For an API reference, a version number and the description of a return value may be key relevance indicators.

But relevance is not just about the subject matter. It is also about the reader's purpose. For a page about a business, the inclusion of a stock price chart may tell you that the page is of interest to an investor rather than a potential customer. Placing that chart at the top of the page helps establish the relevance quickly. For an article about a place, pictures of beaches or nightlife show that the page is relevant to potential tourists but not relevant to residents trying to decide what schools to send their kids to.

Creating content in the subject domain allows you to make sure that writers produce all the pieces of information that make a page relevant to the intended audience. Because the subject domain also factors out presentation order, it allows you to make sure that every page is organized in a way that best shows its relevance.

Better still, because subject-domain content is organized for presentation by algorithms, you can experiment to see if one organization of content works better than another and adapt your presentation algorithm accordingly without having to edit any of the content.

Also, because storing content in the subject domain allows you to use the extract and merge algorithms to pull in content from other sources, you don't have to include the beach pictures or the stock chart with your subject-domain content. Instead, you can have the synthesis algorithm query other content sets or feeds to find the best current tourist shots or to generate the stock chart in real time.

But that, of course, depends on your content having fields that establish relevance and give the synthesis algorithm the information it needs to find relevant pictures or a relevant stock chart. It is not enough to show relevance to readers, you also need to show relevance to algorithms.

Showing relevance to algorithms

Showing relevance to algorithms comes down to breaking information up into clearly labeled fields containing unambiguous values. For example, to query a web service that generates stock charts, you need to provide an unambiguous identifier for the company whose chart you want. The company name is not the best identifier, since there can be duplicate or similar names in different industries (Apple Computer versus Apple Music, for example). A better choice is the company's stock symbol, which is guaranteed to be unique on the exchange where it is listed (though you do have to provide the exchange code: NASDAQ:AAPL, not just AAPL).

This means that the web service needs to index stock prices according to stock symbols (which is what it does, of course). Actually, it is likely that the stock chart drawing service does not hold price information at all but, instead, requests it as needed from yet another web service. The stock ticker symbol is the unambiguous key that identifies the company in each of these transactions.

While humans and algorithms assess relevance in different ways, the foundations of relevance are the same for both: clear identification of the type and subject matter of the information. This means that you need to maintain metadata that clearly identifies the type and subject and also present the content in a way that makes its type and subject evident at a glance.

```
recipe: Hard Boiled Egg
    introduction:
        A hard boiled egg is simple and nutritious.
    picture:
        >>>(image egg.jpg)
    ingredients:: ingredient, quantity, unit
        eggs, 12, each
        water, 2, qt
    preparation:
        1. Place eggs in pan and cover with water.
        2. Bring water to a boil.
        3. Remove from heat and cover for 12 minutes.
        4. Place eggs in cold water to stop cooking.
        5. Peel and serve.
    prep-time: 15 minutes
    serves: 6
    nutrition:
        serving: 1 large (50 g)
        calories: 78
```

Figure 27.1 – Recipe illustrating subject-domain markup for relevance

Figure 27.1 contains structures that both humans and algorithms can use to assess relevance. The document type, clearly identified as `recipe`, establishes basic relevance. Including a picture helps humans visually identify the recipe. Information such as the number of calories could be used by an algorithm to select recipes for a low-calorie cookbook.

It is possible to maintain metadata on type and subject of content in a content management system that hosts content written in the document or media domains. However, in the document and media domains, ensuring that readers can identify the type and subject at a glance is a separate problem from correctly labeling the content in the CMS. In these domains it is possible, with the right authoring discipline, to ensure that content matches the metadata, but it is also possible to have mismatches or even to have a situation where content does not demonstrate its type and subject effectively.

Moving content to the subject domain allows you to show relevance to both humans and machines with the same structures, assuring that the two do not get out of sync.

CHAPTER 28
Composable Structures

Some structured writing algorithms join small pieces of content together to form larger ones. This is *composition*. Composition doesn't occur only in reuse scenarios. It can also occur whenever multiple writers contribute to a single work. In these scenarios, you partition the writing so each piece of content is written only once and is assigned to the person best qualified to write it. Then, of course, you have to stitch the pieces together to create the whole work. This means that you need to create composable structures – structures that can be combined with other structures to create larger structures.

The various methods of stitching pieces of content together are covered in other chapters, in particular Chapter 11, Chapter 13, and Chapter 16. But the mechanics of splicing content together are only half the story. Algorithms can produce reliable results only if they have reliable data structures to work with. Unfortunately, this is often overlooked in designing structured writing systems. Assembling documents out of small easy-to-manage pieces sounds wonderful, and it can be made to work well under the right conditions. However, it will not work for all types of content, and it will work reliably only if you design your content structures to support composability.

Successful partitioning requires that sufficient information be passed between the partitions so that each can do its own work completely without dropping any complexity. When you divide an authoring task among many writers, you need to make sure that the individual pieces come together to form a coherent whole. The definition of what constitutes a coherent whole may vary. The criteria for evaluating whether a collection of Every Page is Page One topics organized as a hypertext is a coherent whole are different from the criteria you would use for a printed product manual designed to be read in linear order. But whatever the criteria, they must be met if you want to distribute authoring to multiple writers or reuse content without compromising rhetoric.

Combining the pieces

The first requirement of composability is that you must be able to combine the pieces at the file level. Many computer file formats won't let you make a bigger file just by appending one file to the next or dropping parts of one file into another. They are not designed to work that way. Many let you add material to a file using cut and paste, but as discussed in Chapter 13, you should avoid that method. You want an algorithm to do the composition, which means you need a format that lets you put the pieces together without turning the files into mush.

Most structured writing formats consist of a hierarchy of structures. Those structures tend to be self-similar in form. For instance, all structures in an XML document are composed of XML elements. This means that you can take an XML document apart at any point in the structural hierarchy and insert, remove, or rearrange the elements at that level. To compose a larger element out of smaller elements, you simply wrap new elements around them. Thus XML (and some other structured writing formats) provides a level of algorithmic composability often lacking in other formats.

Combining structures

The second requirement of composability is that the result of combining markup structures must be a valid document (must conform to the appropriate constraints for that document). The hierarchical nature of XML means that you can take a well-formed chunk of a DocBook file and drop in into the middle of a DITA file. The result would be an XML document that an XML parser could successfully parse, but it would not conform to the constraints of either DITA or DocBook. The tool chains for either language would not be able to process it. You need to do composition in a way that creates a document that meets the constraints of the target language.

The most obvious way to do this it to make sure that all of your pieces come from the same markup language. Thus, DITA has good support for structural composability but only when all of the pieces are DITA.

But using the same markup language is not enough. Markup languages constrain where certain structures can occur, and you must make sure that each piece goes into a place where it is structurally allowed in that language. Just because all the pieces come from the same language does not mean that every possible combination results in a valid document (just as you would not expect components of your car's exhaust system to bolt onto the steering column). Thus you cannot insert a DITA steps structure into a DITA concept or reference topic because steps are not permitted in those topic types. This requires planning and careful management to make sure the combinations you create are valid.

However, it is not essential to composability that all the pieces come from the same language. You can also take content from different sources and with different structures, as long as you can transform their structures to match the structure of the destination document. In other words, you can transform as you compose. This can be a very powerful technique. For instance, you can use it to compose documents from content in a database. (Indeed, all database reporting systems do exactly this: they compose documents in one format from tabular data in another format.)

For this approach to work, however, all of the sources you draw from must have a high level of conformance to their own constraints. If you don't know, or cannot rely on, the structure of the pieces you are drawing in, you cannot reliably combine them with an algorithm. Thus it is often better to focus on strategies for getting the most reliable sources rather than forcing everyone into a common format that they might not use reliably. For more on this, see Chapter 29.

Stylistic and rhetorical compatibility

While composability of structures is vital, it is not always sufficient. You could have pieces in a media-domain language that are structurally compatible but formatted differently. The resulting document would be valid and would publish successfully, but it would be a mess of competing styles and fonts. Similarly you could have two pieces in the document domain that are structurally compatible but have incompatible rhetoric. The resulting document would also publish successfully but might be incomprehensible.

You need to ensure that you have stylistic and rhetorical compatibility between the pieces you are composing, which you can ensure either by imposing stylistic and rhetorical constraints externally or by factoring them out by moving your content to the subject domain.

Narrative flow

Even if you can assemble pieces from different document-domain sources and format them all with a single consistent look, that does not mean that the result will be a complete, correct, and coherent narrative. Creating a coherent narrative is not necessarily a matter of making the document sound like it came from a single person. Many business documents are created by several different writers, sometimes working together, sometimes inheriting and maintaining a document over time. Making such a document sound like it was written by one person is a tall order, but doing so is usually not necessary to achieve a desired business purpose.

What does matter is that the document be cohesive and coherent. The terminology should be consistent from beginning to end. The end should flow logically from the middle and the middle from the beginning. There should be no obvious duplication or omission of content. To achieve this requires constraints on the composition and style of your content.

There are two approaches to narrative composition. One is the information-typing approach that you find in systems such as DITA or Information Mapping. In this approach, content is broken down into a set of broad types, such as procedure, process, principle, concept, structure, and fact (Information Mapping) or task, concept, and reference (DITA – though DITA allows you to define other types though specialization). The idea behind the information-typing approach is that if you keep different types of information (for instance, conceptual and reference) in separate chunks, the chunks will compose more reliably, since you won't duplicate information between chunks of different types. (For more on these mechanisms, see Chapter 33.)

The difficulty with this approach is that these abstract categories don't always make a lot of sense to writers when they are writing about concrete subjects, and different writers may interpret the chunk types or their boundaries differently, resulting in material that does not compose as well as you might hope.

Also, this approach, while it has been shown to improve the quality of writing in some cases, can also impose an artificial clunkiness and lack of flow on the content, leaving it choppy or disjointed. Breaking content into separate chunks of different types, after all, is inherently about weakening the narrative threads that bind them together. And if you intend to reuse chunks in several different narratives, you must keep their attachment to any one narrative thread weak.

If you want to impose a specific rhetorical style or structure, any composition must be subject to the same rhetorical constraints as if the piece had been written as a whole. This is often difficult to achieve, not least because it is difficult for writers to create material that conforms to a rhetorical structure when they can't see the whole structure.

The other approach to composing coherent narratives is to move content to the subject domain. A subject-domain structure does not have to be structured as a collection of abstract chunk types. The structure is specific to the subject matter and, therefore, is much more concrete and less susceptible to varying interpretation by writers. Also, you can use the subject domain to factor out many of the style issues that might otherwise compromise composability. (This is similar to factoring out formatting issues by moving from the media domain to the document domain.) A narrative can then be composed algorithmically by arranging well-identified pieces of information in a predetermined order, relieving the writer of that task.

Obviously, though, this technique can only work with a limited range of content. Not all material fits into obvious, strongly typed subject-domain structures. Content that is more conceptual or theoretical in nature does not have a strong subject-domain structure because it does not approach its subject matter in such a systematic or regular way. Then again, the ability to compose such content out of existing pieces is limited anyway. By its very nature such content requires a continuous flow of exposition that is very hard to assemble from pre-written chunks.

In short, then, composing content from small pieces can help address certain process issues, but it cannot be applied across a broad range of content without doing serious damage to rhetorical quality. As I have stressed before, fixating on any one algorithm is dangerous. You can always eke out more reuse from a system by pushing composition past the point at which rhetorical quality can be maintained. But the point is not to drive up your reuse statistics. The point is to ensure that every part of the complexity of the content process is handled by the person or process that has the skills, time, and resources to handle it properly. When it comes to composing effective narratives, the writer, not an algorithm, is often the one who can do the job best.

CHAPTER 29

Conformance

Structured writing uses constraints to govern rhetoric and to partition and redirect process complexity. In doing so, it transfers decisions from one partition to another. When you transfer a decision, you must transfer the information needed to make the decision, which in turn depends on everyone conforming to the applicable constraints. Every failure to conform means that some piece of information is not transferred and, therefore, some complexity goes unhandled. Since complexity cannot be destroyed, it falls through to a downstream process and ultimately to the reader. Therefore, a key part of developing a structured writing system is designing structures that support and express conformance to constraints.

Conformance is a complex problem. You can't expect to be successful if you just make up structures and systems and demand conformance to their constraints without any thought as to how conformance is to be supported and assessed. It is possible (and all too common) to invent a system that would operate flawlessly if everyone conformed, but which fails in the real world because it is impossible to conform to the constraints.

Constraints have always been part of writing. Style guides and grammatical reference works express constraints that writers are expected to follow. Editorial guidelines tell writers what kind of content a publisher is looking for, at what length, and in what format. If a publisher says that manuscripts must be delivered in DocBook or Microsoft Word, that is a constraint. When the government says that you must submit your online tax return in a particular file format, that is a constraint.

Some constraints are merely statements of requirements. Writers are given no assistance in following them nor is there any verification mechanism (other than perhaps an email from an irate editor). Other constraints are mechanical. Good tax preparation software guides you through your tax forms and checks to make sure that you complete them correctly. It also factors out many of the complexities of the tax code and asks you for information in a way you can understand, thus making it easier for you to conform.

This higher level of conformance checking and support helps make the process easier and the results more reliable. Unless the data passed from one partition to another is reliable, partitioning breaks down and both process and rhetoric suffer. Conformance is the linchpin of structured writing. Without it, none of the other algorithms can work reliably.

How many constraints you need to place on your content depends on your quality and process goals – how and where you want to partition and distribute complexity in your organization. The larger your content set becomes, the more critical rhetorical quality is to your business, the more frequent and dynamic your outputs are, the more your processes rely on algorithms, and the more constraints you need, the more pressing the issue of conformance becomes. For example, content reuse relies on writers conforming to constraints that ensure that reusable parts fit when reused and constraints on how they must assemble reusable parts. Content generation depends on reliable source data. If you want to do any kind of real-time publishing – meaning there is no time to do quality assurance on the output of the algorithm – reliable content is key, and conformance is how you ensure that content is reliable.

Structured writing projects can get into trouble when they introduce constraints to meet management or publishing automation goals without considering how to achieve conformance to those constraints. This can lead to writers being expected to conform to constraints using structures that provide no guidance or validation mechanisms. In some cases, this results in a highly arbitrary approach to conformance, in which writers are trained to implement the constraints, but the structures provide no guidance or validation. The system constraints, in other words, are not reflected in the content structures. If you are creating complex structures and also creating complex constraints that are not reflected or implemented in those structures, you have dumped a huge amount of complexity on your writers, and you are going to have a two-fold conformance problem: conformance will be expensive, and it will be inconsistent.

The best way to ensure conformance with a constraint is to factor out the constraint. For example, if inline citations must be formatted in a certain way, you can factor out this constraint by moving to the document domain and using something like DocBook's `citetitle` element to mark up the titles of works. Now, the publishing algorithm is responsible for conforming to the formatting constraint. You have factored out the formatting constraint.

When you move content creation from the media domain to the document domain you factor out all formatting constraints. When you move content from the document domain to the subject domain you factor out many document design or management constraints and enforce a number of constraints about what information will be captured.

But while this factors out one set of constraints, it creates a new set of constraints in the new domain. When you factor out the formatting constraint for the titles of works cited, you introduced a constraint that requires writers to markup the title of works using `citetitle`. Factoring one constraint into another is useful if it makes the constraint easier to conform to or easier to validate or if it captures additional data that enables other algorithms. A constraint may be easier to conform to if it is simpler, easier to remember, or does not require knowledge that is outside the writer's field. For instance, `citetitle` is a single tag, not a set of formatting instructions, and writers know when they are citing the title of a work.

A constraint may be easier to validate if it has fewer components or can be limited to a narrower scope. For instance, the set of things that are titles of works is smaller than the set of things that are formatted in italic, so validating the `citetitle` constraint requires looking at a smaller and more homogeneous set.

If the constraint you are introducing is not easier to conform to or easier to validate, you should think twice before you introduce it. There are certainly cases where moving content to a more formal document-domain model introduces more constraints than it eliminates without making those constraints easier to comply with or validate. On the other hand, sometimes those additional constraints are required to reduce complexity somewhere else in the process, or to manage previously unmanaged complexity. In other words, you are transferring complexity to writers from somewhere else in the content system.

Inevitably, writers have to make some concessions to the needs of algorithms, but those concessions should not distract them from doing good research and writing quality content. Any complexity that writers can't handle results in poor rhetoric and unreliable data, which in turn results in inefficient processes. If the structures you create for the sake of algorithms prove too complex to conform to or too difficult to validate, consider refactoring those constraints again, perhaps to the subject domain, so you get the precision and detail algorithms need and the ease of use and validation that writers need. In other words, don't address one source of complexity in isolation. Keep moving complexity until every piece of it is handled by a person or process with skills, bandwidth, and resources to handle it.

The recipe examples, such as Figure 29.1, show how a subject-domain structure can transfer complexity away from writers while providing for a high level of conformance checking.

```
recipe: Hard Boiled Egg
    introduction:
        A hard boiled egg is simple and nutritious.
    ingredients:: ingredient, quantity, unit
        eggs, 12, each
        water, 2, qt
    preparation:
        1. Place eggs in pan and cover with water.
        2. Bring water to a boil.
        3. Remove from heat and cover for 12 minutes.
        4. Place eggs in cold water to stop cooking.
        5. Peel and serve.
    prep-time: 15 minutes
    serves: 6
```

Figure 29.1 – Recipe example showing support for constraints

Pulling the prep-time and serving numbers into separate fields makes it easy to validate that the writer has conformed to the constraint to include this information while also making it easier for the writer to supply this information. Putting the ingredients into a record set rather than a list or table factors out any presentation constraint entirely, making it impossible not to comply. And again, it is easier for writers to create content in this form than it is to create, for instance, a presentation-level table.

Completeness

Completeness is an obvious aspect of content quality. Unfortunately, lack of completeness is often hard for writers and reviewers to spot. The curse of knowledge means that omission of information is hard to see unless there is an obvious hole in a predefined and explicit document structure. Defining structures that encapsulate information requirements can significantly improve completeness. In Chapter 4 we saw how calling out the preparation time and number of servings for a recipe helps ensure that writers always remember to include that information, whether or not you decide to present it in fields or as part of a paragraph.

But this is not the only way that subject-domain structured writing helps ensure completeness. Every subject-domain annotation highlights a subject that is important to your business. You can use an algorithm to scan those annotation and build a list of subjects that are important to your business. You can use this list to make sure that all the subjects you need to cover are actually covered.

For example, structured writing allows you to annotate certain phrases such as function names, feature names, or stock symbols.

```
When installing widgets, use a {left-handed widget wrench}(tool)
to tighten them to the recommended torque for your device.
```

Figure 29.2 – Example of inline annotation

Figure 29.2 annotates the phrase "left-handed widget wrench" and records that these words describe a tool. If writers annotate all mentions of tools, you can compile a list of all the tools mentioned in your topics and make sure that you have suitable documentation for each of them. I talk more about this in Chapter 39.

Consistency

Like completeness, consistency can make a big difference to readers, but lack of consistency is hard to spot if the structure of the content is not explicit in all the ways you want it to be consistent.

Being consistent simply means abiding by constraints. You can either enforce the constraint by having writers use a required structure or, preferably, factor out the constraint so that it is handled by an algorithm. We have looked at how you can factor out constraints in both the document domain and the subject domain.

If you annotate the important things in your content set, such as the tool in Figure 29.2, you can use the annotations to check for naming consistency, as described in Chapter 26. For example, if a writer accidentally uses the term "spanner" rather than "wrench," you can catch the error in an audit against a list of approved tool names. This can reveal both incorrect names (consistency) and tools that may be missing from the official list (completeness).

The same applies to values in fields such as the wine match field in the recipe example. You can use the wine match field to compile a list of wines mentioned or check each mention against an approved list. More on this in Chapter 26.

Accuracy

Accuracy problems are often hard to spot. Typos, using old names for things, or giving deprecated examples are all hard for writers and reviewers to see. But there are structured writing techniques than can catch many of these kinds of problems.

For example, if you are documenting an API, you can annotate each mention of a function.

```
Always check the return value of {rotateWidget()}(function)
to ensure the correct orientation was achieved.
```

API function names can be tricky to remember, and typos can be difficult to spot. But if you annotate function names, you can validate all mentions of functions against the API reference or the code base. This technique not only catches misspellings, it can also catch the use of deprecated functions in examples.

Semantic constraints

We can divide constraints into two types: structural constraints and semantic constraints. Structural constraints deal with the relationships between text structures. Semantic constraints deal with the meaning of the content. For instance, consider the structure in Figure 29.3.[1]

```
<person>
    <name>John Smith</name>
    <age>middle</age>
    <date-of-birth>Christmas Day</date-of-birth>
</person>
```

Figure 29.3 – Person structure with badly coded values

Some people certainly describe themselves as middle aged, and Christmas Day is certainly a date of birth, if an incomplete one. The writer has complied with the document structure. But the creator of this markup language was probably looking for more precise information, in a format like Figure 29.4. which an algorithm can more easily read.

```
<person>
    <name>John Smith</name>
    <age>47</age>
    <date-of-birth>1970-12-25</date-of-birth>
</person>
```

Figure 29.4 – Person structure with well-coded values

---

[1] There is redundancy in this example, since age can be calculated from date of birth. I use this example simply to save space. In most cases, you would check information against a different source. But it can be valuable to collect the same data twice, in different forms, as a form of data validation. For instance, a person might remember their age correctly but make a mistake on their date of birth. Asking them to enter both lets you double check.

Some schema languages (a concept I explain in Chapter 35), such as XML Schema (XSD), let you specify the data type[2] of an element. You can specify that the data type of a value in the age field must be a whole number between 0 and 150 and that the date-of-birth field must be a recognizable date format. Figure 29.5 shows an XSD representation of these constraints.

```
<xs:schema
    xmlns:xs="http://www.w3.org/2001/XMLSchema"
    elementFormDefault="qualified">

    <xs:element name="person">
        <xs:complexType>
            <xs:sequence>
                <xs:element name="name" type="xs:string"/>
                <xs:element name="age" type="age-range"/>
                <xs:element name="date-of-birth" type="xs:date"/>
            </xs:sequence>
        </xs:complexType>
    </xs:element>

    <xs:simpleType name="age-range">
        <xs:restriction base="xs:int">
            <xs:minInclusive value="0"/>
            <xs:maxInclusive value="150"/>
        </xs:restriction>
    </xs:simpleType>
</xs:schema>
```

Figure 29.5 – XML Schema markup for data type constraints

The schema in Figure 29.5 uses the built-in types xs:string and xs:date for the name and date-of-birth elements and defines a new type called age-range for the age element. Now, if you try to validate Figure 29.3, the process will fail with data-type errors on both fields.

Applying these kinds of semantic constraints won't work if most of your text is in free-form paragraphs. It is hard to define useful patterns for long passages of text. If you want to exercise

---

[2] The data types referred to in the example above are not data types as they are commonly understood in programming terms (which refers to how they are stored in memory). In XML, as in all major markup languages, the data is all strings. A data type in a schema is actually just a pattern. There is a language for describing patterns in text that is called regular expressions. Regular expressions are a bit cryptic and take some getting used to, but they are incredibly powerful at describing patterns in text. XML schema lets you define types for elements using regular expressions, so there is a huge amount you can do to constrain the content of elements in your documents.

fine-grained control over your content, you must first break information down into individual fields and then apply type constraints to those fields.

In some cases, you can create text structures that exist solely to isolate semantic constraints so that they are testable and enforceable.

This can be particularly effective when you are creating content in the subject domain since you don't have to specify information in sentences, even if you intend to publish it that way. You can break the content out into separate structures and define the data type of those structures to ensure you get complete and accurate information and to ensure that you can operate on that information using algorithms.

```
recipe: Hard Boiled Egg
    introduction:
        A hard boiled egg is simple and nutritious.
    ingredients:: ingredient, quantity
        eggs, 12
        water, 2qt
    preparation:
        1. Place eggs in pan and cover with water.
        2. Bring water to a boil.
        3. Remove from heat and cover for 12 minutes.
        4. Place eggs in cold water to stop cooking.
        5. Peel and serve.
    prep-time: 15 minutes
    serves: 6
    wine-match: champagne and orange juice
    beverage-match: orange juice
    nutrition:
        serving: 1 large (50 g)
        calories: 78
        total-fat: 5 g
        saturated-fat: 0.7 g
        polyunsaturated-fat: 0.7 g
        monounsaturated-fat: 2 g
        cholesterol: 186.5 mg
        sodium: 62 mg
        potassium: 63 mg
        total-carbohydrate: 0.6 g
        dietary-fiber: 0 g
        sugar: 0.6 g
        protein: 6 g
```

Figure 29.6 – Hard-boiled egg recipe

The recipe text in Figure 29.6 is a good example of how content that could be expressed entirely in free-form paragraphs can be broken down in a fine-grained way that allows you to impose a variety of structural and semantic constraints.

This entire recipe could be presented free form. But when structured like this you can enforce detailed constraints such as ensuring that there is always a wine match or that calories are always given as a whole number. The publishing algorithm can stitch all this content into paragraphs if that is what you want. But this format gives writers a huge amount of guidance about the information you want, and you can manipulate and publish the content in many different ways. Readers benefit because every recipe conforms to what you know readers need and want in a recipe.

Entry validation constraints

If algorithms can read the data in your structures, they can check one piece of information against another. For instance, if you have `date-of-birth` and `age` (as in Figure 29.4), you can calculate current age from `date-of-birth` and compare it with `age`. If the values don't match, the writer has made an error, and you can report it. Figure 29.7 shows pseudocode for such a test.

```
if not $age = years-between(now, $date-of-birth)
    error "Age does not match the given date-of-birth."
```

Figure 29.7 – Pseudocode to compare age **field with calculated age**

Referential integrity constraints

In the management domain, there are a set of referential integrity constraints.[3] Referential integrity simply means that if you refer to something, that thing should exist. In the management domain, we often give IDs to structures and use those IDs to refer to those structures for purposes such as content reuse.

If you are going to reuse a piece of content by referring to its ID, there is an obvious constraint that a piece of content with that ID must exist. This constraint is important enough that XML directly supports it (as does SAM). The XML specification says that if you have an element with an attribute of type IDREF, then there must be an element with an attribute of type ID with the

---

[3] The term referential integrity comes from the relational database world, where it has the same meaning.

same value in the same document. This can be useful for checking that things such as a footnote reference corresponds to a footnote somewhere in the document.

Many management-domain algorithms go beyond this constraint and require referential integrity not only within a single document but between documents. Conformance to this type of referential integrity constraint can sometimes only be judged by the publishing algorithm when you publish a particular combination of documents. Thus, it is possible for a document to have referential integrity when published in one collection and to lack it when published in another.

Since it is best to validate a constraint as early as possible, a content management system that is aware of the referential integrity constraints of a system (such as a DITA CMS, for example) may validate the referential integrity of content in all its potential combinations prior to publication.

Nevertheless, referential integrity constraints of this complexity still present a management and authoring headache, even with content management system support. So it is worth looking for a way to factor them out. One example of how this can be done is found in the various approaches to the linking algorithm (see Chapter 18).

Conformance to external sources

Referential integrity constraints can span multiple documents. So can semantic constraints. For example, you may want values in your document to match values in databases or in other documents. A technical writer documenting an API may produce an API reference, much of which can be extracted from the program source code (see Chapter 15), and also a programmer's guide, which is written from scratch. The programmer's guide will obviously mention the functions in the API many times. The writer may misspell one of the names, the API may be changed after parts of the document are written, or the writer may mention a function that no longer exists.

It is clearly a semantic constraint on the programmer's guide that all the API calls it mentions must be present in the API. Since the API reference is generated from the source code, you can express this constrain as: functions mentioned in the programmers guide must be listed in the API reference.

This is an important constraint. When we implemented algorithmic support for this constraint on one project I worked on, it revealed a number of errors:

- The programmer's guide contained misspelled function names.
- The programmer's guide included material related to a private API that was never released to the public.
- The API reference failed to include an important section of the API due to incorrect markup in the source code.
- The programmer's guide documented how to use a deprecated API and neglected to describe how to use the new API.

These errors occurred despite several thorough reviews by multiple people over multiple software releases. Human beings have a hard time spotting these kinds of errors in review, but they have a significant impact on users.

As part of the conformance audit for the programmer's guide, we added an automated check that looked up each reference to an API call, including those in code blocks, in the API reference and reported an error if they did not match. None of the errors listed above would have been detected without this check.

To implement this check, the algorithm had to identify references to API calls in the programmer's guide and find APIs by name in the API reference. For this to be possible, both documents had to be written in a structured format that made the API names accessible to the algorithm. Here is a simplified example. First, a code sample from a programmer's guide (Figure 29.8).

```
code-sample: Hello World

    The Hello World sample uses the {print}(function) to
    output the text "Hello World"

``` (python)
    print("Hello World.")
```

Figure 29.8 – Code example with a function annotated

Next, a function reference listing from the API reference (Figure 29.9).
```

```
function:
    name: print
    return-value: none
    parameters:
        parameter: string
            required: yes
            description:
                The string to print.
        parameter: end
            required: no
            default: '\n'
            description:
                The characters to output after the
                {string}(parameter).
```

Figure 29.9 – Sample API function reference

Because the API reference labels `print` as a function name and the code-sample annotates `print` as the name of a function, you can look up `print` in the API reference to validate the annotated text in the programmer's guide. By adding these structures and annotations to the content, you isolate the semantics of the function call names so that you can apply semantic conformance checks to them.

Conformance and change

Requiring conformance to outside sources means that a document's conformance is neither static nor absolute. A document that was conforming may stop being conforming because of outside events. But this reflects reality. One of the most complex aspects of content management is detecting when a document ceases to be conforming because of a change in the reality that it describes. Using structured writing techniques to validate the conformance of a document against an external source can go a long way to addressing this class of problem. For more on change management, see Chapter 40.

Design for conformance

Mechanical constraints can do a lot to ensure and validate conformance. But conformance is fundamentally a human activity, and you need humans to conform to constraints that cannot be easily expressed or validated in purely mechanical terms.

Schemas and downstream algorithms can verify that writers have created the required mechanical structures and, in some cases, whether they created them with content in the required form. But in most cases there is little or nothing that algorithms can do to verify that the content of a structure is actually what the structure says it is.

If writers don't understand what a structure is for, or if they are not on-board with creating content according to the structure, they are not likely to create content that obeys the constraints that the structure expresses. And no mechanical conformance algorithm can ensure that they did not just write their content the way they wanted to and then wrap the required structured tags around it to make it pass the validation tests.

If there are optional elements to a structure, there is nothing that mechanical validation can do to determine whether the optional parts of the structure were included when they should have been or omitted when they should have been. Only the writers can determine this, and to do so, writers must understand the purpose of those structures and accept that those structures are appropriate for what they are writing.

If you create structures that are difficult for writers to understand (for example, if they are complex management-domain structures that require complex abstract IDs), then they have to think about issues outside of their area of expertise. And when you dump complexity on writers that they are ill-equipped to handle, they have to stop thinking about writing and their subject matter and start thinking about the system. Since most writers don't really understand how complex structured writing systems work, they tend to do whatever it takes to make the system stop throwing errors and accept their work. This often produces dumb compliance to mechanical system constraints rather than intelligent conformance to content constraints. And this dumb compliance dumps the complexity that the writer could not handle on the reader or another part of the organization.

The key to achieving conformance is to create structures that are easy to conform to. For most content, conformance is not about trying to catch evil doers. The writers are on your side, and they try to produce good content. Writers who understand structured content may impose constraints as an aid to their own work, just as a carpenter, for instance, might design a jig to guide their saw. Constraints are a tool for writers, not a defense against misconduct. Constraints may force lazy writers to pull up their socks and do some more research or force inattentive writers to recast their first draft into a more consistent format. But constraints should never prevent a diligent writer from doing good work.

Therefore, the real core of compliance in structured writing is not enforcement; it's creating structures that clearly and specifically:

- Match the subject matter and audience and align with, or at least do not hinder, the rhetoric of the piece
- Supply the information readers need to accomplish their goals
- Communicate to writers what is expected of them in terms they understand
- Either remind writers of what is required or (preferably) factor it out

Of course, writers may disagree about what goals to address, what information readers need to achieve those goals, or how to express that information to a particular audience. When many writers contribute to a common information set, these rhetorical differences must be addressed and resolved professionally, and all writers must be on board with the plan. At that point, well-thought-out content structures can capture the plan and ensure consistency.

For example, suppose you have a set of cooks contributing to a common recipe information set. Once everyone agrees that every recipe should include preparation time and number of servings, you can create a recipe structure that calls out those fields, thus helping cooks remember and comply with this requirement. This will significantly improve compliance compared to merely stating the requirements in a style guide. And, of course, it makes that information available to other algorithms as well.

Auditing and enforcement still have a role to play, not because writers are hostile to the system, but because they are human. But auditing and enforcement are secondary to the main aim of conformance-friendly design. And in that spirit, auditing and conformance should be seen as part of a feedback loop that seeks to improve the design. If you find the same mistakes over and over again, that is not a training problem or a human resources problem, it is a design problem.

Finally, be aware of how your authoring interface affects conformance. How do you know if you are meeting constraints – in any activity? Feedback. With any activity, you need to know when you are done and when the work is correct. In the media domain, there is one form of feedback: how the document looks. With a true WYSIWYG display, if it looks right on the screen, it will render correctly on paper or whatever media you are targeting. The display tells writers when they are done.

That works in the media domain. However, if you are creating content in the document domain but giving your writers a media-domain display – for example, with a structured editor that mimics Microsoft Word – then the feedback they receive will be media-domain feedback, and the conformance you get will be media-domain conformance. There is a risk that content created this way will not work as intended when output to other media, reused, or processed with content from any of the other structured writing domains. Don't expect to get real conformance to any structure that is hidden from the writer. You can't conform to a structure you can't see.

Languages

We have seen examples of markup languages in earlier parts of this book, and I have discussed some of their features. In this section, I discuss markup languages in more detail, looking at the main features of some of the common public languages and also at the way that new languages are constructed to serve new requirements.

CHAPTER 30

Markup

If structures in structured writing consist of nested blocks, the way those blocks are expressed in text is (most of the time) through markup. Markup is, essentially, the insertion of metadata into a text to constrain the creation and interpretation of that text. We have been looking at structured writing examples expressed in markup all through this book. Now let's look at markup itself.

Markup has long been part of partitioning and distributing content creation. For centuries, scribes worked directly in the media domain, using pen and ink to inscribe words and pictures on papyrus or velum. With the printing press, however, came a fundamental partitioning of the publishing process. Writers no longer worked directly in the media domain. While writers still placed ink directly on paper, at first by pen and then by typewriter, they no longer prepared the final visual form of the content. That task was partitioned off and directed to the typesetter.

To tell the typesetter how to create the final visual form, document designers added instructions (metadata) to the writer's manuscript. The designers did this using typesetter's marks, and the process was called "marking up" the document (see Figure 30.1). We still use "marking up" to describe how structured writing is done today and the term "markup language" to describe the languages used for most structured writing.

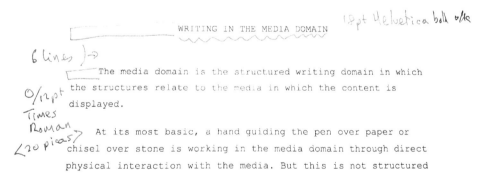

Figure 30.1 – Printer's markup

The writer preparing a manuscript for typesetting worked in the document domain, using basic document structures, such as paragraphs, lists, and titles, without any indication of how they should look in print. The designer then wrote a set of instructions for applying formatting to those structures – a formatting algorithm. Then the typesetter executed that algorithm by setting the type, which the printer used to print final output.

This is almost exactly what you do today when you create an HTML page and specify a CSS style sheet to supply the formatting instructions. A browser then executes those instructions to render the content on screen or paper.

Actually, a better analogy to old-style typesetter's marks is an HTML page with inline styles:

```
<p style="font-family: serif; font-weight: bold;
 font-size: 12pt">
```

You can see that this markup is very similar to the typesetters marks in Figure 30.1.

All writing programs have to store the content you write in files. There are two possible file types they can use: binary and text.

A binary file can only be read or written by a computer program, usually the program that created it. Open up a binary file in a text editor, and you won't be able to make heads or tails of it. Even if parts of it look like plain text, editing those sections and saving the file is likely to result in a corrupt file that the original application can no longer open.

A text file, by contrast, is one that you can open in a text editor and read and write without breaking it. But to express structure in a text file, you need to interpolate information about structure into the text. The way you interpolate structure is with markup – special sequences of text characters that are recognized as defining structure rather than expressing text.

Figure 30.2 contains a snippet of HTML with markup characters shown in bold and plain text shown in regular type.

```
<h1>Moby Dick</h1>
<p>Herman Melville's <i>Moby Dick</i> is a long
book about a big whale.</p>
```

Figure 30.2 – HTML snippet with markup highlighted

A markup parser recognizes the markup characters and builds a structure that has the text and nested structures as content. A processing application then accesses that data and applies rules to the structures defined by the markup, as described in Chapter 11.

Markup vs. regular text

Some markup languages define markup in universal terms. That is, the characters that define markup are recognized as markup no matter where they occur in the file. One example is an HTML tag. Tags are set off by opening and closing angle brackets:

```
<h1>
```

HTML uses open angle brackets (<) to indicate the start of markup and closing angle brackets (>) to indicate the end of markup and a return to regular text. A slash (/) after the opening angle bracket indicates an end tag, which marks the end of a structure:[1]

```
<h1>Moby Dick</h1>
```

Every element in HTML uses the same syntax. Any element with a name that is not recognized as part of HTML is still treated as an element, which the HTML parser will handle as a markup error (usually by ignoring it).

What if you want to enter a markup-start character into the text of your document without having it interpreted as markup? You can't just type it in because the parser will think it is markup. To fix this, markup languages either define an escape character, which tells the parser to treat the following character as text, or they include markup for inserting individual characters in a way that won't be confused with markup characters.

HTML takes the second approach. To include a < character in HTML, you use what is called a *character entity*. A character entity is a code for a character. It begins with another markup-start character, an ampersand (&), followed by a character code, and ends with a semicolon. The character entity for < in HTML and XML is < (lt is short for "less than").

```
<p>In HTML, tags start with the &lt; character.</p>
```

---

[1] Actually, the recognition of markup in languages like XML and HTML is a little more complicated than that. Sequences of characters are recognized rather than individual characters, and therefore, no individual character is universally recognized as markup. But the main point is simply that certain specific sequences in the text trigger a processing program (generally called a *parser*) to recognize when markup starts and ends.

This displays as:

> In HTML, tags start with the < character.

Since & is also a markup start character, you need to replace it with a character entity if you want to include it literally. To include a literal & you use the character entity &.

```
<p>In HTML, character entities start with the &amp; character.</p>
```

This will display as:

> In HTML, character entities start with the & character.

Therefore, to include the literal string & you write &amp;.

```
<p>The character entity for an ampersand is &amp;amp;.</p>
```

This will display as:

> The character entity for an ampersand is &.

Other markup languages do not make a universal distinction between text and markup. For example, in Markdown, you create a numbered list by putting numbers in front of list items:

```
1. First
2. Second
3. Third
```

Here, the numbers are markup. That is, the Markdown processor recognizes them as indicating a list and translates them into an HTML structure like this:

```
<ol>
    <li>First</li>
    <li>Second</li>
    <li>Third</li>
</ol>
```

But numbers followed by a period are only markup in a certain context – the beginning of a line. Elsewhere, they are plain text. You don't need to escape numbers followed by periods when they occur elsewhere in the text. One consequence of this is that you cannot have a markup error in processing Markdown. Anything that is not recognized as markup is simply treated as plain text. (You can certainly make mistakes in marking up up a document in Markdown, but the Markdown processor will not catch them because it will simply treat your incorrect markup as plain text. This is why Markdown editors often use split screen editing, so you can catch your errors visually.)

Thus, the following Markdown file:

```
1. First comes 1.
2. Second comes 2.
3. Third comes 3.
```

translates to HTML as:

```
<ol>
    <li>First comes 1.</li>
    <li>Second comes 2.</li>
    <li>Third comes 3.</li>
</ol>
```

In Markdown, then, markup is not defined in universal terms. It is a pattern recognized in context. Therefore, rather than thinking of markup as being distinct from text, it is better to think of markup as being a pattern within a piece of text that delineates its structure. In some cases the patterns are universal and have the same meaning everywhere, and sometimes they are contextual and have one meaning in some locations and another meaning elsewhere. Sometimes the markup characters are distinct from the text characters, and sometimes a pattern in the text serves as markup.

Markup languages

A set of markup conventions taken together constitutes a markup language. Markdown, DocBook, and reStructuredText are all markup languages. However, each of these languages recognizes markup in a different way. An ampersand (&) is a markup start character in HTML and XML, but it is a plain text character in reStructuredText.

We can usefully divide markup languages into three types: concrete, abstract, and hybrid.

Concrete markup languages

A concrete markup language has a fixed set of markup patterns that describe a fixed set of content structures. For example, Markdown is a concrete markup language that uses a markup that is designed to mimic the way people write plain text email.

Here is the passage about *Moby Dick* written in Markdown:

```
Moby Dick
=========

Herman Melville's _Moby Dick_ is a long book about a big whale.
```

In Markdown, a line of text underlined with equal signs (=) is a level-one heading. A paragraph is a block of text set off by blank lines. Emphasized text is surrounded with underscores or asterisks.

In Markdown, these patterns correspond directly to specific document structures. You cannot invent new structures without inventing a new version of Markdown.

Abstract markup languages

An abstract markup language does not describe specific concrete document structures directly. It describes abstract markup structures, which can be named to represent structures in any domain.

XML is an example of an abstract markup language.[2] The markup in an XML file does not directly indicate things such as headings or paragraphs. Instead, it indicates a set of abstract structures called elements, attributes, entities, processing instructions, marked sections, and comments.

None of these abstract structures describes document structures in any of the structured writing domains. Instead, markup languages based on XML (or its cousin, SGML) indicate subject-, document-, management-, or media-domain structures as named instances of elements and attributes.

---

[2] The formal term for a language like XML is "meta language," a language for describing other languages. In calling XML an *abstract* language, I am focusing on a different property, its use of structures that are not parts of a document but generic containers. A meta language needs such abstract containers, but I find that the term meta language is not helpful to most readers. Therefore, I have chosen to focus on this property of using abstract structures as opposed to the concrete structures of a language like Markdown.

Here is the Moby Dick passage again, this time in XML (more specifically, in DocBook):

```
<section>
    <title>Moby Dick</title>
    <para>Herman Melville's <citetitle>Moby Dick</citetitle>
    is a long book about a big whale.</para>
</section>
```

The structure described by the XML syntax here is an element (`section`), which contains two other elements (`title` and `para`), one of which (`title`) contains text and one of which (`para`) contains a mix of text and another element (`citetitle`). There is no separate syntax for titles or paragraphs as there is in Markdown. Everything is an element, and every element has a name. To define specific document-domain structures, you use named elements. This allows you to create any set of named elements you like to represent any structure you need.

Unlike a Markdown parser, an XML parser does not see paragraphs or titles. It sees elements. It passes the elements it finds, along with their names, to a processing application that is responsible for knowing what the `section`, `title`, and `para` elements mean in a particular markup language, such as DocBook. The parser is common to all XML-based languages, but the processing application is specific to DocBook.

Thus, while processing a concrete language like Markdown is generally a one-step operation, processing an abstract language like XML is a two step-operation. The first step parses the file to discover the structures defined by elements, and the second step processes those structures according to a set of rules applicable to a particular markup language.

Instances of abstract markup languages

This means that DocBook is a instance of the abstract language XML.[3] XML defines abstract structures, and DocBook defines specific document structures by giving names to XML elements. Many common markup languages are instances of XML.[4] XML is virtually the only abstract language used for content these days, so it is the only abstract language I am going to talk about.

Since specific markup languages like DocBook are instances of XML, I need to revise my earlier statement: We can usefully divide markup languages into **four** types: concrete, abstract, instances

---

[3] Or to put it another way, you write in DocBook semantics using XML syntax. Alternatively, since DocBook originated in the days of XML's predecessor abstract language, SGML, you can write DocBook semantics in SGML syntax.

[4] Sometimes also referred to as "applications" of XML, though this usage was far more common in the days of SGML.

of abstract, and hybrid. In fact (spoiler alert), let's revise it again: We can usefully divide markup languages into **five** types: concrete, abstract, instances of abstract, hybrid, and instances of hybrid.

You can't write directly in an abstract language like XML, only in an instance. You can write directly in some hybrid languages, as I'll describe later in this chapter, but not in all of them.

Or to put it another way, as a designer of markup languages you can choose to:

- Design a concrete language from scratch (or modify and existing one)
- Use an abstract language (probably XML) to design a specific instance language
- Use a hybrid language to design a specific instance language

As a writer, you can use:

- A concrete language (like Markdown)
- The concrete parts of a hybrid markup language (like reStructuredText) without extensions
- A specific instance language (like DocBook) based on an abstract language (probably XML)
- A specific instance language (like the recipe markup language used in this book) based on a hybrid language (SAM in this case)

Concrete languages in abstract clothing

The key defining characteristic of an abstract language is the use of abstract named structures such as XML elements. All XML elements share a common markup start sequence followed by the element name. This creates a named block of content. But concrete languages can use named blocks too. For example, JavaDoc, a concrete language for describing Java APIs, uses named blocks using an at sign (@) as a markup start character.

In Figure 30.3, `@param` and `@return` are named blocks. But in JavaDoc, there is a fixed set of named blocks that are defined as part of the language. You can't create a new language by defining your own block names. By contrast, XML itself defines absolutely no element names. Only instances of XML, like DocBook, define element names.

```
/**
 * Validates a chess move.
 *
 * Use {@link #doMove(int theFromFile, int theFromRank,
 * int theToFile, int theToRank)} to move a piece.
 *
 * @param theFromFile file from which a piece is being moved
 * @param theFromRank rank from which a piece is being moved
 * @param theToFile   file to which a piece is being moved
 * @param theToRank   rank to which a piece is being moved
 * @return            true if the move is valid, otherwise false
 */
boolean isValidMove(int theFromFile, int theFromRank,
int theToFile, int theToRank) {
    // ...body
}
```

Figure 30.3 – JavaDoc code fragment

A particularly notable example of a concrete language in abstract clothing is HTML. HTML looks a lot like an instance of XML, but it is not. An XML parser cannot parse most HTML. HTML is nominally an instance of SGML but never did quite conform to it. HTML generally requires a specific HTML parser, such as is found in all browsers. XHTML is a version of HTML that is an instance of XML. HTML5 actually supports two different syntaxes, one of which is an instance of XML and one of which is not, meaning that it has both a concrete syntax and a syntax which is a instance of an abstract language.[5]

The ability to extend

The downside of concrete languages is that their concrete syntax defines a fixed set of structures. If you want other structures, there is no way to create them short of inventing your own concrete language, or a variant of an existing one, and coding the parser and all the other tools required to interpret that language. Designing new concrete languages is non-trivial because you must ensure that any combination of characters that the writer types can be interpreted unambiguously.

---

[5] This is consequence of the long history of HTML and the wild-west approach to both syntax and semantics during the browser wars, leading to significant backward-compatibility problems for today's browsers. This illustrates one of the perils of writing in the media and document domains. Languages in these domains evolve over time to meet new needs and resolve old problems, but this evolution creates backward compatibility problems for documents written in old versions of the format. Formats can evolve in the subject domain as well; however, because subject-domain markup records facts about the subject matter, and not formatting or presentation decisions, it is inherently more stable, and you can pull old subject-domain content into new document- and media-domain formats simply by updating the publishing algorithms.

Some versions of Markdown, including the original, contain ambiguities about how to interpret certain sequences of characters, which detracts from their reliability and functional lucidity.

If you want to define your own structures to express the constraints that matter to your business, you need an easier method. Abstract languages such as XML make this much easier. You create a schema that describes the structures you want and write algorithms to process those structures. (I talk about schemas in Chapter 35.)

The ability to constrain

Extensibility allows you to add structures to a language but does not place restrictions on where they can occur. Constraints let you limit where structures can be used.

Extensibility allows you to have structures called `ingredient` and `wine-match`. Constraints allow you to require that `ingredient` only occurs inside an `ingredients` structure and that the content of the `ingredients` structure must be a sequence of `ingredient` structures and nothing else. Constraints let you say that writers can put `wine-match` inside `recipe` after the `servings` field and before the `prep-time` field but not in the `introduction` or in a `step` in the `preparation`. Constraints allow you to require that every recipe have the full list of nutritional information.

Constraints allow you to partition the content system by creating reliable interfaces between different people and processes. All markup languages have constraints. With a concrete language, you get the constraints that are built into the language. Abstract languages allow you to define your own structures and your own constraints. However, as we shall see in Chapter 34, not all languages that are extensible are also constrainable.

Showing and hiding structure

To reliably create structured content, writers need to see the structures they are creating. In the media domain a WYSIWYG interface shows you the media-domain structures you are creating by visually rendering them on screen. But what about the other domains? The document domain creates abstract document structures that are deliberately separated from their formatting. The subject domain creates subject-based structures that don't have a one-to-one relationship with any organization or formatting of a document. The management domain creates structures that have nothing to do with the representation of content at all. How does the writer get to see these structures when writing in these domains?

This is a big problem with XML, the only abstract language in widespread use today. XML tends to hide structure. As an abstract language, an XML document is a hierarchy of elements and attributes – not the subject-, document-, management-, or media-domain structures the writer is supposed to be creating. Those structures are present in the markup because their names are there, but they are not visually distinguished the way the basic document structures are in a concrete language like Markdown. And XML syntax is verbose, meaning that there is a lot of clutter in the raw text of an XML document, which makes it hard to discern both the structure and the content. XML's verbose syntax and strict hierarchy were designed to make it easy to parse and to help guard against transmission errors. It was not designed to be a format to write in. Thus, it does a poor job of partitioning parsing issues from authoring issues.

To remove that clutter, many writers use XML editors that provide a graphical view of the content similar to that of a word processor. But while graphical XML editors remove visual clutter, they also hide the structure. Even if the writer is supposed to be working in the document domain or the subject domain, the editor is displaying content in the media domain. This greatly reduces the functional lucidity of the document-domain or subject-domain language and encourages backsliding into the media domain.

And then there are the problems that arise when you try to edit the graphical view of an XML document. Underneath is a hierarchical XML structure, but all you can see is the flat media-domain view of the graphical editor. Editing or cutting and pasting structures you can't see can be frustrating and time consuming. You can learn to do it, but even when you learn, the process is still more complicated than it should be. Most editors let you turn on a hybrid view where you see a graphical representation of the XML tags in the WYSIWYG view. Although you can manipulate the tags in this view, this is a complex interface that requires a lot of attention and study to use effectively.

Concrete markup languages like Markdown, on the other hand, show you the structure you are creating and are simple to edit.

Hybrid languages

As we have seen, there are significant advantages and significant disadvantages to both concrete and abstract languages. Hybrid languages try to find a middle way.

By hybrid, I mean a language that combines both abstract and concrete markup in one language. A hybrid language has a base set of concrete syntax describing basic text structures, but it also has abstract structures, such as XML's elements and attributes, that can be the basis of extensibility and constraint.

An example of a hybrid markup language is reStructuredText. Like Markdown, it has a basic concrete syntax for things like lists and paragraphs. But it also supports what it calls *directives*, which are essentially named block structures. For example, a code block in reStructuredText looks like this:

```
.. code-block:: html
   :linenos:

   for x in range(10):
       print(x+1, "Hello, World")
```

reStructuredText provides an extension mechanism that allows you to add new directives. But while reStructuredText directives are similar to XML elements, reStructuredText predefines a core set of directives for common document structures. The `code-block` directive above is not an extension of reStructuredText, it is part of the core language.

Because it defines a large set of document-domain directives, reStructuredText is inherently a document-domain language. You could, of course, add subject-domain directives to it. Most document-domain languages in use today include some subject-domain structures, reflecting the purpose they were originally designed to serve. Nonetheless, reStructuredText is inherently a document-domain language.

Another important note about reStructuredText is that it has no constraint mechanism. You can add new directives, but you can't constrain their use or the use of the predefined directives.

I have developed a hybrid markup language which is designed to be both extensible and constrainable. I call it SAM (which stands either for Semantic Authoring Markdown or Semantic Authoring Markup, as you please). I use SAM for most of the examples in this book.

Figure 30.4 shows the *Moby Dick* passage written in SAM:

```
section: Moby Dick

    Herman Melville's {Moby Dick}(novel) is a long
    book about a big whale.
```

Figure 30.4 – Example of SAM markup

In SAM, as in Markdown and most other concrete markup languages, a paragraph is just a block of text set off by whitespace. Thus, there is no explicit structure named p or para.

At the beginning of a line, a single word without spaces and followed by a colon creates an abstract structure called a *block*. The word before the colon is the name of the block. In the example above, section: creates a block structure named section in the same way that, in XML, an element named <section> creates a block structure named section. Blocks can contain blocks or text structures such as paragraphs and lists.

The hierarchy of a SAM document is indicated by indentation. Thus, the paragraph in Figure 30.4 is contained in the section block. Using indentation to indicate containers helps make the structure of the document visually clear and removes the need for end tags, which reduces verbosity and makes the text easier to read.

Within a paragraph, curly braces markup a phrase, to which you can attach an annotation in parentheses. Here the phrase "Moby Dick" is annotated to indicate that it is a novel. SAM also supports decorations, like the underscores in the Markdown example, so in the media domain "Moby Dick" could have been written _Moby Dick_.

SAM is not intended to be nearly as general in scope as a purely abstract markup language such as XML. It is meant for semantic authoring (which is to say, structured writing). As such it incorporates a number of shortcuts to make writing typical structured documents easier.

In a typical document, a block of text (larger than a paragraph) typically has a title. So in SAM, a string after a block tag is considered to be a title. That means that the markup in Figure 30.4 is equivalent to the markup in Figure 30.5.

```
section:
    title: Moby Dick

    Herman Melville's {Moby Dick}(novel) is a long
    book about a big whale.
```

Figure 30.5 – SAM markup with the title explicitly marked up

However, unlike reStructuredText, SAM does not have an extensive set of predefined named blocks. It only predefines the basic text structures that it provides concrete syntax for. You can write an entire book in reStructuredText without defining any new directives, The most you can write in SAM without defining any blocks is a single paragraph or list.

SAM is designed to have a constraint mechanism, which allows you to write a schema that defines what blocks and annotations are allowed in a SAM document. This includes constraining the use of the concrete syntax as well. SAM represents a different type of hybrid from reStructuredText, which is extensible but not constrainable. This reflects the fact that reStructuredText is intended as an extensible document-domain language, whereas SAM is intended primarily as a neutral starting point for designing subject-domain languages (though you can design document-domain languages in it as well).

Also unlike reStructuredText, SAM is not intended to have its own publishing tool chain. The SAM parser outputs an XML document which can be transformed into an appropriate document-domain language for publishing. This book was written in SAM, using a simple document-domain language I created for the purpose. That language, which includes some subject-domain annotations, was transformed into a semantically equivalent XML document by the SAM parser. That XML document was then transformed into DocBook according to the publisher's specifications (the publisher imposed constraints that are not expressed in DocBook itself). From that point on, the publisher's existing DocBook tool chain took over.

Most concrete markup languages, at least those designed for documents, try to make their marked-up documents look and read as much as possible like a formatted document. SAM is designed to be easy and natural to read, like a concrete markup language, but it is also designed to make the structure of the content as clear and explicit as possible while requiring a minimum of markup. This is why it uses indentation to express structure. Indentation shows structure clearly with a minimum of markup noise to distract the reader.

Because it is meant specifically for authoring, a SAM parser outputs XML, which can then be processed by the standard XML tool chain. Figure 30.6 shows how the SAM markup in Figure 30.5 would be output by a SAM parser.

```
<section>
    <title>Moby Dick</title>
    <p>Herman Melville's <phrase>
    <annotation type="novel"/>Moby Dick</annotation>
    </phrase> is a long book about a big whale.</p>
</section>
```

Figure 30.6 – Example of SAM intermediate XML markup

Instances of hybrid markup languages

I said earlier that this book is written in SAM, but that is not quite accurate. As noted above, you can't directly write anything in an abstract language. You write in instances of those languages. Thus, DocBook is an instance of the abstract language XML. You can write documents in DocBook. (We do say, of course, that we write documents in XML, but that statement is, if not wholly inaccurate, certainly non-specific. Saying that a document is written in DocBook tells you what constraints it meets. Saying it is written in XML merely tells you which syntax it uses, which is a whole lot less informative.) With hybrid languages, it depends on what type of hybrid they are, and whether the concrete part of the language is extensive enough to be used alone. In the case of SAM, it is not. Therefore you cannot write a meaningful document in pure SAM. You have to create your own language using SAM syntax.

So, to be more specific, this book is written in a markup language written in SAM that I created for the specific purpose of writing this book. That markup language was then transformed by a processing application into DocBook, which is the markup language the publisher uses for producing books. From there is was processed through the publisher's regular DocBook-based tool chain to produce print and ebook output.

Choosing the style of markup you will use is very much tied to the choices you make about the structures you want. In the following chapters, I look at a variety of markup languages using a variety of markup styles. As always, of course, your choices should be dictated by what works best for the overall partitioning and redirection of complexity that you are seeking to achieve in your content system.

CHAPTER 31
Patterns

I noted that lightweight markup languages such as Markdown rely on patterns in the text rather than explicit markup characters to delineate basic structures. XML, on the other hand, makes a strict distinction between what is markup and what is text. Markup is always recognized explicitly by markup sequences that cannot occur as normal text in any part of the document.

All the same, even XML markup is defined by patterns. The XML specification[1] defines XML using a series of patterns written in a notation called EBNF. The difference between XML syntax and Markdown syntax is that in XML, patterns are defined absolutely, so a markup start character is a markup start character no matter where it occurs, whereas in Markdown, the patterns are defined in context, so a * character starts a list item only at the beginning of a line, not the end.

Since markup is all about patterns, you are not confined to the patterns defined by the markup language. Many patterns occur naturally in content, and you can recognize those patterns and act on them with algorithms even if they are not formally marked up.

A common example of this is the pattern of a URL. In most web-based editors, if you enter a URL as plain text (with no markup identifying it as a URL), the editor recognizes the pattern and turns the URL into a clickable link in the HTML output.

Recognizing patterns, rather than forcing writers to mark them up explicitly, can increase functional lucidity significantly. Consider something as simple as a date. A date is a complex piece of data consisting of at least three elements, the year, month, and day. (It gets more complex if you include the day of the week or the time.) If you need those individual components of a date for processing purposes, you might be tempted to require dates to be marked up like this:

```
date: {2016}(year)-{08}(month)-{25}(day)
```

or

```
date:
    year: 2016
    month: 08
    day: 25
```

---

[1] https://www.w3.org/TR/xml/

This is fully explicit markup. It places every piece of the data in its own explicit structure. However, it lacks functional lucidity.

A date in the format `2016-08-25` is already a piece of structured markup, which is defined by ISO 8601, an international standard. The only problem with marking up a date using this format is that in another context `2016-08-25` could be a mathematical expression that resolves to `1983`. But you can eliminate this potential confusion if you isolate the context so that a processor can recognize when you use an ISO 8601 date. In other words, all you need to do is create an annotation, as shown here:

```
{1941-12-07}(date) is a day that will live in infamy.
```

If you want to enter the date in another format, you can annotate it with the ISO 8601 format to make the meaning clear to algorithms:

```
{December 7, 1941}(date "1941-12-07") is a day
that will live in infamy.
```

Of course, a sufficiently clever algorithm could recognize `December 7, 1941` as a date as well, because it is a predictable pattern. One of the defining features of patterns is that you can recognize the semantic equivalence between two patterns that express the same information. You can increase the functional lucidity of your markup by recognizing a variety of semantically equivalent patterns, rather than imposing a uniform syntax on writers.

Design implications

XML recognizes the value of patterns. The *XML Schema Datatypes* standard[2] defines a number of common patterns, and the XML schema language (XSD)[3] allows you to define patterns for use in your own markup. It calls these patterns "simple types" (as opposed to "complex types," which are composed of multiple nested elements). Thus, in XSD, you can specify that the contents of a date field must conform to the simple type `xs:date`, which is a pattern based on ISO 8601. If the schema defines the data type for the `date` element as `xs:date`, you can do following:

```
<date>1941-12-07</date> is a day that will live in infamy.
```

---

[2] https://www.w3.org/TR/xmlschema-2/

[3] https://www.w3.org/standards/xml/schema

It is important to understand what this does. If the type of element date in your XML markup language is defined as xs:date, then an XSD schema validator will accept 1941-12-07 as a valid date but reject <date>last Thursday</date> as invalid. It does not break the date down into year, month, and day components. To do that, you need an algorithm. But your algorithm can reliably break down the date because the markup constrains the interpretation, establishing that it is indeed an ISO 8601 date. Most programming languages have library functions that know how to manipulate ISO 8601 dates, so you probably don't have to do any work yourself to get the year, month, and day components of a date.

For other patterns, most programming languages have a library called "regular expressions," which can rip a pattern apart and get at the pieces. XSD goes a step further and lets you define new patterns using regular expressions. Regular expressions let your algorithms decode patterns in your content, as long as those patterns are sufficiently constrained by their context. This frees writers from having to explicitly break down many types of information into separate fields.

For example, in most of our recipe examples, we have explicitly broken down the items in an ingredient list into ingredient, quantity, and unit of measure like this:

```
ingredients:: ingredient, quantity, unit
    eggs, 12, each
    water, 2, qt
```

That format is not terribly onerous to write, but there is still a small functional lucidity penalty. A writer might more naturally write:

```
ingredients:: ingredient
    eggs 12
    water 2qt
```

Ingredient name, quantity, and unit of measure are still distinct. Any human reader can see them instantly. The trick is to get an algorithm to read them the same way. This trick can be accomplished using a regular expression:[4]

```
(?P<ingredient>.+?)(?P<quantity>\d+)(?P<unit>qt|tsp|tbsp)?
```

---

[4] This regular expression is deliberately simple and does not include a full list of all the possible names for units of measures. You would need a more precise expression for production quality code, but it would be distractingly verbose for our present purpose.

To show you how this works, here is a set of ingredient values:

```
eggs 12
water 4qt
salt 1tsp
butter 2tbsp
pork chops 4
```

Here is the result of applying the regular expression above to those values (this is a report generated by regex101.com, a fantastic site for designing and testing regular expressions):

```
Match 1
Full match      0-7      `eggs 12`
Group `ingredient`    0-5      `eggs `
Group `quantity`    5-7      `12`

Match 2
Full match      8-17     `water 4qt`
Group `ingredient`    8-14     `water `
Group `quantity`    14-15     `4`
Group `unit`    15-17      `qt`

Match 3
Full match      18-27     `salt 1tsp`
Group `ingredient`    18-23      `salt `
Group `quantity`    23-24      `1`
Group `unit`    24-27      `tsp`

Match 4
Full match      28-40     `butter 2tbsp`
Group `ingredient`    28-35      `butter `
Group `quantity`    35-36      `2`
Group `unit`    36-40      `tbsp`

Match 5
Full match      41-53     `pork chops 4`
Group `ingredient`    41-52      `pork chops `
Group `quantity`    52-53      `4`
```

As you can see, the regular expression has broken each ingredient down into ingredient, quantity, and unit fields, just like our explicit markup did. The conventional way of writing ingredients, in other words, is just as structured as our formal way. It just takes a line of code to pull them out of the string. Once again we are partitioning and transferring a bit of complexity from the writer to the information architect or content engineer.

Essentially, any expression that you commonly use in text and that follows a well-defined pattern is already structured text, and you don't need to structure it again. All you need to do is to use structure to place it into context so that an algorithm can recognize it reliably. For example, the regular expression for breaking down the components of an ingredient listing only works when applied to text that you are sure is an ingredient listing. Thus, you use normal structured text to delineate the ingredients list and each list item. This provides the necessary context for an algorithm to use the ingredient list pattern to pull out the ingredient, quantity, and unit fields without forcing the writer to make them explicit in the text.

Avoiding unnecessary markup of already recognizable patterns both simplifies your markup design and increase the functional lucidity of your markup languages.

If you were creating this recipe structure in XML and using XML Schema (XSD) as your schema language, you could create a data type for the ingredient field that would verify that the text matched the expected pattern (see Figure 31.1).

```
<xsd:element name="ingredient">
    <xsd:simpleType>
        <xsd:restriction base="xsd:string">
            <xsd:pattern value="(.+?)(\d+)(qt|tsp|tbsp)?"/>
        </xsd:restriction>
    </xsd:simpleType>
</xsd:element>
```

Figure 31.1 – XSD schema type definition for ingredients

This type definition ensures that an error will be raised if the writer enters ingredients in the wrong form or uses the wrong names for units. For instance, these ingredient list items would trigger an error:

```
<ingredients>
    <ingredient>12 eggs</ingredient>
    <ingredient>water 2 cups</ingredient>
    <ingredient>butter 4 lumps</ingredient>
</ingredients>
```

The first ingredient fails because the quantity (12) appears before the ingredient, not after. The next two fail because cups and lumps are not on the list of strings recognized by the pattern, which probably should be extended to allow cups, but not lumps. In Figure 31.1, the names of the fields that were present in the original regular expression are omitted because XML schema is not going to break the values apart and put them in separate fields for us. As far as XML schema is concerned, this is just the definition of a single element. It would be up to a processing application to break the pieces apart into make separate fields.

By the way, though I have used the ingredients listing as an example here, I probably would not use this pattern in a production system. While the `eggs, 12, each` structure is a tiny bit unnatural, its explicitness makes it less error prone. It is less likely that writers will get the order of information wrong and have to pause to deal with an error message. Asking writers to break the information apart explicitly is probably the lesser of two evils. When using patterns to improve functional lucidity, you need to balance naturalness against explicitness.

CHAPTER 32

Lightweight Markup Languages

I commented in Chapter 30 that XML does a poor job of partitioning authoring concerns from parsing concerns. XML is a fully general abstract markup language, and its syntax and its logical model are designed to support the creation of any markup language for any purpose whatsoever. This need to support any possible kind of structure makes XML a heavy, verbose language.

XML's predecessor, SGML, attempted to be fully general and allow you to define specific markup languages that had a light syntax that was easy to author. Unfortunately, the mechanism for creating such languages is complex and difficult to understand. It also made parsing SGML very complicated. SGML is still used in a few niches, but it has never achieved the kind of widespread use that XML has.

Still, XML remains a problem for writers, and a number of languages have been created to try to address the problems created by XML's verbosity. Collectively these are called *lightweight* markup languages. Lightweight markup languages are designed to use a lightweight syntax, that is, one that imposes a minimal burden on the readability of the raw text of the document.

Lightweight markup languages are far less general in their application than XML. In effect, they partition by omission, leaving out capabilities their users don't need in order to make the syntax and structures simpler and easier to understand. This is fine as long as you don't need the capabilities they omit. The key to correct partitioning here is to choose the language that has the best balance between the capabilities you need and the simplicity of authoring you want.

The primary appeal of lightweight markup languages rests on two related properties:

- They have a high degree of functional lucidity at the syntactic level (easy to write) and often at the semantic level (what it means) as well. You can usually read the raw markup of a lightweight language more or less as if it were a conventional text document.
- They can be written effectively using a plain text editor (as opposed to an elaborate structured editor with a graphical editing view). This means that the editing requirements are also lightweight.

Most lightweight markup languages come with a simple processing application that creates output directly in one or more output formats. This means that they have a lightweight tool chain that is easy and inexpensive to implement.

There are a number of lightweight markup languages. Some of the more prominent include Markdown, Wiki markup, reStructuredText, ASCIIDoc, and LaTeX.

Markdown

The most prominent of the lightweight languages, and arguably the lightest weight, is Markdown. Invented in 2004 by John Gruber as a way to quickly write simple web pages using syntax similar to that of text-format email, Markdown has spread to all kind of systems and has multiple variants that have been adapted for different purposes. I have used several examples of Markdown in this book. Here is one of them:

```
Wayne's best yet
================

After tiresome performances in _Rio Grande_
and _Sands of Iwo Jima_, the Duke is brilliant
in _Rio Bravo_.
```

"Adapted for different purposes" mostly means that people have created versions that add to the syntax and semantics of Gruber's first version. For instance, the code sharing site GitHub has adopted "GitHub flavored Markdown" as the standard format for user-supplied information on the site, such as project descriptions and issues, and has added syntax specific to tracking issue numbers and code commits for projects. This allows GitHub to automatically generate links between commits and the issues that relate to them. For example:

```
Issue #135 was fixed in commit
8e8c6a0b4c9c41bd72fab5fd53e3d967e9688110.
```

Markdown is a simple document-domain language. While its semantics are essentially a subset of HTML, it is more squarely in the document domain than HTML because it lacks any ability to specify formatting or even to create tables (though various Markdown flavors have added support for tables).

One of the recurring patterns of markup language development – and technology development generally – is that when a format becomes popular because of its simplicity, people start to add "just one more thing" to it, and it becomes either more complex (and thus less attractive) or more fragmented (and thus harder to build a tool chain for). Markdown is definitely fragmenting at the moment (though a standardization effort, CommonMark, is also under way). There is even a project to add semantic annotation to Markdown as part of the Lightweight DITA project.

None of this is a reason not to use Markdown where its structures and syntax make it an appropriate source. Markdown provides useful constraints on the basic formatting of a web page both by factoring out direct formatting features and by providing a limited set of document-domain features. These constraints help prevent contributors to a site from indulging in extravagant, nonstandard formatting or overly elaborate text structures. It successfully partitions basic web formatting, though not much else.

Markdown is also used in conjunction with static site generators, such as Jekyll, which use Markdown for basic text structures in concert with templating languages such as Liquid.[1] Tom Johnson provides a side-by-side comparison of DITA and Jekyll in a series of posts on his blog (http://idratherbewriting.com/2015/03/23/new-series-jekyll-versus-dita/).

Markdown does not provide any subject-domain structures or constraints. This may be a welcome feature when comparing Markdown with more complex document-domain languages, many of which include subject-domain structures that can be confusing or that writers may abuse to achieve formatting effects.

Because the inspiration for its syntax, text-format emails, has faded to obscurity, it is not clear that everyone automatically knows how to write Markdown, which was the original design intent. However, a lot of it remains obvious and intuitive, meaning that, within its limits, Markdown has good functional lucidity. It works well if you don't need any of the features it partitions by omission.

---

[1] Liquid is essentially a set of management-domain structures, yielding a result that is comparable to a document-/management-domain hybrid in its capabilities, though not in its style.

Wiki markup

Another popular lightweight format is wiki markup, introduced by Ward Cunningham in 1995 as the writing format for WikiWikiWeb, the first wiki.[2] Wiki markup is similar to Markdown in many respects (most lightweight languages share the same basic syntax conventions, based on the imitation of formatted document features in plain text documents).

What makes wiki markup distinct is how it is tied into the operation of a wiki. One notable feature is how it handles linking. In the original WikiWikiWeb markup, any word with internal capitals was considered a *WikiWord* and instantly became a link to a page with that WikiWord as the title. Such a page was created automatically if it did not already exist. This was a simple implementation of a linking algorithm based on annotation rather than the naming of resources.

Different wikis support different markup. Figure 32.1 shows a small example of MediaWiki markup. (MediaWiki is the system that runs Wikipedia).

```
"Take some more [[tea]]," the March Hare said
to Alice, very earnestly.

"I've had '''nothing''' yet," Alice replied in
an offended tone, "so I can't take more."

"You mean you can't take ''less''?" said the Hatter.
"It's very easy to take ''more'' than nothing."
```
Figure 32.1 – Sample of MediaWiki markup[3]

A wiki is a type of simple content management system that allows people to create and edit pages directly from a web browser. A wiki, essentially, is a CMS that partitions and distributes the problem of web content management out to individual contributors, allowing anyone to edit and improve a site. Wikipedia is by far the largest and most well-known wiki. Wiki's are a significant example of a bottom-up information architecture. Anyone can add a page, and that page is integrated into the overall collection by WikiWord-style linking and by including itself in categories (conventionally, by naming them on the page).

---

[2] https://en.wikipedia.org/wiki/WikiWikiWeb

[3] https://en.wikipedia.org/wiki/Wiki#Editing

Cunningham described WikiWikiWeb as "The simplest online database that could possibly work."[4] Like Markdown, its success has led to additional features, fragmentation, and growing complexity. Some commercial wikis are now complex content management systems. Indeed, it is somewhat difficult today to define the boundaries between wikis, blogging platforms, and conventional CMSs.

If wikis have a defining characteristic today it is probably the bottom-up architecture rather than the original novelty of in-browser editing, which is now found across many different kinds of CMS. Cunningham designed wikis to be collaborative platforms – places where people could collaborate with people they might not even know to create something new with no central direction or control. The idea was not only architecturally bottom-up but also editorially bottom-up. However, today, most wikis include features that allow you to exercise a degree of central control. Question-and-answer sites, such as Stack Exchange with its distributed and democratic control systems, may be closer today to Cunningham's idea of a democratic creation space.

What wikis illustrate for structured writing is that simple markup innovations such as the WikiWord can have a revolutionary effect on how content is created and organized. Most wikis today use words between double square brackets for WikiWords, rather than internal capitals, but the principle is the same. You can link to a thing merely by naming it.

WikiWords are also a case of subject-domain annotation. Marking a phrase as a WikiWord says, "this is a significant subject." It does not provide type information – as the subject annotation examples shown in this book do – but merely denoting a phrase as significant says that it names some subject of importance that deserves a page of its own.

This illustrates a critical point about bottom-up information architectures: structured writing, even in very simple form, can create texts that are capable of self-organization and that can be assembled into meaningful collections without the imposition of any external structure. However, the wiki process leaves much of the complexity of content creation and management unhandled. The slack has to be taken up by human effort, which works well for Wikipedia, with its army of volunteer contributors and editors, but is harder to reproduce on a corporate scale.

---

[4] http://www.wiki.org/wiki.cgi?WhatIsWiki

reStructuredText

reStructuredText is a lightweight hybrid markup language most often associated with the Sphinx documentation framework, which was developed for documenting the Python programming language. I looked at reStructuredText briefly as an example of a hybrid markup language in Chapter 30.

Similar to Markdown, reStructuredText uses a plain-text formatting approach to basic text structures. This part of the markup looks natural because it uses characters and patterns that you might use to format a document if you only have a plain-text editor (see Figure 32.2).

```
Hard-Boiled Eggs
================
A hard boiled egg is simple and nutritious.
Prep time, 15 minutes. Serves 6.

Ingredients
-----------

======  ========
Item    Quantity
======  ========
eggs    12
water   2qt
======  ========

Preparation
-----------
1. Place eggs in pan and cover with water.
2. Bring water to a boil.
3. Remove from heat and cover for 12 minutes.
4. Place eggs in cold water to stop cooking.
5. Peel and serve.
```

Figure 32.2 – reStructuredText example

However, reStructuredText also has a feature called *directives*, which you can use to create markup with more complex semantics. Figure 32.3 shows a directive for inserting an image.

```
.. image:: images/harcboiledegg.png
   :height: 100
   :width: 200
   :scale: 50
   :alt: A hard boiled egg.
```

Figure 32.3 – reStructuredText directive for inserting an image

In Figure 32.3, reStructuredText takes the same approach as XML, using characters in a way that they are almost never used in a normal document. This approach simplifies parsing, because there is seldom any question about whether a particular pattern is intended to be markup or text, but it also makes reStructuredText less natural to read and to write. reStructuredText is therefore something of a syntactic hybrid, as well as being a hybrid in the sense that it has both concrete and abstract parts.

If you are looking for a lightweight document-domain markup language of moderate complexity and a degree of extensibility, or if you are interested in Sphinx as an authoring and publishing system, reStructuredText is an option to consider.

ASCIIDoc

ASCIIDoc is a lightweight markup language based on the structure of DocBook. It is intended for the same sort of document types for which you might choose DocBook, but it allows you to use a lightweight syntax. In appearance it is similar to Markdown, as shown in Figure 32.4.

```
= My Article
J. Smith

http://wikipedia.org[Wikipedia] is an
on-line encyclopaedia, available in
English and many other languages.

== Software

You can install 'package-name' using
the +gem+ command:

 gem install package-name

== Hardware

Metals commonly used include:

* copper
* tin
* lead
```

Figure 32.4 – ASCIIDoc example[5]

---

[5] https://en.wikipedia.org/wiki/AsciiDoc

However, while Markdown was designed for simple web pages, ASCIIDoc was designed for complex publishing projects with support for a much wider array of document-domain structures, including tables, definition lists, and tables of contents.

If you are looking for a lightweight document-domain markup language of medium complexity that is compatible with DocBook (meaning you are interested in creating books rather than web pages), ASCIIDoc is an option to consider.

LaTeX

LaTeX is a document-domain markup language used extensively in academia and scientific publishing. It is based on the syntax of TeX, a typesetting system developed by Donald Knuth in 1978.[6] Figure 32.5 is an example of LaTeX.

```
\documentclass[12pt]{article}
\usepackage{amsmath}
\title{\LaTeX}
\date{}
\begin{document}
  \maketitle
  \LaTeX{} is a document preparation system for
  the \TeX{} typesetting program. It offers
  programmable desktop publishing features and
  extensive facilities for automating most
  aspects of typesetting and desktop publishing,
  including numbering and  cross-referencing,
  tables and figures, page layout,
  bibliographies, and much more. \LaTeX{} was
  originally written in 1984 by Leslie Lamport
  and has become the  dominant method for using
  \TeX; few people write in plain \TeX{} anymore.
  The current version is \LaTeXe.

  % This is a comment, not shown in final output.
  % The following shows typesetting  power of LaTeX:
  \begin{align}
    E_0 &= mc^2                                   \\
    E &= \frac{mc^2}{\sqrt{1-\frac{v^2}{c^2}}}
  \end{align}
\end{document}
```

Figure 32.5 – LaTeX example

---

[6] https://en.wikipedia.org/wiki/LaTeX

Figure 32.6 shows how that markup is rendered.

$$\LaTeX$$

$\LaTeX$ is a document preparation system for the $\TeX$ typesetting program. It offers programmable desktop publishing features and extensive facilities for automating most aspects of typesetting and desktop publishing, including numbering and cross-referencing, tables and figures, page layout, bibliographies, and much more. $\LaTeX$ was originally written in 1984 by Leslie Lamport and has become the dominant method for using $\TeX$; few people write in plain $\TeX$ anymore. The current version is $\LaTeX\,2_\varepsilon$.

$$E_0 = mc^2 \tag{1}$$

$$E = \frac{mc^2}{\sqrt{1 - \frac{v^2}{c^2}}} \tag{2}$$

Figure 32.6 – Output from LaTeX[7]

The equation markup shows why LaTeX is popular for academic and scientific publishing. While not exactly transparent, the markup is compact and functionally lucid for anyone with a little experience with it.

Wikipedia offers a comparison of various math markup formats which shows how big a difference syntax can make to the lucidity of a markup language.

For the equation:

$$x = \frac{-b \pm \sqrt{b^2 - 4ac}}{2a}$$

1

The LaTeX markup is:

```
x=\frac{-b \pm \sqrt{b^2 - 4ac}}{2a}
```

Whereas the XML-based MathML version looks like Figure 32.7.

---

[7] The original uploader was Bakkedal at English Wikipedia - Own work, CC BY-SA 2.5, https://commons.wikimedia.org/w/index.php?curid=30044147

```
<math mode="display" xmlns="http://www.w3.org/1998/Math/MathML">
 <semantics>
  <mrow>
    <mi>x</mi>
    <mo>=</mo>
    <mfrac>
      <mrow>
        <mo form="prefix">&#x2212;<!-- - --></mo>
        <mi>b</mi>
        <mo>&#x00B1;<!-- &PlusMinus; --></mo>
        <msqrt>
          <msup>
            <mi>b</mi>
            <mn>2</mn>
          </msup>
          <mo>&#x2212;<!-- - --></mo>
          <mn>4</mn>
          <mo>&#x2062;<!-- &InvisibleTimes; --></mo>
          <mi>a</mi>
          <mo>&#x2062;<!-- &InvisibleTimes; --></mo>
          <mi>c</mi>
        </msqrt>
      </mrow>
      <mrow>
        <mn>2</mn>
        <mo>&#x2062;<!-- &InvisibleTimes; --></mo>
        <mi>a</mi>
      </mrow>
    </mfrac>
  </mrow>
 </semantics>
</math>
```

Figure 32.7 – MathML markup

Clearly MathML was not designed with the idea that anyone would ever try to write it raw. It is intended to be the output of a graphical equation editor.[8] While you could use a graphical equation editor to create LaTeX math markup, it is certainly possible to write and read raw LaTeX.

LaTeX is not as lightweight as Markdown. Its markup is almost entirely explicit – except for paragraphs, which are delineated by blank lines just as in Markdown. But its syntax is certainly lighter when compared to XML-based languages, and it has much greater functional lucidity.

---

[8] Interestingly, MathML comes in two different flavors. Presentation MathML is a media-domain language that describes how an equation is presented. Content MathML is a subject-domain language that describes what it means.

The hallmark of a lightweight design is sufficient functional lucidity that you can write in raw markup rather than needing a graphical editor. But LaTeX's structures are barely out of the media domain, which limits its usefulness for structured writing.

Subject-domain languages

So far I have looked at languages that are primarily document domain in design. The document domain is an obvious choice for a public language since the use of common document types such as books and articles is widespread. But there are public subject-domain languages as well.

```
/**
 * Validates a chess move.
 *
 * Use {@link #doMove(int theFromFile, int theFromRank,
 * int theToFile, int theToRank)} to move a piece.
 *
 * @param theFromFile file from which a piece is being moved
 * @param theFromRank rank from which a piece is being moved
 * @param theToFile   file to which a piece is being moved
 * @param theToRank   rank to which a piece is being moved
 * @return            true if the move is valid, otherwise false
 */
boolean isValidMove(int theFromFile, int theFromRank,
int theToFile, int theToRank) {
    // ...body
}
```
Figure 32.8 – JavaDoc example

JavaDoc (Figure 32.8), which I looked at in Chapter 16, not only has subject-domain tags for parameters and return values, it effectively incorporates the Java code itself (computer programs are a kind of structured text), pulling information from the function header into the output.

Many other languages for documenting programming languages exist.[9] However, it is difficult to find public subject-domain lightweight markup languages outside the realm of programming language and API documentation. This is probably because only programmers are likely to write their own parser to create a markup language. Most other people are going to choose an extensible language as a base, which today usually means XML. Part of my motivation for creating SAM is to provide a way to create subject-domain languages with lightweight syntax.

---

[9] Wikipedia maintains a list at: https://en.wikipedia.org/wiki/Comparison_of_documentation_generators.

CHAPTER 33
Heavyweight Markup Languages

I use the term *heavyweight* as a contrast to the commonly used term *lightweight*, even though heavyweight is not used commonly for markup languages. Nonetheless, it fits. Both the abstract language XML and specific instance languages such as DocBook and DITA are heavyweights in the sense that they provide a lot of capability at the expense of a large footprint.

Having said that, there is an important distinction between the heavyweight syntax of XML and the heavyweight semantics of a DITA or DocBook. You can represent the semantics of DITA or DocBook using a more lightweight syntax. And you can certainly create simple markup languages with lightweight semantics using the heavyweight syntax of XML. Despite this, there is a definite connection between heavyweight syntax and heavyweight semantics, perhaps because languages with more heavyweight semantics have a greater need for the heavyweight capabilities of XML and the processing tools that go with it.

This chapter contains a brief survey of some heavyweight languages. Heavyweight languages often contain structures from more than one domain. Their core is usually in the document domain, but they typically contain some media-domain structures for things, such as tables, that are hard to abstract from the media domain in a generic way. They often also contain some subject-domain structures, typically related to technology, since many heavyweight languages originated for documenting technical products. Finally, most contain some management-domain structures, particularly for capabilities such as conditional text.

Why is the structured writing landscape dominated by a few large and loosely constrained markup languages? Here are a few reasons:

- Big, loosely constrained document-domain markup languages are the smallest step into the document domain from word processors and desktop publishing applications. Thus, they represent a relatively small conceptual change for writers. However, this also means that they do relatively little to partition and redirect complexity away from writers. They exchange some formatting complexity for structural complexity, which may or may not be an overall win, but they do nothing to enhance rhetorical repeatability or drive information architecture.

- Too often, reducing the amount of complexity that goes unhandled in the content system is not the main driver in the adoption of structured writing. Instead, adoption is typically driven by a desire to improve some aspect of content management, particularly content reuse. While you can do these things without resorting to structured writing, structured writing formats make it easier to integrate individual pieces of content and help ensure that your content doesn't get locked into a vendor's system.

- Constraints are onerous if you don't get them right, and the benefits of getting them right are often under-appreciated, especially in content management applications where the consequences of a lack of constraint show up years down the road, when it becomes all too easy to blame any problems on human failure rather than poor system design.

For all these reasons, it is worthwhile to look at where the big public languages fit in the structured writing picture.

This chapter is not a buyer's guide. A detailed evaluation of large systems such as DocBook, DITA, and S1000D would take too much space in this book to do them full justice. Rather, this chapter, and the book as a whole, provides a framework to help you understand your structured writing requirements independently from any system and to evaluate, compare, and contrast systems in more or less neutral terms.

DITA

You can look at DITA in two ways: as a complete structured writing system that can be used more or less out of the box[1] or as an information typing architecture (as its name proclaims).[2]

Out of the box, the DITA specification gives you a fixed set of topic types. The DITA information typing architecture gives you the capability to create a wide variety of information types (though not necessarily all the types you might want). In Chapter 34, I discuss DITA as an information typing architecture. Here I look at out-of-the-box DITA.

---

[1] I say "more or less" because even packaged applications such as Microsoft Word or FrameMaker are not used completely out of the box for serious content creation: you need to customize styles and output formats to some extent, and the same is true of DITA.

[2] The acronym DITA stands for Darwin Information Typing Architecture, with the word Darwin representing DITA's approach to the extensibility of markup: specialization.

Out-of-the-box, DITA comes in two main forms:

1. **The DITA Open Toolkit:** You can download the open-source DITA Open Toolkit for free and use it to produce content. The formatting style sheets that come with the toolkit are basic, but they can be customized, which you will likely want to do.
2. **Packaged DITA tools:** There are a variety of tools that package DITA. Some are XML editors that include support for processing DITA. Others are content management systems. These tools may add capabilities beyond what is supplied by the DITA Open Toolkit and also may hide the underlying DITA structures. Evaluating these tools is beyond the scope of this book.

If you are evaluating how well DITA fits your needs, the key features of out-of-the-box DITA to consider are its topic model and its focus on the reuse algorithm. The description of the document-domain/management-domain approach to reuse in Chapter 13 is based on the DITA model, which provides comprehensive support in those domains.

As described in Chapter 22, the DITA topic model is based on the idea that you can break down information into different abstract types and that there is value in separating the different types. One of the problems with this theory, and consequently with the application of DITA's topic model, is that it is not clear how big an information type is and whether an information type constitutes a rhetorical block or just a semantic block. Specializing DITA allows you to be more specific on this point, but with out-of-the-box DITA, you are probably using three topic types: concept, task, and reference.[3] These topic types do not unambiguously constitute semantic blocks or rhetorical blocks; in practice they are used both ways.

The principal thing that sets out-of-the-box DITA apart from other approaches to structured writing is its map and topic architecture. In most other systems, the unit that the writer writes and the unit that the reader reads are the same. For long works, you may choose a mechanism for breaking up and assembling pieces, but the content is still described by a single schema. For instance, in DocBook, you can write a book using a `book` document type that includes various `chapter` document types to create a complete book out of multiple files. In other words, a Doc-Book `book` is a single document structure that happens to be made up of individual files.

However, DITA partitions this document assembly function into a separate file type, the map. A DITA map file is an independent structure. It does not create a single logical document structure.

---

[3] Current versions of out-of-the-box DITA include other topic types, such as machine-industry and troubleshooting tasks, for more specialized uses.

It does not usually contain any actual content, and you can't write an entire book in a single map file. Instead, a map file is a set of instructions to a publishing tool chain about how to assemble a larger work out of component pieces.

This distinction is critical. In the DocBook model, there is a continuity of constraint between the book and its chapters. In DITA, the constraints on the map and the constraints on the topics in the map are completely separate. This means that in DITA, the topic is the largest unit of content to which content constraints can be applied (at least in the conventional way). This partitioning leaves the responsibility for the rhetoric of larger units entirely with the writer.

Maps are structured like a tree, so you can construct arbitrarily deep hierarchies. This means you can choose which parts of your structure you create using a map and which parts you create inside a topic. If you have a list of four items, each of which needs two or three paragraphs of description, you can create one topic with the list of four items in it, or you can create one topic for each item and then tie them together using a map. Since the topic is the largest unit of content constraint in DITA, if you break content down to this level, you lose the ability to apply constraints to the content as a whole. On the other hand, you can reuse each item separately.

This presents a dilemma. In Chapter 22, I characterized structured writing as dividing content into blocks, and I made a distinction between semantic blocks and rhetorical blocks. In the design of a markup language, you build rhetorical blocks from semantic blocks, which in turn may be made up of smaller semantic or structural blocks. This works fine for developing the structure of a rhetorical block, which is the work that will be presented to the reader. In that scenario, writers write rhetorical blocks. The semantic blocks are just elements of the model.

But things become more difficult when you attempt to do fine-grained reuse of content. In that case, you may need to write independent semantic blocks and combine them to produce rhetorical blocks. DITA lets you do this in two ways. The first (which is frequently discouraged) is to nest one topic inside another. The second is to combine topics using a map, with the map representing the rhetorical block. However, DITA does not provide a high-level way to constrain the structure of a rhetorical block built this way.

If you want a constrained rhetorical block, you have to model it as a single DITA topic type. You can certainly do this by specializing from the base `topic` topic type. However, in doing so you probably need to move away from the information typing idea of separating different types of information, because a full rhetorical block, such as the recipe example, often requires different types of information.

This leads to the confusion about whether a DITA topic is a semantic block or a rhetorical block. For people who use out-of-the-box DITA this can be a problem because, by default, the DITA Open Toolkit creates a separate web page for each topic. This is not an appropriate presentation if your individual DITA topics are not rhetorical blocks. If your rhetorical block is made up of multiple DITA topics strung together by a map, and you want that rhetorical block to appear on a single page, you need to use a procedure called *chunking*, which is not as straightforward as it should be.[4]

The idea that blocks are reusable is attractive, but you need to think through exactly what the reusable unit of content is. It is one thing to reuse complete rhetorical blocks (perhaps with some variations in the text). It is quite a different thing to reuse semantic blocks below the level of the rhetorical block, particularly if you need to constrain the rhetorical block or apply any of the other structured writing algorithms at the level of the rhetorical block.

Rhetorical quality should be a serious concern with this model. If you assemble rhetorical blocks out of smaller reusable units without paying attention to the rhetorical integrity, completeness, and consistency of the result, quality can be seriously affected. It is difficult to maintain rhetorical quality if writers no longer see, think, or work in the context of the rhetorical block and if the structure of the rhetorical block is not constrained.

DITA, as a technology, does not prevent you from working in whole rhetorical blocks or from constraining your blocks in any way you want – DITA's information typing capabilities make this possible. However, the block-and-map model, whether implemented by DITA or any other system, creates an inherent tension between creating smaller semantic blocks to optimize for reuse and creating constrainable rhetorical blocks to optimize for rhetorical quality. And because it provides no facilities for imposing constraints above the topic level, DITA makes it difficult to partition and redirect this complexity away from the writer.

However, you do not require either document-level or topic-level constraints to reuse content blocks. When used correctly, constraints may improve quality and reliability, but you don't need them to compose larger blocks out of smaller blocks. This has led many organizations to select DITA for its reuse capabilities without taking advantage of its constraint capabilities or its information typing roots. People taking this approach will sometimes write their content in the base

---

[4] Chunking is on the agenda to be fixed in DITA 2.0.[http://docs.oasis-open.org/dita/dita-1.3-why-three-editions/v1.0/cn01/dita-1.3-why-three-editions-v1.0-cn01.html#future-of-dita]

`topic` topic type rather than a more constrained specialization. Quite simply, they don't find DITA's information typing to be useful, so they ignore it.

The growing popularity of this approach to reuse has led to the development of alternatives to DITA that provide the same reuse-management capabilities but remove the constraint mechanisms. One example is Paligo, a reuse-focused component content management system that uses DocBook as its underlying content format, specifically for the purpose of minimizing constraints on the content.[5] Such systems can reduce the up-front complexity of component-based content reuse, though possibly at the expense of costs down the road if the failure to manage and constrain rhetoric, and thus to capture subject metadata, result in poor rhetoric and inefficient processes.

DocBook

DocBook is an extensive, largely document-domain language with a long history and an extensive body of processing tools and support. As I have noted, DocBook is not a tightly constrained language. Instead it provides broad capabilities for describing document structures. In other words, it is mainly concerned with partitioning the document domain from the media domain in as comprehensive manner as possible. It does not constrain the rhetoric of content in any significant way and, therefore, makes no attempt to partition and distribute any part of the creative aspects of the content system.

Unlike DITA, DocBook was not created to support any particular information typing theory. It does not reflect any methodology for writing or organizing content. DocBook supports the structure of books but does not constrain the rhetorical structure of a work; in fact, the developers of DocBook made every attempt to avoid constraining rhetorical structure, leaving that task to the writer. The downside of this hands-off approach is that DocBook is a complex system, and writers must deal with this complexity without the benefit of having other aspects of their task, such as information typing, partitioned and directed away.

Of course, this is not to say that using DocBook is any more complex than using a publishing tool such as FrameMaker. It may well be less so. Nonetheless, DocBook and its tool chain can be challenging to learn and use. Because of this, writers often use simplified subsets of DocBook. (Where DITA is sometimes customized by the addition of elements, DocBook is often customized

---

[5] http://idratherbewriting.com/2016/08/01/paligo-the-story-xml-ccms-in-the-cloud/

by their subtraction.) However, DocBook remains popular with many for its lack of constraints combined with its rich feature set.

Because of its lack of constraints, DocBook does not fit well with the idea of structured writing as a means to partition and redirect complexity nor does it fit the model of intimate connection between process and rhetoric explored in this book. However, it can play a useful role in a structured writing tool chain as a language for the presentation algorithm. This is exactly how it was used in the production of this book. The book was written in a small, constrained language based on SAM and transformed by the presentation algorithm into DocBook, which was then fed into the publisher's standard DocBook based publishing tools. The publisher's requirements are far more precise than the generic DocBook specification, but the DocBook created by this method matches the publisher's specifications exactly.

In using this approach, I partitioned the complexity of meeting the publisher's DocBook requirements from the task of writing the book and gave myself a much simpler authoring interface, both syntactically and semantically. Of course, I transferred that complexity to myself in another role (information architect or content engineer), since I had to write the algorithm that transformed my SAM source into the publisher's preferred form of DocBook. But this was a big win for me, because writing that algorithm was a separate activity that did not impinge on my attention while I was writing the book. Another benefit was that because my lightweight book-writing language is highly constrained, maintaining conformance was much easier and the DocBook produced by my algorithm was more consistent and conformed more closely to the publisher's requirements than if I had written in DocBook directly.

I wrote my previous book, *Every Page is Page One*, directly in DocBook (an experience that contributed to my decision to develop SAM), but it took a lot of revision to get my DocBook content into the form that the publication process required. The publishing process has a set of constraints that are not enforced by DocBook itself. Instead, they have to be imposed by the writer while writing. But in the SAM-based markup language I used for this book, all those constraints were factored out, which enabled me to translate content reliably into a DocBook instance that met the publisher's requirements. In addition, I didn't need to remember any of those constraints.

S1000D

S1000D is a specification developed in the aviation and defense industries for the complex documentation tasks of those industries. It supports the development of the Interactive Electronic

Technical Manuals (IETMs) that are typically required in that industry. Although S1000D has subject-domain structures, it also has media-domain structures targeted at the production of IETMs and extensive management-domain structures designed to support the common source database (CSDB), the content management architecture that is part of the S1000D specification. S1000D is much more than a structured writing format. It specifies a complete document production system for a specific industrial sector. In other words, S1000D represents a particular partitioning of the content system designed for a particular industry.

HTML

HTML is widely used as an authoring format for content. For the most part, HTML is used in the media domain: people writing for the web in its native format, often using a WYSIWYG HTML editor.

Efforts have been made over the years to factor out the media-domain aspects of HTML and leave the formatting to CSS style sheets. This makes HTML a legitimate document-domain markup language. People interested in using HTML this way often use XHTML, a version of HTML that follows the constraints of XML. Being an instance of XML means you can write XHTML in an XML editor and process it with XML processing tools. This means that you can potentially publish content written in XHTML by processing it into other formats or by modifying its structure for use in different HTML-based media such as the web or ebooks.

Subject-domain languages

There are hundreds of public subject-domain languages written in XML. Most of these are more data oriented than content oriented, but you might be able to derive content from some of them using the extract and merge algorithms. Wikipedia maintains an extensive list.[6] However, most subject-domain languages used for content are developed in house to suit a particular subject matter and reader base. Examples of XML-based subject-domain languages include the following:

- Auto-lead Data Format: for communicating consumer purchase requests to auto dealers.
- BeerXML: for exchanging brewing data
- Keyhole Markup Language: for geographic annotations
- RecipeML: for recipes

---

[6] https://en.wikipedia.org/wiki/List_of_markup_languages

CHAPTER 34

Extensible and Constrainable Languages

The languages I have looked at to this point are publicly specified and have existing tool chains. Some are more constrained than others, and some support different structured writing algorithms and different ways of partitioning and redirecting complexity. Choosing one of them makes sense if the constraints they express and the algorithms they support partition content complexity in a way that is right for your organization. If not, you need to create your own structures to improve how complexity is partitioned and distributed in your organization.

You have three options for doing this:

- Create your own concrete language entirely from scratch, creating both the syntax and the semantics. (This is what John Gruber did when he created Markdown.)
- Use an existing abstract markup syntax, such as XML or SAM, and create your own semantics by defining named structures using that syntax (as described in Chapter 30).
- Take an existing markup language with extensible and/or constrainable semantics, such as DITA or DocBook, and extend and/or constrain it to meet your needs.

Each of these approaches has merits and drawbacks. For instance, creating a new concrete language may enable you to achieve exceptional functional lucidity for a particular type of information; extending/constraining an existing language may save tool development costs; defining your own semantics based on an existing syntax may enable you to find the right balance between functional lucidity and development costs.

This chapter looks at extensible and constrainable markup languages.

XML

The X in XML stands for eXtensible, but, as noted in Chapter 30, XML is an abstract language that does not define any document structures itself. Therefore, with XML, you extend from zero. Syntactically, everything is defined for you. Semantically you start from scratch.

You can define the structure of a new XML markup language using one of several available schema languages. I look at schema languages in Chapter 35, but the mechanics of defining a markup language in XML are out of scope for this book.

ReStructuredText

ReStructuredText defines a base set of structures – such as paragraphs, titles, and lists – with a concrete syntax. It defines other blocks using directives. Figure 34.1 shows a directive that includes an image and a set of attributes for rendering that image.

```
.. image:: images/biohazard.png
   :height: 100
   :width: 200
   :scale: 50
   :alt: alternate text
```

Figure 34.1 – reStructuredText directive

You extend reStructuredText by adding new directives. However, there is no schema language for reStructuredText. To create a new directive, you must create code to process it.

There is an important distinction between languages that are extensible by schema and those that are extensible only by writing code to process the extension.

If a language is extended by writing processing code, the only way to know if the input is valid is by processing it. If it raises a processing error, it is invalid. If you have only one processor for a language, you can treat that processor as normative. That is, the definition of a correct file is any file that can be successfully processed by the normative processor. The language is defined by the processor. But if you have multiple processors, how do you determine who is at fault when one of those processors fails to process a given input file? Is the processor incorrect or the source file?

A schema creates a language definition that is independent of any processor. That is, it partitions and redirects the complexity of validation. The schema is normative, not any of the processors. If the source file is valid per the schema, the processor is at fault if it does not process that file correctly. If the source file is not valid per the schema, the blame lies with the source file.

In the case of reStructuredText, the capacity of the processor to be extended using directives is built into the processor architecture. There is a specific and well-documented way to extend the language. Although reStructuredText allows you to extend the language by adding new directives, it does not have a constraint mechanism. There is no mechanism (other than hacking the code) to restrict the use of existing or new directives and structures.

TeX

TeX (pronounced "Tek") is a typesetting system invented by Donald Knuth in 1978. As a typesetting language it is a concrete media-domain language. But Knuth also included a macro language in TeX, which allows users to define new commands in terms of existing commands. (I say commands because that is the term used in TeX. Markup in the media domain tends to be much more imperative than markup in the subject domain, which is entirely descriptive, so commands is an appropriate name for TeX's tags.) This macro language has been used to extend TeX, most notably in the form of LaTeX, a document-domain language described in Chapter 32.

As with reStructuredText, the ability to extend is not the same as the ability to constrain. Introducing new commands does not create a constraint mechanism.

SAM

Although lightweight languages provide great functional lucidity, they suffer from limited extensibility (which generally requires writing code) and a lack of constraint mechanisms. I believe that a fully extensible, fully constrainable lightweight markup language would be a valuable addition to the structured writing toolkit. This is why I developed SAM, the markup language used for most of the examples in this book and for writing the book itself.

SAM is a hybrid markup language that combines concrete markup similar to Markdown with an explicit syntax for defining abstract structures called blocks, record sets, and annotations. It also has concrete markup for common features such as insertions, citations, and variable definitions.

SAM, like XML, is for defining specific markup languages. However, unlike XML, languages defined in SAM share a small common base set of text structures for which SAM provides concrete syntax. This allows SAM to combine lightweight syntax for the most common text structures with the ability to define constrained markup languages for specific purposes, particularly subject-domain languages. In other words, SAM represents a different partitioning of the markup design process from both the common lightweight languages and XML.

SAM is designed to be extensible and constrainable through a schema language (this is not complete at time of writing, but hopefully will be available by the time you read this). My intent is that the schema language should be able to define and constrain new block structures, constrain the use of existing concrete structures, and constrain the values of fields using patterns.

SAM is not designed to be as general as XML in its applications. As a result, its syntax is simpler and more functionally lucid, and its schema language should also make it simpler and easier for writers to develop their own SAM-based markup languages.

I use SAM for the majority of the examples in this book because SAM is designed to make structure clear. All of the examples could have been expressed in XML, but that would have made them harder to follow. Naturally, to write in SAM you need to know more about the rules of the language, but you should be able to read a typical SAM document and understand its structure with little or no instruction (see Figure 34.2).

```
examples: Basic SAM structures

    example: Paragraphs
        The is a sample paragraph. It is inside
        the {block}(structure) called `example`.
        It contains two {annotations}(structure),
        including this one. It ends with a blank
        line.

        This is another paragraph.

    example: Lists

        Then there is a list:

        1. First item.
        2. Second item.
        3. Third item.

    example: Block quote

        Next is a block quote with a {citation}(structure).

        """[Mother Goose]
            Humpty Dumpty sat on a wall.
```

Figure 34.2 – Basic SAM structures

This objective is similar, but not identical, to the aim of mainstream concrete and hybrid languages, such as Markdown and reStructuredText, which is to have the source file be readable as a document. Those languages are document-domain languages, and they strive to make the document structure clear and readable from the markup. SAM has the same goal, except that SAM was designed primarily for creating subject-domain languages. As such, SAM is designed to make the subject-domain structure of a document clear to the reader.

A SAM document may not look as much like a finished document as a Markdown or reStructuredText document. For example, it does not use underlines to visually denote different levels of header. Instead, it focuses on creating a hierarchy of named blocks and fields. In doing so, it uses the kind of markup people commonly use to create named blocks of text and to express a hierarchical relationship between them. As you can see in Figure 34.2, blocks are introduced with a name followed by a colon, and hierarchy is expressed through indentation.

SAM is an open source project. A description of the language and a set of associated tools are available from https://github.com/mbakeranalecta/sam.

DocBook

DocBook is not really extensible in the same sense as the other languages mentioned here, but it still deserves a mention. DocBook provides a deliberately modular construction that makes it easy to create new schemas that include elements from DocBook. DocBook takes full advantage of the extensibility features built into XML schema languages.

DocBook has a large tag set, so if you want a small constrained document-domain markup language, you can often create one by sub-setting DocBook. DocBook provides just about any document structure out there, so if you are building a document-domain language, DocBook probably has the pieces you need. You can even add additional constraints.

The great advantage of creating a new language as a subset of DocBook is that the result is also a valid DocBook document and, therefore, can be published by the DocBook tool chain. You don't have to write any new publishing algorithms if you take this approach (other than to customize formatting, which is required with any system). Creating a subset of DocBook allows you to impose more constraints and improve functional lucidity significantly, compared to standard DocBook, without having to write any processing code at all.

Technically speaking, any XML-based markup language is extensible in the same way. However, DocBook's structure, and the implementation of its schemas, was designed to support both extension and sub-setting, which is not true for many markup languages.

DITA

DITA's approach to extension is unique among markup languages. In fact, it is something of a misnomer to call DITA a markup language. The DITA standard calls it an information typing architecture. What is an information typing architecture?

The conventional way to define a document type is to use a schema language. That is how XML and SAM do it. Schema languages describe constraints on markup structures. So what does an information typing architecture provide over and above what a schema language provides?

An architecture is a set of principles for addressing a certain kind of problem. For instance, there are several different architectures for building bridges, including the beam bridge, the truss bridge, the arch bridge, and the suspension bridge. Figure 34.3 shows four basic architectural types, which you can see in hundreds of different bridges.

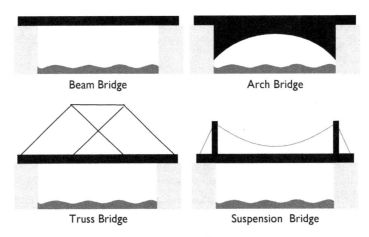

Figure 34.3 – A comparison of bridge architectures[1]

An architecture, by itself, is just a set of ideas. The virtue of an architecture is that it assembles elements in a way that can be tested and that can support the development of tools and/or prefabricated components that you can use to build specific bridges. Therefore, if you decide to build an arch bridge, you can draw on existing tools, practices, and data to make design and construction faster and more reliable.

---

[1] A combination of images, copyright © 2013 Themightyquill [https://commons.wikimedia.org/wiki/User:Themightyquill], CC BY-SA 3.0 [https://creativecommons.org/licenses/by-sa/3.0/deed.en]. Wikipedia [https://en.wikipedia.org/wiki/Bridge]

Architectures partition the design problem for a structure. They divide the design space into pieces and provide design and construction guidance within each piece. A single architecture cannot address all cases. The suspension bridge is a great solution for crossing some gaps, but a beam or truss bridge is simpler and more economical for most common gaps. An architecture does not tell you what type of bridge to build; it helps you build a particular type of bridge once you know which type you need.

At the heart of each architecture you usually find one simple idea or principle. The arch bridge depends on the strength of the arch to support the load. The truss bridge depends on the rigidity of the triangle to provide strength, and the suspension bridge uses cables to transfer weight to towers. At the heart of the DITA architecture is a similarly simple idea, which I call the block-and-map architecture. That is, DITA treats information products as being constructed of blocks of information (which it calls *topics*) according to a hierarchical map. The assemble-from-pieces reuse algorithm discussed in Chapter 13 is an example of a block-and-map architecture.[2]

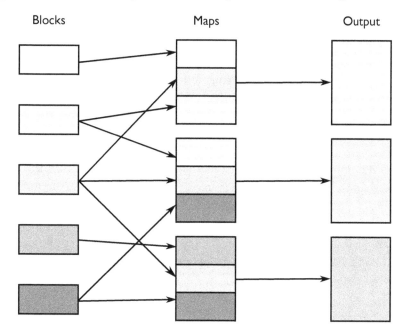

Figure 34.4 – A block-and-map architecture used to implement content reuse.

---

[2] Note that DITA is not the only block-and-map architecture, and maps are not the only way to assemble blocks.

An architecture does not solve the entire design problem for a bridge or for a content system. It provides a starting point, but you still need to design, build, integrate, and maintain the bridge or system to address a specific need. Even out-of-the-box systems require some customization to make them fit your needs. When you choose an out-of-the-box system, choose the one that comes closest to meeting your needs, so you can live with the fewest limitations and do the least customization.

If you choose an architecture, rather than an out-of-the-box system, as the basis for building a custom system, you will have more work to do, but you will be able to eliminate more limitations and partition your system to more closely meet your needs. However, the principle remains the same: start with the architecture that comes closest to meeting your needs. Choosing an existing architecture should require less work than if you designed and built from scratch, but this is only true if the architecture is a good fit for your organization. If not, trying to adapt an architecture to meet a need it was not designed for can be more work than designing and building something from scratch or from lower-level components.

Just as every out-of-the-box tool has passionate advocates who will tell you it is right for every problem, there are passionate advocates of architectures like DITA who will tell you that it is right for all content systems. But the very nature of an architecture is to make it easier to address a particular class of problems by implementing a particular kind of solution. A universal architecture would have no content. Architectures get their usefulness by being specific and limited, not by being general or all-encompassing.

Although DITA is unique in calling itself an information typing architecture, there are other content architectures. For example, Adobe FrameMaker is based on an architecture that builds a document out of a set of nested frames. Wikis are based on an architecture that connects pages to each other using references – wikiwords and categories – embedded in the pages themselves. (In other words, a wiki is a block architecture where assembly is not based on maps.) Blog platforms implement a block-based architecture based on temporal sequence supplemented by categorization and tagging.

Even though all of these examples are architectures, they have not been described or formalized as architectures outside of the tools that implement them. The only other architecture I know of that is being formalized in this way is the one I am developing myself, SPFE, which I describe briefly later in this chapter.

Of course, there is more to an architecture than just its core idea. For example, each of the bridge architectures can be divided into different sub-architectures. The Bailey Bridge (Figure 34.5) is an sub-architecture of the truss bridge architecture. Sub-architectures specify more details, which makes them easier to implement but further limits their applicability.

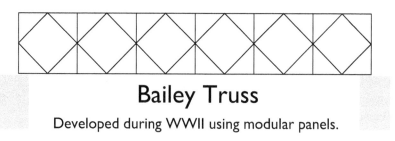

Bailey Truss

Developed during WWII using modular panels.

Figure 34.5 – The Bailey bridge architecture[3]

DITA is a sub-architecture of the block-and-map architecture, of which there are several other examples. As such, DITA comes with a lot of built-in design and implementation detail. As with the Bailey Bridge, this additional detail makes DITA easier to implement in applicable cases but more limited in its scope of applicability.

Of course, this added ease of implementation applies only if you use DITA within its scope of applicability. If you try to build something outside of that scope, you may succeed, but the further you depart from that scope, the more difficult you make the implementation.

DITA has grown into a large and complex architecture over the years, and it is out of scope for this book to describe its architecture in full.[4] Instead, I attempt to map key features of the DITA architecture to the structured writing concepts explored in this book.

---

[3] Copyright © 2012 Fmiser CC BY-SA 3.0 [https://creativecommons.org/licenses/by-sa/3.0] Wikimedia Commons [https://commons.wikimedia.org/wiki/File:Bailey-truss.svg]

[4] Obviously I am not a fan of the DITA architecture or I would not be developing my own architecture in SPFE. For a much fuller and more sympathetic treatment of the DITA architecture, see *DITA for Practitioners Volume 1: Architecture and Technology* by Elliot Kimber, http://xmlpress.net/publications/dita/practitioners-1/

Basics of the DITA architecture

The fundamental block-and-map architecture of DITA consists of these pieces:

1. A collection of schemas for a set of base block types ("topics" in DITA parlance)
2. A specification for a map mechanism for assembling blocks into information products
3. A specification for a specialization mechanism for creating specialized types of blocks from the base types and a parallel specification for writing processing algorithms to work with specialized content.

There is a great deal more in the current DITA specification, but those elements are the core of the architecture.

Therefore, information typing in DITA can be defined as creating specialized versions of the base information types using the specialization mechanism. I have already looked at several ways to do information typing, so DITA's information typing ability is not unique. What distinguishes it is the use of specialization as a tool for information typing. Like any feature of an architecture, specialization is designed to make a particular class of information typing easier.

This approach seeks to provide thought-out and tested information-type designs. This means you don't start from scratch. When you specialize from an existing DITA type, you don't have to do all the design work from scratch, you only have to design the parts that are specific to your specialization.

The limitation is that DITA's base types are not universal. They are designed to work with a specific architecture. Some applications of XML, such as recording transfers between banks or storing the configuration options of an editor, are not instances of DITA topics. DITA lets you design a range of information types more easily, but it limits the range of information types you can create.

This is the proper role of any architecture. It gives you a head start in its own domain, but it is only appropriate for use in that domain and only if the choices baked into the architecture are the choices that make sense for your application. These limitations are reflected in the way DITA defines information typing.

The DITA specification defines information typing as follows:

> Information typing is the practice of identifying types of topics, such as concept, reference, and task, to clearly distinguish between different types of information.[5]
> —Darwin Information Typing Architecture (DITA) Version 1.3 Part 3: All-Inclusive Edition

Unfortunately this definition is largely circular – information typing defines information types. But it does help establish a scale. Information typing is about defining a unit of information called a topic.

The specification goes on to define the purpose of information typing as follows:

> Information typing is a practice designed to keep documentation focused and modular, thus making it clearer to readers, easier to search and navigate, and more suitable for reuse.

DITA, then, is about information typing at a particular scale – the topic – and is designed to support particular algorithms such as reuse. DITA information typing is not as general as structured writing. That does not make it impossible to work at other scales or implement other algorithms in DITA, it just means that the architecture provides better support for certain areas.

Out-of-the-box DITA is commonly associated with the idea that there are just three information types: task, concept, and reference. The DITA specification makes it clear that this is not the intention of DITA as an information typing architecture.

> DITA currently defines a small set of well-established information types that reflects common practices in certain business domains, for example, technical communication and instruction and assessment. However, the set of possible information types is unbounded. Through the mechanism of specialization, new information types can be defined as specializations of the base topic type (<topic>) or as refinements of existing topics types, for example, <concept>, <task>, <reference>, or <learningContent>.

---

[5] http://docs.oasis-open.org/dita/dita/v1.3/csd01/part3-all-inclusive/dita-v1.3-csd01-part3-all-inclusive.html#information-typing

As I have noted, many structured writing algorithms partition complexity best when given more specific markup, particularly markup in the subject domain. The ability to create other information types is, therefore, relevant to getting the most out of structured writing.

Specialization

What is DITA specialization? XML syntax defines abstract structures that do not occur in documents: elements, attributes, etc. To create a markup language in XML, you define named elements and attributes for the structures you are creating. This is a type of specialization.

For example, in DocBook, `para` is a type of element. `para` has what is called an *is-a* relationship to elements: `para` is an element, but it is a special type of element. An XML parser will process a `para` generically as an element, reporting its name to another algorithm whose job is to interpret and act on that structure. This algorithm must supply a rule that processes just this specialized `para` element (and not the also specialized but different `title` element). All of the algorithms in this book act on structures reported to them by a parser.

DITA specialization follows the same principle, but moves it up a level. The base `topic` topic type is the abstract structure. You can create a more specific type, such as `knitting-pattern` or `ingredients-list`, as a specialization of `topic` (or of other topic types that are themselves specializations of `topic`). Each of these specialized types has an *is-a* relationship with the type it was specialized from. So `knitting-pattern` *is a* `topic`.

DITA specialization differs from simply creating new named elements in XML in that the base DITA topic type is not an abstraction like an XML element. You cannot create an XML document without inventing a set of elements and giving them names. The base DITA topic type, on the other hand, is a fully implemented topic type that you can use directly. You can, and people do, write directly in the base topic type without inventing anything new. I noted in Chapter 7 that it is sometimes easy to treat what is intended as a meta model as a generic model. This is the case here. All topic types in DITA are derived by specialization from the generic `topic` type. They all have an *is-a* relationship to this generic type.

One consequence of this is that a specialization-aware algorithm (one written to recognize and act on the specialization mechanism) can successfully process a specialized element as though it were the base element. The result may not reflect all aspects of the specialized element, but the processing will not fail. This is an attractive quality because it is often easier to modify an existing piece of code than it is to write new code from scratch, particularly if most of the rules in the base code can remain unchanged.

To specialize a topic type, you specialize the root element and any child elements or attributes that you need. Each specialized element or attribute should have an *is-a* relationship to the element it specializes. Thus a procedure element might be a specialization of an ordered-list element and its step elements might be specializations of a list-item element. In this case, processing a procedure as an ordered list would produce meaningful output: a conventionally labeled numbered list. However, you would probably want to specialize the output of steps in a procedure, perhaps by prefixing each step with "Step 1:" rather than just "1." Modifying the ordered-list code to handle procedures would probably take less code than writing a procedure processing algorithm from scratch.

The second way in which DITA specialization differs from giving names to abstract elements is that specialization is recursive. For example, suppose you have created a topic type called `animal-description`, which is a specialization of `topic`. You want to impose additional constraints on the description of certain types of animal, so you create `fish-description` and `mammal-description`, which are specializations of `animal-description` (and which would be processed like an `animal-description` if no other processing is specified for them).

Then you might decide to impose still more constraints to describe different kinds of mammals, so you create the type `horse-description`, which is a specialization of `mammal-description`. This type would be processed as a `mammal-description` if no specific processing is provided for `horse-description`; as `animal-description` if no specific `mammal-description` processing is provided; and as `topic` if no specific `animal-description` processing is provided.

The third way in which information typing in DITA differs from creating a language from scratch is that DITA information types share a common approach to processing and to information architecture. In particular, they inherit a common set of management-domain structures and their associated management semantics.

A DITA topic, then, comes with a lot of built-in functionality and structure that you don't have to reinvent when you create a specialized topic type. But the corollary is that not all content types are specialized instances of a DITA generic topic. This means that while the set of information types that you can create using specialization is unbounded in number, it is not unbounded in type. There are types you can't create by specialization, at least if you want to maintain the *is-a* relationship to the base type (without which specialization isn't specialization at all).

When we factored the ingredients of a recipe out of a document-domain list into a subject-domain record structure it was, in part, to make them independent of the decision to format them as a table or as a list.

```
ingredients:: ingredient, quantity, unit
    eggs, 3, each
    salt, 1, tsp
    butter, .5, cup
```

Figure 34.6 – Subject-domain ingredients list

The ingredient record set in Figure 34.6 no longer has an *is-a* relationship to a table or a list. The point was to break that relationship so you could present this content any way you wanted to. Because any structure specialized from a DITA topic should maintain an *is-a* relationship to a DITA topic, which is a generic document-domain structure, you can use specialization to create subject-domain structures that intersect with the document domain, such as the introduction, ingredients, and preparation sections in a recipe document. However, you can't take the next step of factoring out the presentation, as I've done in Figure 34.6, since factoring out presentation breaks the *is-a* relationship with the document domain.

In other words, the specialization mechanism lets you create any element name you like as a specialization of any other element name you like. As such, it can create any structure you like. But the entire justification of the specialization mechanism rests on the maintenance of the *is-a* relationship between the specialized element and the base element. Once this is broken, code inheritance and fallback processing no longer work, and specialization simply becomes an unnecessarily complex and error prone way of writing a new and unrelated schema.

Labels are a big part of document-domain content. When we created a subject-domain structure for recording nutritional information for a recipe, we factored out all of the labels by putting the content in named fields. In other words, as shown in Figure 34.7, the labels went from being data (in the content) to metadata (part of the structure).

```
nutrition:
    serving: 1 large (50 g)
    calories: 78
    total-fat: 5 g
    saturated-fat: 0.7 g
    polyunsaturated-fat: 0.7 g
    monounsaturated-fat: 2 g
    cholesterol: 186.5 mg
    sodium: 62 mg
    potassium: 63 mg
    total-carbohydrate: 0.6 g
    dietary-fiber: 0 g
    sugar: 0.6 g
    protein: 6 g
```

Figure 34.7 – Subject-domain structure for nutritional information

Although the structure in Figure 34.7 looks superficially like a list, it is really a data record. If this structure were created as a specialization of a list and then published using a generic list publishing algorithm, the result would look like Figure 34.8.

- 1 large (50 g)
- 78
- 5 g
- 0.7 g
- 0.7 g
- 2 g
- 186.5 mg
- 62 mg
- 63 mg
- 0.6 g
- 0 g
- 0.6 g
- 6 g

Figure 34.8 – Generic rendering of the subject-domain structure in Figure 34.7

In short, there is no *is-a* relationship between a pure subject-domain structure and a generic document-domain structure. The subject-domain structures are simply data fields with no presumption about presentation attached to them.

Does this mean that you cannot create a subject-domain structure using DITA's specialization mechanism? No, you can usually create the structure you want, and in a way that meets the syntax requirements of a DITA specialization. However, it won't be an actual specialization of its base type. Because it is a subject-domain structure, it breaks the *is-a* relationship with the document domain. As I noted at the end of Chapter 18, subject-domain algorithms work completely differently from those of the document domain, so inheriting some of the processing of a base document-domain structure is moot once you move to the subject domain. Your new subject-domain structure will be an essentially unrelated structure for which you must create completely new processing algorithms, just as you would if you had defined your structure from scratch.

Generally, the fewer pieces of an architecture you use, the less value there is to basing your work on that architecture, both because you have more work to do and because you take less advantage of the infrastructure, tools, and expertise surrounding that architecture. If you get too far away from the intention of architecture, you will create a system that is less understandable to people versed in the architecture. All this betrays a poor fit between the system partitioning you need and the partitioning the architecture provides. All architectures come with overhead, and even if you don't use their features, you have to live with the overhead, which adds cost and complexity to your system. Thus, while you can use DITA and depart from the default DITA way of doing things, the value of using DITA diminishes the further you depart from the DITA way. The same is true of any information typing architecture, just as it is true of any content development system.

Limits on rhetorical constraint

Throughout this book I have stressed the value of constraining rhetoric, not only to improve rhetoric itself, but also to improve process. One of the limits imposed by DITA's topic and map architecture is that it limits the size of unit to which you can apply rhetorical constraints.

For instance, if you regard a recipe as being made up of three topics – one concept, one reference, and one task – DITA lets you constrain the rhetoric of a specialized ingredient list reference topic, or a preparation task topic, but not of the recipe as a whole. DITA lets you write a map to combine a concept topic containing an introduction, a reference topic containing a list of ingredients, and a task topic containing preparation instructions. But it does not let you specify that a recipe topic consists of one concept topic, one reference topic, and one task topic in a particular order. In other words, DITA does not provide any direct way to define larger types or the overall rhetorical structure of documents.[6]

You can define a recipe topic type in DITA, but to do so you need to adopt a different view about how atomic a DITA topic is. Some DITA practitioners might say that a recipe is not a map made up of three information types, but a single task topic. In this view, a task topic is much more than what Information Mapping would call a procedure. It allows for the introduction of a task, a list of requirements, and the procedure steps all within the definition of a single topic. However, this approach abandons DITA's principle of segregating information types into separate files.[7]

One of the reasons for this uncertainty about how to define an atomic topic in DITA is DITA's focus on content reuse. DITA topics are not only units of information typing, they are units of reuse. Making a recipe a single topic leaves you with fewer, larger units of content, which makes individual topics harder to reuse. The atomic unit of content that is small enough to maximize potential reuse is much smaller than the atomic unit of content that contains a complete rhetorical pattern. The atomic unit of reuse is smaller than the atomic unit of use.

Because DITA has no direct mechanism for describing models larger than a topic, a DITA practitioner must choose between modeling for maximum reuse and modeling to constrain a topic type to rhetorical structure. In practice, DITA users make different decisions about how atomic their topic types should be based on their business needs.

SPFE

SPFE is another project of mine. It is an architecture for implementing structured writing algorithms. Its structure follows the model I laid out in Chapter 19. Every structured writing publishing chain uses a similar model, since every publishing chain has to get content from the subject domain or the document domain into the media domain. But how that processing is partitioned is key to which structured writing algorithms you can implement in the publishing process and how cleanly you can insert them. SPFE specifies a division of responsibilities in a publishing tool

---

[6] Actually, it might be technically possible to create a map that was constrained in this way through specialization. That is, you could create a recipe map that is allowed to contain exactly one recipe-intro topic, one ingredients-list topic, and one preparation topic in that order. But there is no higher level way of specifying this kind of constrained map and the functional lucidity of such an approach is clearly low. But the real point here is not the technical limitation but the limits of the block-and-map architecture itself to support effective rhetorical constraint.

[7] I have asked a number of DITA practitioners how a recipe should be modeled in DITA and have received each answer from multiple people.

chain that is intended to allow for the efficient partitioning of the content process and the development of structured writing algorithms.

However, that is a little too general a description. As I said, no architecture can be optimized for all cases. The SPFE architecture, just like the DITA architecture, is optimized to favor certain algorithms and certain information designs. For instance, DITA is optimized for the reuse algorithm and hierarchical information architectures, while SPFE is optimized for the linking algorithm and a hypertext or bottom-up information architecture. This does not mean you can't do reuse in SPFE or linking in DITA. It simply means that these things are at the heart of the design of each. Architectures are just a starting point for building systems. They don't determine the features of a specific system, they provide guidance and tools for building systems of a particular type.

SPFE and DITA are also optimized for working in different structured writing domains. DITA is based in the document domain, since its base topic types are document domain types, and like all document-domain systems, it makes heavy use of management-domain structures. SPFE does not include any base types in the architecture, so it is not tied to one domain in the same way DITA is, but it is intended mainly to support the subject domain. (The subject domain requires an additional processing stage compared to document-domain systems and the SPFE architecture supports that additional stage.) Since the SPFE architecture is not based on any particular content model, it supports pulling in content from diverse sources and source formats.

Another big contrast between DITA and SPFE is that DITA is a block-and-map architecture, whereas SPFE is designed to support automatic collection and linking of content based on metadata, meaning that no maps are required. Of course, this means that information architects or content engineers have to write the algorithms to do the desired collection and linking for a specific information set, as opposed to using a generic mechanism as in DITA. The SPFE architecture is designed to support this work.

One of the features of the subject domain is that it allows you to greatly reduce the amount of management domain features you need. Generating information architecture from metadata rather than specifying it with maps, means writers don't need to know how to link or assemble the pieces they write. Together, these features mean that writers don't need to know much if anything about the publishing system. While writers working in DITA typically require extensive DITA training, individual writers working in a subject-domain SPFE system need to know little or nothing about how SPFE works, as long as they follow the constraints of the markup language they are using.

I have noted that the DITA architecture allows some ambiguity about whether a DITA topic is a rhetorical block or merely a semantic block and that DITA cannot constrain any unit larger than a topic. However, the real source of this problem for DITA is the block-and-map architecture. SPFE places no inherent constraint on the size of units processed by a SPFE system, so it does not place limits on the rhetorical constraints you can apply. However, if you try to implement block-and-map reuse below the level of a rhetorical block with any system, including SPFE, you will encounter the same problem. There are forms of reuse, particularly in the subject domain, that avoid these issues, though they are not as general as block-and-map reuse in the document domain.

SPFE does not define a base set of content structures the way DITA does, nor does it develop information types by specialization the way DITA does. In SPFE, you can do information typing from scratch, as long as you follow certain guidelines. But SPFE also supports a higher level of information typing that uses a technique called composition.

As with DITA specialization, SPFE composition includes both content structures and the code that processes them. But rather than creating one topic type from another by specialization, SPFE lets you assemble an information type, of any scale, by assembling existing definitions of semantic blocks and their associated algorithms, like building a model out of Lego blocks. Because a semantic block can also serve as a structural block in assembling a larger semantic block, this process also works iteratively. The result is that you can take advantage of existing design work and coding in the creation of your information types without the restrictions on size and type that come with DITA specialization.

To create a subject-domain markup language in SPFE, therefore, you define the key subject-domain fields and blocks that are essential to your business. You can include all the other semantic blocks you need, such as paragraphs, lists, tables, and common annotations, using pre-built components, along with their default processing code.

By strictly segregating the presentation and formatting layers, SPFE reduces the effort required to process custom markup formats. Custom formats are processed to a common document-domain markup language which is then processed to all required media-domain output formats. The SPFE Open Tool Kit includes a basic document-domain language for this purpose, but you can also use DocBook or DITA in this role, allowing you to take advantage of their existing publishing capabilities. This also allows you to install SPFE as an authoring layer on top of an existing DITA or DocBook tool chain.

I developed SPFE because I am not a fan of the block-and-map architecture, of which DITA is the most widely known example today. Block-and-map architectures provide good support for ad hoc reuse, but at the expense of poor support for many other structured writing algorithms, for the subject domain, and for functional lucidity. I believe that many organizations would benefit from a different partitioning of their content systems supported by a different architecture. But picking a architecture is all about meeting your business goals efficiently and if extensive ad hoc reuse is the overwhelming driver of your content system redesign, then a block-and-map architecture like DITA (but not just DITA) is something you should look at.

Both SAM and XML are supported as markup syntax for SPFE, and you can freely mix and match SAM and XML content.

SPFE is an open source project available from https://github.com/mbakeranalecta/spfe-open-toolkit.

CHAPTER 35
Constraint Languages

Structured writing is about applying constraints to content and recording the constraints that the content follows, both to constrain what writers write and to constrain how algorithms interpret the content. This requires some way to express constraints in a formal and machine readable way. Schema languages partition this problem and redirect it to a common validation algorithm expressed by a standard piece of software that everyone can use. Schema languages are, quite simply, languages for expressing constraints.

For concrete markup languages, such as Markdown, the constraints are established in the code of the processor. A Markdown file is validated when the processor parses that file. (In practice, though, Markdown does no meaningful validation. Anything it does not recognize as markup, it simply outputs as text.)

For abstract markup languages, such as XML, you define structures yourself. Basic XML syntax is validated by the parser, but the definition of constraints is the business of a schema language. The validation of those constraints is the business of a piece of software called a *validator*.

A schema language is a structured language for defining structured languages. The schema for a markup language says what structures are allowed and in what order and relationship. A given document either conforms to those constraints or it does not.

Figure 35.1 contains a schema fragment in a schema language called Relax NG, which is one of several schema languages available for defining XML-based markup languages.

```
<element name="book" xmlns="http://relaxng.org/ns/structure/1.0">
   <oneOrMore>
      <element name="page">
         <text/>
      </element>
   </oneOrMore>
</element>
```

Figure 35.1 – Relax NG schema language

Figure 35.1 defines two elements and three constraints. The first element is called book and the second is called page. The constraints are:

■ The page element must occur inside the book element. (Because the page element is defined inside the book element structure.)

■ There must be at least one page element inside the book element, and there can be more. (Because the page element is defined inside a Relax NG oneOrMore element.)

■ Text can occur inside the page element, but not directly inside the book element. (Because the Relax NG text element occurs inside the definition of the page element, but not as a direct child of the book element definition.)

Thus, if you write the passage in Figure 35.2, a validator will report an error because the words "Moby Dick" are directly inside the book element and text is not allowed in that position.

```
<book>Moby Dick
    <page>Call me Ishmael. Some years ago- never mind how long
    precisely- having little or no money in my purse, and nothing
    particular to interest me on shore, I thought I would sail
    about a little and see the watery part of the world.</page>
</book>
```
Figure 35.2 – Passage from Moby Dick, marked up in XML

There are several different schema languages for XML, each of which is capable of expressing and enforcing different sets of constraints. It is not unusual to combine different schema languages to more completely constrain a markup language. In particular, many schemas, including the DocBook schema, use a schema language called Schematron in concert with another schema language, such as Relax NG or XSD.

While most schema languages work by modeling the structure of a document, as in the Relax NG example (Figure 35.1), Schematron works by making assertions about the structure in a language called XPath. A Schematron schema would make a lousy guide for authoring, but it uses assertions to enforce constraints that no other schema language can.

Figure 35.3 is a simple Schematron example that defines one of the constraints listed above, namely that a book element must contain a page 240, meaning it must contain at least 240 pages.

```
<schema xmlns="http://purl.oclc.org/dsdl/schematron"
 queryBinding="xslt2">
    <pattern>
        <title>Book constraint</title>
        <rule context="book">
            <assert test="page[240]">A book must contain at
            least 240 pages.</assert>
        </rule>
    </pattern>
</schema>
```

Figure 35.3 – Schematron rule that constrains book **elements to contain at least 240** page **elements**

The rule says that in the context of the element book, the assertion that there is an element page that is number 240 in the sequence of pages must be true. In other words, there must be at least 240 pages in a book. If not, the message in the assert element is displayed.

In the Relax NG schema example (Figure 35.1), the schema is essentially a template that models the hierarchy of the document. The constraints are consequences of the structure of that hierarchy. For example, the book element cannot contain text because there is no place for text in that part of the hierarchy. Only documents that conform to the template are valid.

In the Schematron example (Figure 35.3), the schema is a series of constraint statements. The hierarchy of the document is a consequence of meeting all the constraints. Any document is valid as long as it conforms to all the stated constraints.

There are two ways to describe constraints:

- Nothing is allowed unless a specific rule allows it.
- Everything is allowed unless a specific rule forbids it.

Schematron is based on the latter doctrine. It says anything is valid as long as it passes a given set of tests. Other schema languages take the former approach. For example, if you validate an XML document against an empty XSD schema, the validator will say the document is invalid because it can't find the definition of the root element. An XSD or Relax NG schema describes constraints as a set of permissions. The writer is constrained to stay within the boundaries of what is permitted.

Starting with the doctrine that everything is allowed is not a good model for structured writing. To write algorithms that can handle a document, you have to know what elements and attributes are allowed and in what combinations. This is why Schematron is seldom used as a standalone schema language. It is more frequently used to express constraints that cannot be expressed in the main schema language.

The wider point here is that in the course of writing and publishing content you often need to express and enforce constraints. Some constraints – for example resolving key references – can be expressed only during the publishing process. You can match a key to a value or resource only after the synthesis algorithm is complete and all the elements of the published content set have been assembled and resolved.

Constraints, therefore, do not exist at just a single point in the writing and publishing process. You must consider them across the entire publishing process. In particular, when an error occurs in the publishing process, trace it back to where the constraint that could have prevented it was violated and ask how you could better implement or enforce that constraint.

Two of the most basic design principles of any system are to catch errors as early as possible and to design any error-prone process to make errors less likely. The earlier you catch an error, the less work you waste before the error is detected. And if you can remove the possibility of the error occurring at all, you can avoid all the costs of detecting and fixing the error as well as the greater costs of not catching the error before the content is in the hands of the reader.

Schema languages are excellent tools for expressing constraints in structured writing. However, in partitioning your content system you also need to determine where each constraint should be enforced. The content process is about getting ideas from the writer's head to dots on a page or screen. Therefore, you need to enforce constraints and catch errors as close as you can to the beginning of the process: the ideas in the writer's head. Rhetorical constraints can play a large role in improving not just quality but also process, because they impose constraints and catch errors nearer to ideas in the writer's head, rather than trying to catch them as dots on the page.

This means moving your content closer to the subject domain. When you sit down to write a schema, you should not simply think about how to express constraints; you should also think about whether you can factor them out, possibly by moving the content to a different domain. Subject-domain structures tend to be simpler and stricter than document- and media-domain structures. This means that it is easier to write schemas for the subject domain than for the other domains. Schema languages are powerful tools, but they should be used in conjunction with a correct partitioning of the content system.

Management

Structured writing is a tool for partitioning the content development process. Now that we have looked at the fundamentals of structured writing – algorithms and structures – it is time to look at how to use these algorithms and structures to partition and manage the content creation system.

CHAPTER 36

Content Management

Managing content assets is fundamental to managing your content process. Any large collection of content needs to be managed, and when you manage the work of many contributors to produce an integrated information architecture, managing content assets becomes key.

To make content management decisions, you must be able to locate and examine the content. These functions can be performed by algorithms if you attach sufficient and accurate metadata to your content. Attaching metadata to content to allow algorithms to make decisions is, of course, precisely what you do in all forms of structured writing. Therefore, content management is a structured writing process. In this section, I add a set of management algorithms to the content manipulation algorithms described in Part III. Like the other structured writing algorithms, these algorithms work differently depending on which domain your content is written in.

But what if your content does not contain sufficient metadata to allow management tasks to be performed by algorithm – or even by hand? In that case you can use a content management system (CMS) to attach additional metadata to the content as external labels. Most commercial CMSs support this capability.

Of course, you can choose to manage your content by hand. This means that everyone who has a role in managing content – that is, everyone on your content team – must have the knowledge, skills, and time needed to perform the management tasks. Because these conditions are hard to meet with perfect consistency, content management complexity can fall through the cracks, resulting, as unmanaged content complexity always does, in inferior rhetoric and inefficient processes. But managing content by hand does avoid introducing a lot of complex tools into your system. Introducing tools is never neutral. New tools always bring complexity and unless they provide or enable a better partitioning of complexity, they can cause more problems than they solve.

Some content management systems attempt to be all encompassing – to be the only tool anyone in the organization ever uses to create and deliver content. These often involve forcing all users to use a single source format for all content. Others are designed to work in concert with other tools. Some are more frameworks than tools – platforms on which you can construct your own content management functionality. All of them implement some form of structured writing for at least some functions in the media and document domains, with some from the management domain mixed in. Some work with or, at least, support certain management functions for the subject domain. Some content management systems attempt to manage the entire content creation process, from design and authoring to workflow and publishing. Others focus on more limited aspects of the process. Some focus on a single media platform (usually the web), while others support multiple media.

A CMS does exactly what I describe throughout this book: it partitions and distributes the tasks of the content process. Different CMSs partition tasks differently, and there is, of course, no guarantee that a particular system's partitioning is the best fit for your organization or that it will direct all complexity to people or processes that have the skills, knowledge, and resources to handle it. And if not, unhandled process complexity will result in poor rhetorical quality.

As tempting as it may be to simply go shopping for a content management system, buying a CMS means buying a complete process and partitioning, which may or may not be the best fit for your organization. A better approach is to design the partitioning of your system first and then look for tools that fit the process you have designed. Depending on how you partition your processes and how and where you direct complexity, you may not need a conventional CMS at all.

Many of the decisions you have to make in a large content system involve examining large volumes of content. Thus the main interface to many content management systems is a file system or database view of the content repository. Whatever specific tasks the system is performing, the essence of its interface is that it allows you to view and apply metadata to large volumes of content. However, if your content already contains the metadata needed to make these decisions, the need for such an interface diminishes (thought the need for the algorithms themselves remains).

Structured writing and content management systems both work to make metadata available to algorithms. But the algorithms work differently – and require a different class of metadata – in each structured writing domain. Content management systems tend to supply subject-domain and management-domain metadata for algorithms that process document-domain and media-domain content. But subject-domain content supplies the subject-domain metadata itself, and, as we have seen, you can often factor out the need for management-domain metadata by moving content to the subject domain. Just as a single set of subject-domain structures can often serve the needs of multiple content manipulation algorithms, the same set can often also serve the needs of multiple management algorithms.

This does not mean that if your content is in the subject domain you will never need a CMS, but it does mean that you may not need one or that you may require a less elaborate system. Certainly it means that if you do need a CMS, you should choose one that partitions tasks in a manner that supports the subject domain.

If your content does not contain the metadata necessary for management, the CMS must gather and store it separately. This adds complexity to the CMS interface, which can be a major source of pain for users. For CMSs that are designed to be the only tool you use, this generally means that the structured writing format is baked in and can only be changed in limited ways, if at all. Transferring metadata from the CMS to the structures of your content would require a major reconfiguration of the CMS. If you buy this kind of system, you need to look at the total picture – how the CMS partitions and redistributes all of the complexity of the content systems and what complexity, if any, it neglects and lets fall through. There will likely be little you can do to change the partitioning or distribution after the fact, so make sure it is what you want going in.

Most off-the-shelf content management systems are designed for media-domain or basic document-domain content. This makes sense from a commercial point of view because it allows CMS vendors to develop their own metadata scheme and management algorithms independent of the content that will be stored. This means a vendor can sell its CMS to a wider variety of clients and advertise that it has simple editors or works with the editors that people already have.

The problem with this model – in terms of developing a comprehensive solution for managing complexity across the content system – is that it draws a hard line between the management tasks and constraints supported by the CMS and any management tasks or constraints associated with the rhetoric of your content. This means that algorithms that depend on the consistency of content or its relationships with other content are largely unsupported by the CMS, and there is no integration between those algorithms and the algorithms provided by the CMS. Due to this lack of integration, complexity gets dropped, with the usual consequences for rhetoric and for the reader.

Other CMSs are built for more complex document-domain languages. These CMSs typically support management-domain features specifically for those languages and are sold as such: a DITA CMS or an S1000D CMS. The most common systems of this type today are based on DITA.

You might expect that a CMS would primarily record management-domain metadata. After all, the management domain is an intrusion into the structured writing world, since it does not actually describe the structure of content. The reason for the intrusion of the management domain into content is to allow for the management of the content below the level of whatever file or chunk size you store in the CMS.[1]

But while you will rarely find much media-domain or document-domain metadata stored at the CMS level, CMSs often contain a great deal of subject-domain metadata. If you manage a large volume of content, you will need some way to find content on a particular subject. If you are doing content reuse, for example, writers will constantly be asking if content already exists on the subject they are preparing to write about. If your CMS is managing the delivery of content dynamically to the web, it will need to respond to queries based on subject matter. And if you are optimizing your content for search you will need to provide the search engine with subject metadata in the form of keywords or microformats. All of this depends on subject-domain metadata. Subject-domain metadata is therefore central to CMS operations. This is also why the use of subject-domain structures in your content can lessen your reliance on content management systems.

To figure out what type of content management system you may need, if any, it helps to understand how content management systems work with metadata and where the metadata in question resides in your content. Essentially structured writing and content management systems both work to make metadata available to algorithms, so structured writing and content management systems are both working towards the same goal, and depending on your process you may want to assign more of that task to one or to the other.

The location of metadata

Content management system do their job largely through the collection and management of metadata. Partitioning and redirecting complexity requires a method to pass information between partitions in a reliable way, and that is what structure and metadata do. Metadata provides a record of the identity and status of content. Management actions are actions on metadata: either creating and updating metadata or performing actions (running algorithms) based on metadata.

The location of the metadata that records the identity and status of the content and the constraints it obeys differs from one structured writing domain to another. The media domain captures virtually no metadata that is useful for content management, the document domain captures some,

---

[1] In some CMSs, this distinction between the chunk stored in the CMS and the structures expressed inside that chunk is moot. A CMS based on a native XML database, for instance, makes no distinction between the chunk and the structure of the chunk, but treats the entire repository as a single XML resource that it can query and manage down to any level of granularity. Even with such a system, however, this distinction remains for writers, who have to deal with the structure of whatever sized chunks of content they are asked to write.

but not enough, and the subject domain often captures almost everything you need except, perhaps, for workflow information.[2]

It is a common pattern for a CMS to store document-domain or media-domain content and attach subject-domain metadata to it as an external label. For instance, a CMS might store recipes written in Markdown and attach separate metadata records to each recipe listing the key recipe metadata needed for retrieval and sorting of recipes. One of the things that writers often complain about with CMS systems is that they are not allowed to submit content to the system without filling out complicated metadata records.

An alternative is to write recipes in a subject-domain format in which all the recipe metadata is included in the content from the beginning. The CMS then requires no external metadata labels, though it does require a way to access and query the embedded metadata. (CMSs based on XML databases often have this capability as a natural consequence of the XML database architecture.)

Which approach is preferable? The conventional CMS approach arises because most CMSs are based on relational databases, which are good at storing metadata records and attaching them to blobs of text but are not good at storing or querying the hierarchical structure of structured content. The conventional approach has several disadvantages, all of which introduce complexity that is often not handled well. The disadvantages of this approach include the following:

1. It can only record the characteristics of a chunk of content as a whole. It cannot look down into the content to find more fine-grained metadata. One of the advantages of writing a recipe in the subject domain is that it allows you to do things like querying the collection of recipes for all those with a calorie count below 100. But unless the metadata record for the recipe includes that level of detail, the CMS cannot respond to that query. And if you want to store that level of detail, you are effectively asking the writer to write the entire content twice, once in the document-domain document and once in the subject-domain metadata label. This requires more work, and the two versions are likely to fall out of sync.

2. It provides no support for subject-domain validation of the content, which means it does nothing to help improve content quality. Requiring document-domain content as the storage format precludes the use of the subject domain for writing and cuts writers off from all the advantages the subject domain provides.

---

[2] Workflow information is management-domain data, and there is nothing to prevent you from adding workflow information to document-domain or subject-domain content if you want to.

3. The system has no way of telling if the content conforms to its constraints. It records the content constraints in a separate record without ever validating that the content meets them.

4. It separates the metadata from the content it describes. This allows for drift between the content and the metadata.

Storing metadata in the content presents challenges, too. Having each piece of content stored in the subject domain makes sense from a rhetorical point of view and makes it easy to submit content, since writers don't need to fill out additional metadata forms. The problem is how to retrieve the metadata.

A CMS is essentially a database, and the way you retrieve information from a database is to write a query. A query is different from a search. A search is fuzzy. A search engine takes a plain-text question or search phrase and tries to figure out which documents are the best match. Search engines may be powerful and sophisticated, but their results are a sophisticated, mechanical guess, and sometimes they get it wrong. Ask a search engine for a list of recipes with less than 100 calories, and it will give you a bunch of guesses based mostly on the plain text of those documents. Chances are it will catch some, miss others, and give you some false hits.

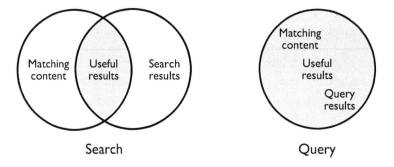

Figure 36.1 – Search results vs query results

A query, on the other hand, is a request for items whose metadata precisely matches specified criteria. A query to return recipes for which the field `recipe/nutrition/calories` has a value less than 100 returns all the results, misses none, and gives you no false hits. However, it works only for content that contains this field, and to write this query, you need to know exactly how the system stores recipes (Figure 36.1 illustrates the difference between a search and a query).

If you have many different content structures in your repository, you need to know how each of them is structured in order to create queries that return the correct results. Although this is a complicating factor, it is not the end of the world. Information architects and content engineers can create and save queries that writers can run as needed, saving them from having to remember the details of every structure.

In the end there is no way around this. Accurate and reliable queries depend on precise and consistent metadata that is specific to the object it belongs to. There is no such thing as a generic metadata record. They are always specific to the things they describe. Subject-domain metadata is specific to its subject. If you want to be able to find all recipes with calorie counts less than 100, you need recipe-specific metadata that records the number of calories in the recipe. If you want to find a used car listing for a blue convertible, you need metadata that specifically records the car color and body style. There is no generic metadata format that supports both of these queries. The inherent variability of content means you cannot create a generic query system. The trick is to find the right balance and make your content as easy to query as possible.

Hybrid approaches

There is an alternative to the two approaches described above, which is a hybrid of document and database. In the subject domain, you can create fielded data that an algorithm can use to create content. The relational database on which most CMSs are based is great at managing fielded data, and this allows you to create a hybrid system in which the fielded data parts of a subject-domain document type are stored as database fields and the narrative content is stored in text fields, using document-domain markup such as markdown or HTML. Hybrid systems are common in custom web CMS solutions, where they are not looked on as hybrids so much as logical extensions of the database model.

The advantage of this model is that writers no longer need to add metadata after the fact, because it is part of their normal authoring environment, which is essentially a database form. This is a perfectly legitimate implementation of subject-domain structured writing, since there is no requirement that structures have to be expressed and captured as markup. Nonetheless, many content strategists have come to realize the limitations of this model, which lacks any constraints on the content of the narrative fields, thus making it hard to create the consistent rhetoric that content strategy tries to foster.

Alternate repositories

At the heart of any content management solution is a repository: a place to put the content assets you are managing. The simplest form of repository is a file system, either local or shared on a network. The amount of metadata you can associate with content on a file system is limited. Some people use a separate metadata store, such as a spreadsheet, but there are limits to how far such a system can scale, and it is hard to apply content management algorithms in such an environment.

Virtually every full-fledged CMS comes with and is built around a repository that is set up to capture the metadata used for content management. Having the repository baked into the CMS provides obvious integration benefits for the CMS itself. However, this means that access to your content by any algorithm not provided by the CMS itself comes only by permission of the CMS. Many systems do not make access to their content easy for outside processes. This is in part because vendors want to present their systems as providing everything you need, but also in part to preserve the integrity of the repository, which could be corrupted by a poorly written algorithm.

Even if you don't need the services of a full-fledged CMS, or you simply can't find one that supports the partitioning you want for your content system, you may still need two content management features for your content repository: access control and version control. Both of these services can be provided by the version control systems used by software developers.

Access control gives you the ability to permit or deny access to resources on an individual or group basis. This improves the security of your system by helping to prevent deliberate or unintentional damage to your content by unauthorized individuals.

A version control system (VCS) stores all of the changes writers make to content, allowing them to find and restore previous versions. Most version control systems allow you to maintain content in an ordinary file system, which means that tools and algorithms do not have to know anything about the VCS in order to work on the content. The VCS stores all changes that are made to files under its control, storing each iteration along with labels and commentary that allow you to find any version you might want to go back to.

VCSs also allow you to back up content, along with its version metadata, to a central server that can be accessed by many different contributors. Each contributor maintains a local copy of the files and has the ability to upload files to the central server, where they can be pulled down to the workstations of other contributors, thus ensuring that everyone has access to the latest content.

VCSs have various mechanisms for preventing people from accidentally overwriting each other's work, including the ability to create a branch of the repository where you can make changes independent of other contributors and merge those changes into the mainline of the repository when they are ready for release.

A VCS cannot match the complex facilities of a typical CMS, but they are used successfully for very large software projects involving hundreds of developers. How is it possible to support collaboration on this scale with such simple repository features? The key is that code is by its nature highly structured text to which all of the same algorithms I have described for structured writing can be applied.

Software code can be strongly partitioned to minimize the collaborative overhead, and a rich set of tools is available to ensure conformance and to perform audits to make sure that code is complete and conforms to its constraints. In short, software projects do not need the kind of services provided by a typical CMS on the content side, because the code itself contains the metadata necessary to perform these same functions with external algorithms and because developers enforce the right kind of discipline among contributors. The open nature of a VCS is ideal for software development because it does not impose any barriers to implementing new processes or new algorithms to manage the content.

There is a movement within the documentation community called Docs as Code[3], which advocates using the same management processes and tools for content as developers use for code, including VCSs and scripted content builds, rather than conventional CMS solutions.

Multiple repositories

As I have noted before, it is often preferable to be able to draw content from multiple sources and this often means multiple repositories. It is, of course, possible to store content in many different source formats in a single repository (unless that repository is designed to only work with one format) but in many cases it will also make sense to store content in different formats in different repositories, particularly if those repositories also serve different purposes for different groups.

Actually, even talking about multiple versus single repositories can be misleading. As far as your algorithms are concerned, if there is one interface they can query to retrieve content, they are dealing with one repository. Behind that interface there may be different systems using different

---

[3] http://www.writethedocs.org/guide/docs-as-code/

internal formats and residing in different physical locations. The web is a perfect example of such a system. You can retrieve any resource on the web by issuing an HTTP request. that one request interface can retrieve all kinds of different resources from all kinds of different systems but the web architecture, based on HTTP, makes the entire web look like one repository.

The digital world is not like the physical world. You don't need to put all you stuff in the same cupboard in order to organize and retrieve it. You can build a interface to any data source you care about and treat all your diverse sources as one repository to feed your shared publishing infrastructure.

Making content manageable

If you want to manage your content, you need to make your content easy to manage. The danger is that if you start with the principal aim of making your content manageable, you may find your management goals at odds with your rhetorical goals. It is easier to manage content (or anything else) if its structure is uniform. The constraints you impose to make content more manageable make it more uniform. Thus a system, such as DITA, which starts by proposing just three types of content (concept, task, and reference), has an obvious appeal from a management point of view. However, remember that it is your rhetorical goals that serve your readers. Your management goals should be subservient to your rhetorical goals.

The constraints that you impose to improve content quality are those which make sure that a piece of content does just the job it is supposed to to. Such constraints ensure that a recipe contains everything a recipe needs and is presented in the way a recipe should be presented. They are specific to the subject matter and the audience. Three generic content types cannot provide all the constraints you need to effectively manage content quality. Indeed, some of the constraints that are designed to facilitate content management may be positively damaging to content quality.

If improved content quality was not one of your business goals, you might naturally choose uniformity in implementing a CMS. However, I would argue that a system based on subject-domain content can lead to better content management in the long run.

In any system that relies on constraints, on data that is known to meet certain constraints, you need to make sure that the constraints are actually being met. As we have seen, it is often easier to provide effective guidance and perform effective validation in the subject domain. Also, the subject domain allows you to factor out many constraints, which is the most effective way of making sure they are obeyed. The document domain provides far fewer opportunities for factoring out constraints and providing effective guidance. In addition, it is much more difficult to audit constraints correctly in the document domain.

Thus, while a simple document-domain system of concept, task, and reference topics meets the content manager's desire for uniformity, it provides little opportunity for ensuring that the full range of constraints necessary to make content management and reuse work are actually followed. The result can be a deterioration in the quality of the content set over time, a process that tends to be self-perpetuating because disorder in current content makes it harder to impose order on new content, just as you cannot put things away neatly in a messy drawer.

The variety of constraints and formats found in a subject-domain system may present a greater content management challenge initially, but it can go a long way toward ensuring that the necessary constraints are met. And, as we have seen, you can often use subject-domain structures to factor out management-domain concerns (see Chapter 5), which can go a long way toward removing the conflict between rhetorical structures and management structures in content. This not only leads to more effective management, but also to a simpler writing experience.

The key to conformance is to create structures that are easy to conform to. The content management algorithms rely heavily on conformance to constraints and structures. To manage content successfully, you must know exactly what assets you are managing. The more you know about each asset, and the more you can rely on what you know, the more confidently you can manage those assets and the less likely it is that your management system will descend into chaos.

This means that the content management algorithms depend on content that is easy to manage. It may seem like generic units of content would be the easiest to manage, but the problem with such generic units is that you know very little about them, and what you do know is often unreliable. Generic units may be easy to create and easy to store, but they are not easy to manage. It may require more initial thought and planning to manage highly specific, well-constrained content units, but such units will be the easiest to manage over time, especially as your content set grows.

Managing structured writing assets

Finally, consider that structured writing requires you to record content in media-domain, document-domain, or subject-domain structures, factor out invariants into separate files, express constraints, and create algorithms that translate the content to the media domain for publishing. All of this creates artifacts that you need to track and requires processes both for keeping track of artifacts and for running the structured authoring and publishing tool chain.

Thus, you need to manage both the artifacts and the process. Don't fall into the trap of assuming that these are generic processes. Every artifact results from the partitioning of your content system, and how they should be managed is determined by how the partitions communicate with each other. This is specific to the overall partitioning strategy of your particular content system. You need to find a tool that fits your partitioning strategy.

One class of tool commonly used for document- and management-domain structured writing systems, particularly DITA, is a specialized type of content management system called a component content management system or CCMS. A CCMS is designed for managing small fragments of content, rather than whole documents, and content reuse is the primary design objective. Most DITA tools are build around a CCMS.

However, if your content is in the subject domain, you may not need all the services of a CCMS and may be able to do just fine with a relatively simple version control system.

CHAPTER 37

Collaboration

Creating an effective information architecture for a large content set requires collaboration. If your work has to be integrated with the work of others, you have to make decisions that affect or are affected by the work of others. This means you need information about their work, and you need the skills required to integrate your work with theirs. In other words, your work becomes more complex when you collaborate with others.

As Tom Johnson writes:

> So often we place the bar for contribution at whether someone can write. In reality, it's not just whether someone can construct clear, grammatically correct sentences. It's whether the person can integrate the information into a larger documentation set.
> —http://idratherbewriting.com/2016/12/14/higher-level-technical-writing/

And, of course, the larger the documentation set becomes, the more complex the task of integration becomes, to the point where it can quickly come to exclude the participation of anyone except full-time writers who have specific training. How does each collaborator know what others are doing? How do they know which parts of the wider work they are responsible for? How do they integrate their work with the work of others? How do you manage the overhead created when collaborators have to be aware of other people's work as well as their own? How do you maintain process efficiency and rhetorical quality when writers can't work without knowing what everyone else is doing all the time?

The fundamental constraint on collaboration is the amount of time that collaborators have to spend on collaborative activities – orienting themselves to the work of others and integrating their work with the work of others – as opposed to creating new work. Unless you can ease this burden you quickly reach the point where adding more collaborators actually slows the project down because every new person you add increases the total collaboration overhead by more than the amount of work time that person can add to the project. Once this state is reached, either the volume and pace of work plateaus (or declines if even more people are added to the project) or some of the required coordination is abandoned, with the inevitable impacts on process and rhetoric.

Before the web, organizations handled the overhead of collaboration largely by assigning different books, pamphlets, and other publications to different writers and issuing some basic style guidance for language, physical appearance, and layout. All the other aspects of collaborative complexity – such as making sure that everyone was saying the same thing at the same time in the same way, that people were not creating the same content over and over again, that at least one person was saying everything that needed to be said, and that the best rhetorical models were being followed – were ignored, and the results were dumped on the customer in the form of information that was inconsistent, incomplete, incorrect, contradictory, hard to understand, and hard to find.

With the advent of the web, the same content, produced using the same non-communicative approach to collaboration, got dumped onto the company website. It wasn't any better or worse than it was before, but because it was now all searchable in one place, the quality problems became much more obvious. In response to this mess, the discipline of content strategy was born.

One of the challenges of collaboration between different roles and departments in an organization has always been that writers used different tools or, if they used the same tools, those tools didn't support collaboration or content integration. One approach to addressing this problem has been to adopt a universal platform. In most cases, this universal platform has only been for web content, though there are some that attempt to be a universal platform for every piece of content that the organization creates.

Some of these universal platforms are designed for simplicity, aiming to create a platform that anyone can use with minimal training. Examples include wikis and certain web CMS platforms. These platforms cannot support capabilities such as more sophisticated publishing, content integration, content reuse, consistent rhetoric, or any kind of systematic content management. Other platforms attempt to provide a full range of content management, reuse, integration, and publishing capabilities, but they generally prove too difficult for many contributors. Finding a happy medium can be difficult.

Partitioning collaboration

The central problem of collaboration is how much collaborators have to know about each other's work and what they have to do to integrate their work with content from other collaborators. The way you reduce the collaborative overhead is to reduce what each collaborator has to know and do to ensure that content works together.

This same need to reduce overhead occurs in the programming world. There, the solution to this problem is the Application Programming Interface (API). An API is a standardized, structured, published method for calling code written by someone else. With an API in place, programmers have to know far less in order to write code that works with other people's code. They just call the API.

An API acts as a set of constraints. It tells programmers exactly what they have to do to get the result they want. If they meet the constraints, they should be confident that the code will produce the promised results.

Structured writing does something similar for the collaboration problem in content. It is not exactly the same, because content does not call other content – at most it links to it. But structured writing languages provide constraints that writers must follow. If those constraints are well designed, they can tell writers all they need to know to ensure that their content will integrate with other contributors' content in the information architecture.

For example, if you have many writers contributing to a cookbook and they use a subject-domain structured writing format such as the one in this book, each writer does not have to know how the other writers format their recipes or whether they include serving counts or nutritional information. The subject-domain recipe format becomes a kind of content creation API that allows writers to write with confidence and without having to know what everyone else is doing. If they use some of the techniques discussed in Chapter 38, they also don't have to worry about accidentally duplicating an existing recipe.

More than this, however, recipe writers want readers to have the ability to access information about cooking implements and techniques mentioned in the text. Providing these abilities is the very stuff of information architecture and what sets it apart from mere cataloging.

Subject-domain annotation of tasks and tools (as described in Chapter 18) allows writers to pass on to readers the information required to deal with these matters. These annotations enable information architects to use the linking techniques described in Chapter 18 to construct and maintain the architecture and the audit techniques described in Chapter 39 to monitor coverage and discover new content requirements. Thus, writers don't have to communicate with their collaborators by any means outside of the content itself, greatly reducing the collaborative overhead of the process and giving them more time to write and to focus on accuracy and rhetoric.

Dealing with diverse collaborators

Part of the complexity of collaboration is the difference in skills, knowledge, and background between contributors. Some of your contributors may be full-time professional writers, while others may be engineers, marketers, field personnel, or support people for whom communication is an important, but not central, part of their jobs. Asking all these people to use the same tools and structures imposes far more difficulties on one group than another.

It makes perfect sense in these situations to design a collaborative system that distributes authoring complexity from one group of contributors to another; from your occasional contributors to your full-time writers, for example, or, better still, to your information architects and content engineers. People can handle far more complexity in their central task than they can in any of their peripheral tasks. Asking occasional contributors to fill in a form that captures all the elements needed to create an effective document according to a tested rhetorical pattern is far more effective than presenting them with a blank wiki page and much less complex than asking them to create a DITA or DocBook document from scratch.

A structured writing approach in which different contributors use different structured writing languages, each best suited to their contributions, can be very effective. Of course, any use of structured writing distributes some degree of complexity towards writers, since they now must know and follow the structure. On the other hand, structured writing can also distribute a lot of complexity away from the writer. While a blank page may seem like the simplest possible interface, it gives no task guidance at all. It is an interface without affordances that forces every contributor not only to write but also to design the information, decide what needs to be said, and determine how to say it.

General document-domain structured writing languages don't help; they offer no rhetorical guidance at all. Yet, they still demand that writers conform to structures dictated by the needs of the publishing process – needs that are meaningless to anyone other than a full-time professional writer.

Structured writing can give writers a rhetorically specific authoring interface, thereby distributing the design complexity to an information architect who designs the structured writing language. What is vital here, however, is that the structured writing language not distribute any other complexity to the writer. Any language that requires writers to master publishing or content management concepts, for instance, is not going to work well for this purpose. What works is a simple subject-domain language that addresses writers in terms they already understand and asks for annotations using concepts and ideas that they already know.

Wikis

One of the more common reasons for introducing content management and/or structured writing to an organization is to improve collaboration. However, these techniques are certainly not the only way to facilitate collaboration. In fact, the more common approach is to create simple and largely unstructured tools such as message boards and wikis. One of the largest collaborative projects in the world – Wikipedia – runs on a wiki using a fairly simple document-domain markup language that is often hidden behind a simple WYSIWYG editor.

This is powerful model for collaboration because it is inexpensive, readily available, and requires minimal technical skills. But it is also a model that distributes almost all of the rhetorical and management complexity to people rather than algorithms. There is simply not enough structure in this model to allow you to transfer very much complexity to an algorithm. Nor is there anyone making sure that all of the complexity gets handled. The system relies on the uncoordinated work of volunteers for auditing and content management as much as it does for content creation.

Why, then, turn to a more complex structured writing system for collaboration? In a word: integration. The kind of collaboration supported by message boards and wikis chiefly means that everybody can see what everyone else is doing. Any connections between the pieces created by different people are loose and non-critical, mostly taking the form of ad hoc hypertext links. And such connections are managed by large-scale community efforts. Wikipedia is full of links between articles largely because anyone can edit an article and add a link to an article on a related subject.

But not all collaboration can rely on such loose and uncoordinated activity. Often you need to bring the pieces created by collaborators together to form an integrated and cohesive whole. You can do that by hand, of course, but that can be cumbersome and time consuming. It may be hard for any one coordinator to keep up with all the content being created, especially if new content is continually being written without any freeze period to allow integration to take place.

Equally important, the collaborators on a message board or a larger wiki such as Wikipedia are largely ignorant of each other and each other's activity. Duplication of effort and even outright contradiction may be frequent. If you have huge numbers of volunteers constantly looking for duplication and contradictions, as Wikipedia does, you can live with this (though there will always be parts of the system that are in error at any given time). However, an organization that pays its writers and editors may not be able to afford this labor-intensive approach. It may need a more efficient way to coordinate the activity of its collaborators to avoid duplication and error. This means using structured writing to transfer a lot of that complexity and effort to algorithms.

Management-domain structured writing

Languages, such as DITA, that rely on the management domain, move a lot of the complexity of content management and collaboration into the source files in the form of management-domain structures. That complexity is transferred to any writer who works directly in DITA. In a collaborative environment, this management-domain burden is transferred to the collaborators, which means that it takes more time, discipline, skill, and knowledge to be an effective collaborator.

This is not to say that adopting this model is not a step in the right direction. This approach may impose a management system that is complex and demanding to use, but it often replaces a system in which management – and the tools and capacity to manage – was non-existent. Difficult and complex management techniques may be better than no management at all.

A well-designed management-domain system backed by well-designed tools may reduce the collaboration overhead compared to ad hoc methods, but it still leaves a lot of the collaborative overhead on the writer's plate. Therefore, the same limits on the effective scale of collaboration apply, although they have been pushed out a little. Alternative approaches have a greater potential to reduce overhead at all scales.

Bridging silos

There is a lot of talk in content management circles about breaking down content silos.[1] The word *silo* is used to mean a closed system in which neither the tools nor the participants communicate to others – or even make available for discovery – what they writing. It is difficult, of course, to create an integrated and effective information architecture when different content groups operate in complete isolation from one another.

---

[1] Actually, attitudes to silos are beginning to change with members of the content strategy community increasing realizing both the difficulty of dismantling silos and the advantages the silos provide in providing reliable semantically rich content and data. For example, see *Don't Dismantle Data Silos, Build Bridges* by Alan J. Porter at https://www.cmswire.com/customer-experience/dont-dismantle-data-silos-build-bridges/.

It is certainly possible to overstate the case here. Not every piece of content produced by an organization needs to be fully and seamlessly integrated with every other piece, and asking any current silo to open itself up to the wider organization likely means adding significant complexity in the form of additional knowledge and skills required to integrate content and systems with the rest of the organization. Not every silo must be torn down, but if silos are compromising your content strategy and information architecture, then you need to open up their content and processes to integrate better with the wider organization.

The naive way to do this it to have everyone use a single system and a single markup language. There are two ways to do this:

- Adopt a simple document-domain language, such as Markdown, that everyone can easily learn. The disadvantage is that such languages do not offer enough structures to meet everyone's needs or to support process and rhetoric.
- Adopt a large, complex document-domain language, such as DITA or DocBook, that offers structures to meet most needs. The disadvantage is that such languages have poor functional lucidity, especially for part-time contributors.

Each approach focuses on one problem and pays no attention to where the complexity that is directed away from that problem gets dumped.

The most pernicious myth about collaboration is that everyone must use the same tools and must understand each other's work. In fact, this is the worst way to collaborate, because it creates a huge amount of overhead that can make it difficult get anything done. Efficient collaboration is achieved by limiting how much collaborators have to know about each other's work and each other's tools. This allows each group or individual to work efficiently while still creating a product that can be integrated successfully with the work of others.

This approach to collaboration is seen throughout the worlds of engineering and computer programming. The secret ingredient that allows workers to collaborate with minimal knowledge of each other's work is the interface. A structured writing language is an interface to content creation that works by partitioning and redirecting the complexity of the content system.

To look at it another way, structured writing is a *tool* for partitioning and redirecting complexity in a content system. The examples we have examined so far have looked at ways in which content decisions can be partitioned away from writers towards document designers, content strategists, information architects, and content engineers. But structured writing can also be used to partition decisions differently for different types of contributors, so that the interfaces they are asked to use are a good match for the kinds of decisions they are qualified to make and shield them from those they are not qualified to make.

Part of a well partitioned system is limiting what people in different partitions have to know about each other's work. This has three benefits:

- It means there is less collaboration and communication overhead between partitions, making for a more efficient and reliable process. (The less information flow your processes require, the less vulnerable you are to breakdowns in the information flow.)
- It means people working in the partition can focus on their own work with the minimum of distractions, allowing them to do more work of higher quality.
- It reduces the complexity of working in the partition, which reduces the number of decisions that people working in the partitions have to make, which reduces the knowledge and skills they need to work successfully, which means you can recruit a wider range of people to work in the partition, including those with greater skills and knowledge in the core business of the partition and occasional contributors from other disciplines and other departments.

Silo is just another name for partition. If silos are a problem in your organization, this means that your system is partitioned incorrectly. The worst possible response to this is to simply break down the walls of all the silos, since this exposes every contributor to the full complexity of the content creation process with massive negative impacts on process and rhetoric. The correct approach is to re-partition your process to make sure that every part of the complexity of your content creation process is handled by a person or process with the knowledge, skills, and resources to handle it.

CHAPTER 38

Avoiding Duplication

Duplicate content is a significant source of complexity in any content organization. Not only is it expensive, but different versions of the content may not agree with each other, which creates complexity for readers. When the subject matter changes, only one of the duplicates may get updated, causing further drift between the two versions (and updating two versions is twice the work). Duplicate and near-duplicate content may also affect search results.[1] The need to detect duplicate content is therefore a major source of complexity in the content system.

Writers don't hold the entirety of the organization's content collection in their heads, so when they decide to write something, they may not know if that content already exists somewhere in the content set. Obviously, writers should look for duplicate content before they write, but it can be difficult and time consuming to find out if content on a particular subject for a particular audience already exists.

Indeed, even defining what constitutes duplication, let alone detecting that it exists, is not easy. Structured writing techniques can help, to a point, but don't get carried way in the attempt to eliminate duplication; you can easily do more harm than good and inject more complexity into the content system than you remove.

To avoid creating duplicate content, writers need to determine as quickly as possible whether a piece of content already exists or not. This check takes place countless times, so even if the cost of a single check is small, the cumulative cost can be large, significantly slowing down your content system. Small costs with high repetition rates are often the hardest to detect and least satisfying to fix, whereas the wins from detecting duplication are often more visible and more satisfying. But it is important to make sure that your efforts to limit content duplication don't cost more than they save.

This problem obviously affects any attempt at content reuse, since every time you set out to reuse content, you have to determine if reusable content exists. The longer that effort takes, the longer, and therefore more expensive, each instance of content reuse becomes. And bear in mind that

---

[1] The issue of how search engines treat duplicate content is complex. A lot of that complexity has to do with how duplicate is defined and how the search engine interprets the intent of the duplication it finds. For example, a search engine may filter duplicate or near duplicate content and list only one of the duplicated pages, which may not be the page your reader is looking for. For more information see https://www.hobo-web.co.uk/duplicate-content-problems/

you incur this cost every time a writer looks for duplicate content, whether it exists or not, but you realize any savings only when reusable content is found. Indeed, failed attempts are often more expensive, since the writer has to exhaust all possibilities, whereas successful attempts end as soon as relevant content is found.

To implement a system for detecting and eliminating duplicate content, you need a clear plan for exactly how writers are going to detect duplicate content or find content to reuse, and you need to ensure that the overhead of using such a system does not outweigh the benefits it provides.

Given the cost of finding duplicate content by hand, handing the task over to algorithms is highly desirable. But to hand the task over to algorithms, you need to define what constitutes duplication in precise terms and preserve the information that reflects those terms.

Creating a formal system for ensuring that content only exists once is sometimes called establishing a "single source of truth." Having a "single source of truth" does not mean that there is only one place or system from which all truths come; it means that every significant truth you manage is stored only once. Different truths can certainly be stored in different places, but you need to make sure that the same truth is not stored in two different places or two times in the same place.

A formal system for detecting duplication essentially means establishing a set of constraints that allow you to define and detect duplication. In other words, you need a set of rules that says that if item X matches item Y in aspects A, B, and C, then X is a duplicate of Y. (These are rhetorical rules – rhetoric is at the heart of process here as elsewhere.) These rules constitute an algorithm for detecting duplication. However, for a machine to execute this algorithm, aspects A, B, and C must either be a defined part of the content model for X and Y or be stored with X and Y. The content needs to retain this information in explicit rhetorical structures.

Where no such formal constraints exist, writers can use a search engine to look for existing content. Although a search engine can find existing content on a subject, search engines are not precise enough to detect duplication reliably. They may miss an actual duplicate because of a variation in terminology or return possible matches that are not duplicates. The writer must assess each match to determine if a duplicate exists. That takes a lot of time, and to ensure there is no duplication, the writer must perform this task before writing anything. This is a tremendous amount of overhead to impose on the content system.

How exactly do you construct a set of constraints for detecting duplicate content? First you need to define duplicate content. A reasonable general definition is the following:

> Duplicate content is content that describes the same subject matter to the same audience for the same purpose.

All three aspects of this definition matter. Subject matter obviously matters. You communicate to achieve an objective, so purpose matters. You communicate with a variety of people, so audience matters. Ignoring any one of these three aspects could result in serious quality problems.

Let's consider two potentially duplicate pieces of content and look at how to establish a formal system for deciding if they are duplicates. Consider two movie reviews, written in Markdown:

```
Disappointing outing for the Duke
=================================

After a memorable outing in _Rio Grande_
and _Sands of Iwo Jima_, John Wayne
turns in a pedestrian performance
in _Rio Bravo_.
```

and:

```
Wayne's best yet
================

After tiresome performances in _Rio Grande_
and _Sands of Iwo Jima_, the Duke is brilliant
in _Rio Bravo_.
```

Let's examine these according to our three criteria: First, do we have two pieces on the same subject? The subject of each is a movie. But is it the same movie? A human reading the text can easily tell that the subject of both pieces is the movie *Rio Bravo*, but an algorithm would have no way to tell, since nothing in the markup of either review directly identifies which movie is being reviewed.[2] Even if it could recognize the names of movies in the text, it would have no way to tell which one was the subject of the review.

---

[2] I am talking here, and throughout this book, about conventional algorithms. The question of whether an AI algorithm could tell the difference is out of scope here. When and if AIs become sophisticated enough to read and write content effectively in human language, the partitioning of the content system and the role of structured writing in that system will become very different. So, algorithm here means the kind of algorithm that a content engineer or information architect with the appropriate skill set could design and code in a reasonable amount of time with ordinary programming tools.

But suppose these same reviews were written in SAM in a subject-domain movie review language:

```
movie-review: Disappointing outing for the Duke
    movie-title: Rio Bravo
    review-text:
        After a memorable outing in {Rio Grande}(movie)
        and {Sands of Iwo Jima}(movie),
        {John Wayne}(actor) turns in
        a pedestrian performance
        in {Rio Bravo}(movie).
```

and:

```
movie-review: Wayne's best yet
    movie-title: Rio Bravo
    review-text:
        After tiresome performances in {Rio Grande}(movie)
        and {Sands of Iwo Jima}(movie),
        {the Duke}(actor, "John Wayne") is brilliant
        in {Rio Bravo}(movie).
```

Now it is easy for an algorithm to tell that these two pieces of content are both movie reviews and that they are both reviews of *Rio Bravo*.[3]

What about audience? There are multiple audiences for movie criticism. There are moviegoers and Netflix subscribers who simply want to decide what movie to watch on a Saturday night, but there are also academic film students who may want a detailed analysis of *Rio Bravo* according to some school of film criticism.

With the two Markdown versions, it is impossible for an algorithm to tell whether one of these reviews is written for moviegoers and the other for film students. The structured versions do identify the audience, but not in the form of an `audience` field in the markup. Instead, the entire `movie-review` markup language is intended for writing reviews for moviegoers. To write a review for film students, you would not use the same markup language, you would create a separate document type, which you might call `film-study`. Thus, you can tell from the document type alone whether two reviews are meant for the same audience.

---

[3] If the two Markdown examples were stored in a CMS that kept movie-review specific metadata, including the name of the movie reviewed, an algorithm could do the same check using the CMS metadata. For more on the choice between locating subject-domain metadata in the document vs. in a CMS, see Chapter 36 and Chapter 25.

The subject domain is concerned with the rhetoric of a piece of content: which pieces of information need to be presented to the reader in order to achieve its purpose. Since the purpose of any piece of content is to serve the needs of a particular reader seeking a particular goal, the definition of the reader is inherent in the definition of the subject-domain document type itself. It is the type of document that supports this person achieving this goal. If you were recording your content in the document domain, however, you might want to include a field to identify the intended audience of the piece, since audience is not implied by a document-domain structure.

But what about purpose? Although they review the same movie, the two reviews express two different opinions. Does that constitute a difference in purpose? You could argue that one review encourages readers to see the movie and the other encourages them not to. Or you could argue that having the two reviews together gives readers a more complete picture and more options to inform their viewing choices.

Ultimately, the decision comes down to whether keeping both versions serves a purpose for the organization that is publishing them. For example, does the organization want to present itself as having a single firm opinion on every movie or does it want to present itself as a neutral arbiter that presents a variety of opinions?

Therefore, whether or not you keep both reviews is a business decision. You will seldom find two independently written pieces of content that are word-for-word identical, and you will seldom find reusable content that is exactly what you would have written yourself.

The definition of duplicate content is not based on identical text; it is based on whether or not the content serves an identical rhetorical purpose, which is to say an identical business purpose. The constraints that determine if two pieces of content are duplicates of each other, therefore, are business rules.

Let's say you are willing to have multiple reviews of the same movie in your collection as long as they give different opinions. To accommodate this you can change your business rule for detecting duplicate movie reviews by adding a grading system to your review structure. Now you can rewrite your duplicate detection rule for movie reviews to say that two reviews duplicate each other if the `movie-title` fields have the same value and the `5-star-rating` fields have the same value.

By that rule, these two reviews are not duplicates, because their `5-star-rating` values differ:

```
movie-review: Disappointing outing for the Duke
    movie-title: Rio Bravo
    5-star-rating: 2
    review-text:
        After a memorable outing in {Rio Grande}(movie) and
        {Sands of Iwo Jima}(movie), {John Wayne}(actor) turns
        in a pedestrian performance in {Rio Bravo}(movie).
```

and:

```
movie-review: Wayne's best yet
    movie-title: Rio Bravo
    5-star-rating: 5
    review-text:
        After tiresome performances in {Rio Grande}(movie) and
        {Sands of Iwo Jima}(movie), {the Duke}(actor, "John Wayne")
        is brilliant  in {Rio Bravo}(movie).
```

However, in other cases, the business rules for determining duplication can be more difficult to express. For example, consider a recipe for guacamole. Is guacamole a single dish for which there can only be one recipe? If so, then detecting duplication is easy. If the type of the item is `recipe` and the value of the dish field is `guacamole`, then the content is duplicate.

But there are many different ways to prepare guacamole. Some differ only slightly from one another, but some present welcome variations that people might like to try. Clearly a recipe site would not want eight essentially identical guacamole recipes, but neither would they want to pick one variation to the exclusion of all others. So the question becomes, how do you decide when a recipe is an effective duplicate of an existing recipe and when it is a welcome variation? If you decide that a particular recipe is a welcome variation, how do you differentiate it from other guacamole recipes in your collection? In some cases, for example spicy versus mild, you could add another data field, but in other cases, for example different secondary ingredients, the choice may have to be a human editorial decision.

Clearly, the business rules for detecting duplication are not universal. The method you use to detect duplication for recipes is not the same method you would use for API reference topics, used car reviews, movie reviews, or conceptual discussions. Duplication detection happens in the subject domain and is specific to a particular type of content about a particular subject serving a specific business purpose. Whatever constraints you choose, the business processes and systems that ensure that those constraints are followed are specific to each function and organization.

The scale of duplication

So far, we have looked at detecting duplication of whole documents. But what if you want to avoid duplication that occurs below the level of a document? In Chapter 5 I looked at the example of a warning that was to be attached to all dangerous procedures. The duplication of that warning occurred at a much smaller scale. It was just a single structure within a procedure. Not only could it occur in many different documents, it could also occur multiple times within a single document.

Although you generally want to eliminate duplication of whole documents, you often need to duplicate parts of documents. For example, you want the identical warning to occur in every dangerous procedure so that readers are duly warned when attempting that procedure. In fact, all forms of content reuse are methods for deliberately duplicating content in multiple places in the content set.

Reuse techniques are about eliminating duplication on the writing and content management side of the content system, while creating duplication on the publishing side. However, although reuse techniques give you a mechanism for inserting duplicate content in various locations, these techniques can easily go off the rails if you don't clearly define what is and is not duplicate content.

It helps greatly if the duplicate content plays a consistent role. That is the case for the warning for dangerous procedures described in Chapter 5. That warning has a clear purpose and a clear anchor point in any document it appears in.[4] Without a clear purpose and anchor point, the chances of writers remembering to include the existing content, rather than creating a duplicate, goes down substantially. A subject-domain structure – the mandatory `is-it-dangerous` field – can fully define both the purpose and the anchor point, thus making it impossible for writers to forget or neglect to include the warning. This is a clear example of rhetorical structure supporting reliable process.

Finding a reliable anchor for detecting duplication gets more difficult the smaller the content unit you try to apply it to. For example, should you try to remove duplicate sentences that occur frequently but in different contexts? The phrase "Press OK" occurs frequently in technical document, and it always means the same thing. However, would replacing it with a variable reduce complexity or make any part of the content system more reliable? The number of repeated words, phrases, and sentences in the average technical document is very high and treating them all as

---

[4] An *anchor point* is a defined location in your content where reused text is inserted or referenced. Using the warning example, in a document-domain structure, the anchor point is the position in the text where you would place markup that identifies where the warning should be inserted.

duplicate content is obviously not feasible, nor would it solve an obvious problem. So, when should you regard a piece of content below the document level as duplicate?

Duplication of information versus duplication of text.

One reason for not wanting to factor out "Press OK" is that anything you replace it with would probably be longer and certainly more abstract. But the most important reason lies in the distinction between duplicating *information* and duplicating *text*. The same text can occur in multiple places without actually being duplicate information. Each instance of "Press OK" refers to a button in a different dialog box. Those buttons all have the same name, and thus, the instruction to press them is identical. However, they are different buttons. It is entirely possible that a redesign of one of those dialog boxes could result in the OK button being renamed to something more specific to the function of the dialog box, such as Print or Send. Thus, each instance of "Press OK" is a different piece of information, even though it is expressed with the same text. (See Chapter 17 for an approach that eliminates the duplication of "Press OK" without running into this issue.)

By contrast, the warning for a dangerous procedure is a single piece of information occurring in multiple contexts.[5] It applies with equal force to all procedures that are dangerous. Of course, a procedure could go from being dangerous to not being dangerous. (A new version of the product may include a safer design.) In this case the warning should be removed. But the value of the warning remains the same for all procedures to which it applies, whereas the name of the make-it-go button for a dialog box can change independently of other dialog boxes, all of which still have make-it-go buttons. That is, it is a difference in applicability, not a difference in content.

If this distinction seems a little hard to get your head around, that is a good indication of how difficult it can be to detect true duplication in content. And when you eliminate duplicated text that is not duplicated information, you introduce complexity that will either make change management more difficult down the road or get missed and damage rhetoric.

It is probably better, therefore, to stick to cases where you are certain that the duplication you are detecting is genuinely duplication of information and not merely duplication of text.

---

[5] In the example we have been looking at. Of course, some procedures are dangerous in unique ways and require unique warnings.

Duplication and the level of detail

Another problem with identifying duplication deals with the level of detail with which a subject is treated. For example, most Wikipedia articles on countries contain a section on the economy of that country. At the beginning of that section there is a link to an entire article describing the economy of that country followed by a brief summary, which is less detailed. There may also be a brief mention of the highlights of the country's economy in the four or five context-setting paragraphs that lead most Wikipedia articles. These different levels of detail serve different user needs, and therefore, each is a valuable contribution to the content set. In other words, they are not duplicates, because, although they address the same subject, they serve different audiences and different purposes.

Everything we know about effective rhetoric tells us that you need to address different audiences and different tasks differently. Taking a piece of content designed for one audience and using it for all other audiences – or attempting to write generic content that takes no account of any audience's needs or tasks – is certain to produce content that is significantly less effective.

The question, then, is whether differences in level of detail reflect different audiences and purposes. While it may seem like this is a distinction that any writer should be able to make via inspection, it is often quite a difficult distinction, because it relies on the writer understanding the audience and purpose of the content being examined.

A writer who thoroughly understands one audience and purpose may look at a piece of content designed for a different audience and purpose and fail to recognize the difference. The writer may not even know that the audience and purpose for which the document was written even exist and, therefore, may not understand its rhetorical purpose and context. To that writer, the document may look badly written, unnecessarily verbose, incorrectly ordered, or too brief. In other words, it can be difficult for a writer to tell whether a potentially reusable piece of information is badly written and, therefore, in need of editing to be reused or an excellent piece written for a different audience and purpose and therefore not reusable at all.

Here again, subject-domain structured writing can come to the rescue by making the rhetorical purpose and context of the content explicit. As we saw with movie reviews, treatments of the same subject for different audiences and purposes have different structures. If you find a piece of content on a similar subject, but with a completely different structure, you can then look up the documentation for the document structure to determine what its audience and purpose is. This will help prevent accidental editing and inappropriate reuse.

Duplication in less structured contexts

Since duplication detection rules define when a piece of content is unique, they make it easy to determine if a piece of content exists. If you know which fields of a proposed content item define it as unique, you can query for a topic that has the same values in those fields. If you find one, you can be confident that the content already exists; if you don't, you can be confident that it does not exist and needs to be written.

At least in theory you can. The problem is that not all content can be structured to the same degree. For example, while we can determine with a high degree of certainty whether the API reference contains only one entry for the `hello()` function in the `greetings` library, it is much harder to detect if a writer has inappropriately inserted a full description of the `hello()` function into the programmer's guide.

API references typically contain a clearly identified entry for each API they document. Programmer's guides typically deal with the relationships between different APIs and other parts of the system and how to accomplish real-world tasks that require programmers to use several APIs together. This focus lends itself to a reasonably strict content type for programming topics, but that does not help you detect duplication of material on related subjects such as API functions and libraries. Therefore, detecting that the writer of a programming topic has duplicated information provided by the API reference can be difficult.

Also, the programming guide writer may have had a good reason for duplicating information from the API guide. For example, to explain why someone might choose the `hello()` function from the `salutations` library rather than from the `greetings` library, the writer may have needed to explain the differences between the `hello()` functions in each library, using information taken from each library's API reference. Simply referring readers to the two API reference entries to compare and contrast for themselves would eliminate the duplication, but at the expense of dumping the complexity of detecting and understanding the differences onto the reader.

Content, by its nature, deals with the complex and irregular aspects of the world, and you cannot expect to fully remove all duplication or everything that might be duplication from your content set without creating far more complexity in the content system than you have redirected. However good reuse looks on paper, attempting to remove all duplication is likely to leave you with more unhandled complexity than it removes, which always results in compromised rhetoric.

But while you should be cautious, don't throw up your hands and abandon the attempt to tackle duplication in your content set. There are effective strategies, particularly on a smaller scale.

Localizing duplication detection

If detecting duplication in the general case often introduces more unhandled complexity than it removes, most of the duplication that really matters occurs locally. The risk that your movie review collection will accidentally duplicate content in your recipe collection is low. It is duplication within each collection you need to worry about, not duplication between them.

Therefore, subdividing your total content set can make for a much more practical approach to duplication detection. You can subdivide your content set in several different ways. Dividing it by document type is the most obvious. However, there are cases where you may want to detect and remove duplication between different document types covering the same subject.

For instance, some organizations try to minimize duplication between technical, training, and marketing content for each product. Although these three types of content obviously have different content models, their content models may include key fields that can be used to define a duplication detection rule (they address the same subject, so logically they will have certain subject-domain structures in common). Where the purpose of the content is similar, therefore, you can define anchor points for inserting the same information in multiple places. Of course, you have to keep in mind that technical, training, and marketing content do not always address the same audience for the same purpose, so an over-zealous approach to removing duplication could do more harm than good.

Finally, the more local the content set you are dealing with is, the more likely it is that everyone who creates and maintains the content set will know, or be able to guess, what content exists.

Reducing duplication through consistent content models

Using specific content models based on sound rhetorical models can help reduce duplication of content at all scales. A tightly constrained subject-domain content model, in particular, makes sure that there is a place for every piece of information and that every piece of information stays in its place. There is far less scope for incidental duplication between different content types if each content type is appropriately constrained. (And this works equally well to combat the opposite of duplication, which is omission.)

Strongly defined content types tend to be more cohesive – meaning that each instance of that content type for a different subject still covers the same ground and covers it more consistently. Without strong types, different writers may chunk ideas and information differently, which means that topics from two different writers can partially overlap. Not only is partial overlap hard to detect, since there are fewer points of similarity, it is also hard to fix because each item contains different information that the user needs. Eliminating one of the duplicates means finding a place for all of the extra information it contains, a process that can affect other content items and raise other duplication detection questions.

Using consistent content types ensures that when writers have a question about whether a certain subject has been covered or not, they know where to look.

Reducing duplication through consistent quality

Duplication can occur even when writers know that a piece of content on the same subject and with the same purpose already exists. A writer may think that the existing content is inadequate but not be willing or able to track down the original writer to discuss the situation or to determine how other uses of the content would be affected by editing it to bring it up to standard. Instead, the writer may simply write a new version.

There are two basic issues here. First, the quality of the existing content may not be good enough. A focus on creating consistent quality across the content system will go a long way towards avoiding this issue, because it will help ensure that when writers find content they would like to reuse, that content will be good enough to reuse.

Second, the two writers may have different views about the appropriate style or content to use to describe a particular subject to a particular audience for a particular purpose. Differences of opinion about how things should be written are common and, thus, a common source of complexity in any content system. However, you can manage these differences if you set up well-defined constraints that define the appropriate rhetorical strategy for a particular subject, audience, and purpose. Here again, structured writing can help enormously by allowing you to set rhetorical standards that are clear to writers and enforce those standards through the constraints built into the structured writing languages they use. This helps to avoid content duplication caused by disagreements over rhetoric and style.

Reducing the incentive to create duplicate content

Of course, merely knowing that content already exists is no guarantee against duplicate content being created. Even if there are down-the-road benefits to avoiding duplication, it may still be easier for a writer on a deadline to create duplicate content if the means for finding and reusing the existing content are cumbersome or difficult to use and understand.

Many of the reasons for avoiding duplication and for reusing content have to do with downstream savings in change management and translation. It is not a given that these techniques make life easier for the writer. If the system is too cumbersome or difficult to use, writers will rewrite rather than reuse, despite any downstream problems this causes.

It is important to understand that content reuse does not always save writers a lot of work. For instance, reuse does not necessarily reduce the amount of research they have to do. After all, how can you determine if a piece of existing content adequately describes your subject to your audience for your purpose if you have not done your research? Only systems that abstract the entire question away from the writer actually remove the need to research the topic. Once the research is done, writing the content can be a fairly straightforward and technically simple operation. There is no particular incentive for writers to undertake the complex tasks of looking for existing content, assessing it, and inserting it into their work, especially when the search may turn up nothing, leaving them with the writing work still to do.

In other words, if creating duplicate content is less complex and takes less time and energy than reusing existing content, it is likely that you will get lots of duplication, even if you have solved all the technical challenges of identifying and reusing content. Solving the technical challenges alone is never enough. You must remove complexity from key players in the system, so it is easier for them to do the job the right way.

A technically less sophisticated solution that is simpler for people to use will almost always win out over a more sophisticated system that is more complex to use. Or, to put it another way, complexity is not real sophistication. Real sophistication is not about adding functionality; it's about directing complexity to where it can best be handled. Real technical sophistication in content systems is always about partitioning and directing complexity to the person with the skills, time, and resources to handle it. A simple subject-domain markup language can provide a simple (and therefore sophisticated) content creation interface that loses none of the functionality you need.

Less formal types of duplication detection

Having hard and fast rules that define duplication as two pieces of content having the same values in the same set of fields works well when the subject matter lends itself to that degree of structure. But not all content can be structured this precisely. Fortunately, structured writing techniques can be used to implement some less formal approaches to detecting duplication. These approaches are more probabilistic than certain and may be more appropriate as audit tools rather than tools you would expect writers to use before starting to write. Still, they can be useful, and they can detect duplication that other methods might not find.

For instance, if your content contains topic-level subject annotations – like those we looked at in Chapter 18 – that list the subjects they cover by type and term, you could have an algorithm compare those annotations across your topic set and flag pairs of topics on the same subject. Often it is appropriate for topics of different types, such as a programming topic and an API reference topic, to have similar or identical subject annotations, since these topics treat the same API functions in different ways (same subject but different audience or purpose). However, if you find topics of the same type with matching or near-matching sets of subject annotations, those topics may contain duplicate content.

You can do a similar check based on inline subject annotations on phrases. This is less precise than comparing topic-level annotations, because inline subject annotations annotate subjects the phrase refers to rather than subjects the topic describes.[6] However, if two topics contain subject annotations that refer to many of the same subjects, those topics may contain duplicate content.

There are tools available that claim to detect duplication using natural language methods. I'm not going to comment on the specific capabilities of such tools, but there are a few obvious downsides to these methods. First, they can only compare two completed texts to see if they are duplicates. They can't detect whether the text you are thinking of creating already exists. Second, these methods look for the same or similar text, which is not the same as looking for the same or similar information. Similar information can be expressed using very different words, and similar words can convey very different information.

---

[6] That is, phrase-level annotations tell you what subjects are mentioned in the course of describing the subject of a topic. For example, John Wayne may be mentioned in a review of *Rio Brave*, but the review is about the movie, not John Wayne.

Reducing duplication by merging sources

Sometimes you want to deliberately duplicate content with variations for different publications. But while you want duplication on the output side, you don't want it on the input side.

One way to reduce duplication in content is to merge information that is intended for different publications into a single structured source file from which the different variations can then be created algorithmically. We looked at an example of this in Chapter 5 when we combined alcoholic and non-alcoholic beverage matches for a recipe into a single subject-domain recipe document, allowing us to produce variations of the same recipe for *Wine Weenie* and *The Teetotaler's Trumpet*.

Because a true subject-domain document is simply a collection of information about a subject from which you can select items to create many different documents, you can use the subject domain to eliminate duplication on the input side of a project. For instance, if you have a product with multiple versions, you can merge the information about all of the versions into a single source file and then publish only the blocks that apply to a particular version.

Duplication and the structure of content

As we have seen, placing content in well-defined structures makes it easier to detect duplication because it allows algorithms to compare values in a structure. When you can determine the common information requirements of a rhetorical pattern, you can create data points that can be compared to detect duplication before, during, and after writing content.

But using consistent, well-defined structures has a deeper importance for detecting and avoiding duplication. Dividing content into consistent blocks makes it easier to compare. Reuse is difficult when two pieces of content use different structures to describe the same content. You can't simply replace one with the other because they don't cover the same ground. Only units of the same size, shape, and scope can be effectively compared to see if they are duplicates. Using well-defined content structures ensures that each piece of content on a particular subject – each movie review, recipe, API reference, feature description, or configuration task – has the same shape, size, and scope, meaning they are comparable.

But the benefit of structure is not simply that units are comparable. If content is written in well-defined structures, each of which has a specific job to do, the chances of creating duplicate content, even by accident, are greatly reduced. In addition, writers and managers will have a good sense of what has been created and what has not, because well-defined structures ensure that different approaches to describing a subject won't result in partial overlaps (or omissions).

Detecting duplication before and after authoring

Detecting and removing duplication is an important part of keeping your information collection tidy and should be a part of your regular audit practices. However, detecting duplicate content after it has been created means that you have paid writers to write the same content twice. Ideally, you would like to detect the duplication before the writer creates the duplicate content.

The extent to which you can do this depends on how well-structured your content types are. Let's look at some scenarios:

- The ideal scenario is that the content model factors out repeated content altogether. If your `procedure` structure contains a compulsory `is-it-dangerous` field, then the warning text has been completely factored out. All the writer has to do is correctly fill out the `is-it-dangerous` field.
- Merging sources is also a powerful approach, since the material for both versions is now created by the same writer at the same time.
- The next best scenario is to have the authoring or content management system refuse to let writers create a duplicate content record in the first place. For instance, if writers must supply values for the movie name and the five-star rating before they enter content for a movie review, the CMS can raise an error before they waste any time writing a new review. However, although this kind of interface is common in the database world, it is not common for content, in large part because of the number of content types that you would have to create this kind of interface for. Also, writers are not used to working in this type of system.
- The next best thing after that is for the system to allow writers to query the current collection to determine if there is already a piece of content with those values in those fields. The main difference here is that writers must initiate the query, rather than having the CMS check for duplication before it allows content to be created. This approach requires an interface that allows writers to run queries. However, a content engineer or information architect can provide pre-defined queries that match the organization's business rules for determining duplication, thus relieving writers from having to learn how to create queries. This approach can be implemented using a much wider variety of tools and does not necessarily require any form of database or content management system.

- After this the next best thing is to use a search engine. This is time consuming and uncertain, so it should be supplemented by using structured writing techniques and regularly auditing the content set to catch and remove accidental duplication.

- Then there is the case where writers create content outside of the CMS but must complete a metadata record when submitting the content to the CMS. The CMS can detect duplication in the metadata fields and refuse to accept the content. The downside, of course, is that by the time the CMS detects the duplication, you have already paid for creating the content. Technically, though, the mechanism here is exactly the same as the case in which the writer queries before writing. In practice the difference is a matter of discipline rather than the technical implementation.

- Finally, it may be appropriate to accept that certain types of duplication are just too hard to define and detect and that your content system will be simpler overall if you simply tolerate those types and/or handle them with periodic auditing, perhaps assisted by structured writing techniques.

Optimize the whole, not the parts

Elimination of all duplication from a content set is not feasible. The variations on how, when, and why you mention particular facts in content, and the various purposes and audiences for which you describe different subjects, make it impossible to have hard and fast rules about what is and is not a duplicate for every type of content. If you pursue a single source of truth too zealously, you can eliminate valuable differences in information and presentation, which will damage rhetoric by creating excessively generic content that is hard for readers to understand and may not give them the information they need.

However, as we have seen, there are structured writing techniques that you can use to detect certain kinds of duplication and that make duplication less likely to occur, while reducing the amount of unhandled complexity in your content system.

CHAPTER 39
Auditing

Although the appropriate structured writing techniques can help keep your content set from falling into disorder, maintaining a healthy content collection and a well-integrated information architecture still requires constant monitoring to find and fix errors and to ensure that your processes are working as well as possible.

In Chapter 29, I looked at how structured writing techniques can improve the conformance of individual pieces of content. In Chapter 25, I looked at how they can help you maintain metadata and taxonomy across your information set. Now, let's look at how these techniques can help you audit a content set to ensure that it meets its constraints and that its constraints are consistent with your goals.[1]

Even if every item in a collection meets its individual constraints, that does not mean that the whole collection meets its constraints. For instance, even if every item in your collection conforms, that does not mean that the collection is complete, that all the links that should exist do exist, or that your links point to the best resources. For issues like these, you need both a sound strategy for creating and supporting your information architecture and a sound audit process to make sure everything is in its place.

Auditing is about making sure that:

- The content set is defined correctly (you know what types of content it should contain and which instances of each type)
- The content set is complete (it contains all the items of each type that it should)
- The content set is uncontaminated (it does not contain any items or types it should not)
- The content set is integrated (it expresses all of the relationships between items that it should)
- Each item in the content set conforms to its constraints

Auditing a large content set is difficult, and many CMS solutions are deficient in audit capabilities. The main reason they are deficient is that most content is recorded and stored in media-domain

---

[1] Content strategists often use the term "content audit" to mean a current-state analysis performed at the beginning of a website redevelopment project. A content strategy content audit is about cataloging, and possibly categorizing, the content you already have. Here, I use the word audit to refer to an ongoing activity to ensure that your content set continues to meet its goals.

or generic document-domain formats, which make it difficult to mechanically assess what content you have and what state it is in. It is hard to know if you have all the pieces you should have if you can't tell exactly what each piece is.

One of the biggest, and least appreciated, benefits of structured writing is that it makes content more auditable. When content management systems fail or become unmanageable, the root cause is often either an incorrect distribution of complexity from day one or a failure to audit. Failure to audit may mean a lack of attention to regular audits or an inability to audit effectively. Without effective audits, your content set can end up incomplete, corrupt, and poorly integrated, which reduces quality, increases costs, and creates a body of unmanaged complexity that every down-stream process and person has to deal with, including, of course, the reader. A vicious cycle can develop in which writers, frustrated with the difficulties of the system, create workarounds that further corrupt the information set. Unmanaged complexity breeds more complexity. Whatever expenses you may incur to implement a more structured writing approach could well be offset or even exceeded by the savings associated with more effective auditing.

Auditing the definition of the content set

Content strategists spend a great deal of time and effort developing a content plan (usually this is for a website, but the same principle applies to any content set). How they do this is beyond the scope of this book, but the result should be a definition of the content set: which types of information it is supposed to contain and which instances of those types. This definition is based, of course, on the goals for the content, which the content strategist needs to define.

The definition of a content set is not necessarily static. It is not a fixed list of topic types or specific topics to be developed. First, the subject matter may change during content development, which would change the content pieces needed and, perhaps, require new content types or modifications to existing types. Second, the exact set of pieces or types may not be knowable at the outset. Only during content development can you fully explore and understand the complex set of relationships between subject matter and the needs and background of readers.

It is hard to be disciplined and deliberate in evolving the big picture model of the content set if you are not disciplined and deliberate in how you create the pieces. If you ask writers to assign CMS metadata after they have written content, they will tag that content using the terms that seem like the closest fit to the content they have already written. However, they will probably not revise the content to fit the labels. Their view will not be that the content is wrong, but that the labels don't fit the content. And since the labels won't fit the content, you won't really know what type of content you have in your collection.

If you don't know what type of content you have, you can't tell if you have defined your content set correctly. Content may perform poorly because it does not fit the type definition properly or because the type definition is wrong. Unless you can determine what the problem is, you won't know what to fix.

Of course, a writer may come up with a better type definition. This is a good thing. If it really is better, you will want to change all similar content to match this improved model. However, if your models are not formally defined and the writer executes this new model without defining it as a formal structure and then just tags it using the current CMS tags, you will never know about the new model, never have the chance to test it to see if it is better, and never have the chance to update the official definition so that new content follows the new model. Unless your content types are codified and auditable, you won't detect improvements in the types and they won't carry over to other content. See Chapter 41 for more on this.

Having strong well-defined content types makes it easier to audit your content types to make sure they do the job they were designed to do. Similarly, having strong, well-defined content types helps ensure that each item does the job it is supposed to do and that you cover all the subjects you should cover.

Assessing completeness

Structured writing can also help you assess the completeness of your content set. If you create content in the subject domain, including annotating the subjects that you mention in the text, you can use algorithms to extract a list of the types and subjects that your content talks about. In your initial top-down plan, you may not have thought about the need for content that covers certain subjects or that supports certain activities, but if those subjects or activities start showing up in your content, that is a strong indication that they are related to the purpose of your content set and should probably be included in the definition of the content set.

Subject-domain structured writing lets you know what your content is actually talking about and what writers are discovering or think needs saying. Content needs are ultimately driven by subject matter, and it is your writers, who work with the subject matter every day, who are on top of what the subject matter is and how it is changing. Bottom-up content planning distributes the responsibility for discovery outward and for coordination inward, which keeps you in touch with evolving content needs. Without this information flow it is difficult to determine whether your content set is meeting its coverage goals. (We saw the same pattern of information flow with bottom-up taxonomy development in Chapter 26.)

The ability to compile lists of subjects your writers are writing about attacks two audit problems. If writers are writing about things outside your current coverage definition, either your coverage definition needs updating, or writers are polluting the content set with irrelevant material.

Avoiding contamination

Subject-domain content structures and annotations can help prevent contamination of the content set by irrelevant or poor quality material. But more important than catching writers in the act is catching flaws in content types that allow contamination to creep in.

A major form of contamination is redundant content. As I noted in Chapter 38, you have to carefully define what it means to avoid redundant content, because it is not simply a matter of addressing a subject only once. Avoiding redundancy means addressing an audience need only once, and that may require several topics on the same subject addressed to different readers.

It is all too easy for duplicate content to sneak into a content set. Some of it comes in because the same functionality is repeated in many products or in content delivered to different media. Some comes in because writers simply don't know that suitable content already exists.

Content reuse is a major motivator for structured writing for exactly this reason. But the content-reuse algorithm addresses only the problem of how to reuse content. It provides a method to reuse content you are aware of. It does not prevent you from duplicating content when you did not look for or did not find existing reusable content. You need to audit your content regularly to make sure that writers can find potentially reusable content and to make sure that duplication does not creep in.

Even when you find redundancies, they can be difficult to consolidate if they don't have similar boundaries within their respective documents. This is an example of the composability problem, which I discussed in Chapter 28. Strongly typed subject-domain content – content that conforms to a model that breaks down and enforces constraints on the pieces of information required to cover a topic – enables you to detect duplication in a much more formal and precise way. Duplicate subject matter is much easier to detect when content is captured in the subject domain.

When you consult a repository to see if there is content you can use, you need to be able to query the repository in a sensible way for the type of content you are looking for and recognize appropriate content when you see it. And you need to be able to rely on that content conforming to its type in order to use it with confidence. Subject-domain topic typing helps with all of these things. Subject-domain labeling of document- and media-domain content can help as well, but only if it conforms to the appropriate constraints, a problem discussed in Chapter 29. The easier it is to correctly identify reusable content and use it, the less corruption of the repository will occur.

Maintaining integration

A content set is never a collection of wholly independent pieces. The pieces have relationships with each other that matter to readers. Whether you express those relationships through links or cross references, or whether you rely entirely on tables of contents and indexes, it is still important to understand and manage them.

Relationships matter for management reasons as well. If you have documentation for multiple releases of a product, the relationship between the documentation for feature X in version 3 and that for feature X in version 2 matters. It may matter because the feature has not changed, and you can reuse the item. Or it may matter because you found an error in version 2 and want to fix it in version 3. And if you put this content online, the relationship matters for readers. You don't want a search from a reader using version 3 to return information for version 2.

You can describe relationships between items externally. Items are related whenever they have the same value in any one of their metadata fields. Which field it is tells you what the relationship is. Finding the relevant metadata field allows you to manage the relationship. But the same problem exists here as with all external metadata (see Chapter 36) – the content may not conform to the metadata, and, without structured metadata in the content itself, it is hard to audit how well content conforms to its metadata. In-band information, such as subject-domain annotations, is always more reliable than out-of-band information, such as external metadata.

But the bigger problem is that external metadata does not map the important relationship that can exist between a part of one item and the whole of another item. Are the function names mentioned in the programming topics all listed in the API reference? Are the utensils mentioned in a recipe all covered in the kitchen tools appendix? These are important content relationship questions, but these relationships cannot be mapped using external metadata. You need subject-domain markup inside the content that identifies function calls and the names of kitchen tools.

Structured writing, particularly in the subject domain, helps you discover and manage these relationships by clear identifying the subjects on which these relationships are based.

Making content auditable

I have talked all through this chapter about how using strong content types makes content easier to audit. What is a strong content type? Fundamentally, a strong content type makes explicit what the content is supposed to say and how it is supposed to say it. Or, to put it another way, a strong content type captures, enforces, or factors out the major constraints of the content, including its major rhetorical constraints. A strong content type constrains the interpretation of content as well as its composition, and the more reliably content can be interpreted, the more reliably it can be audited. Strong content types are almost always in the subject domain.

You can create content that conforms to its rhetorical constraints without using structured writing techniques. But as we saw in Chapter 29, strong content types provide explicit guidance to the writer and facilitate the use of conformance algorithms. Thus, you design content structures to support conformance. The same holds true for auditing; strong content types make your content easier to audit, so you design content structures to help you meet your auditing goals.

Facilitating human review

Auditing is sometimes not as straightforward as conformance, even with structured writing techniques in place. Auditing often requires human review, not only to make sure that all subjects have been covered but also to discover new issues or subjects that need to be addressed. Human review of a large content set is difficult due to the sheer volume of content. An algorithm can simplify this work by creating different views of the content set that humans can review more easily. This is an application of the content generation algorithm for internal purposes.

Suppose, for instance, that you are using subject-domain annotations to drive linking, as described in Chapter 18. Every topic in the collection is supposed to be annotated to state the type and name of each subject it covers. Every mention of a significant subject is supposed be annotated with its type. The linking algorithm uses these annotations to link the content without requiring you to create or manage links in the source text. But that does not guarantee that all the right links get made. There could be errors in annotation that are impossible to detect when conformance testing individual topics. But you can do an lot to catch these kinds of errors when you audit the content set as a whole.

These are some of the audit functions you can perform based on annotations and index entries:

- Create a sorted list of all annotated phrases and see if they are annotated consistently. This will tell you a lot about your subject types, including how well they are understood and what instances of each type you should cover.
- Create a list of all annotated topics and check it against your content plan, using a taxonomy if you have one. This will tell you a lot about whether your coverage is complete, whether your writers are on track, and whether your content plan or taxonomy matches reality.
- Create a sorted list of all the annotated terms and check it against the list of annotated phrases to find phrases that are indexed but not annotated or annotated but not indexed. This can identify subjects that are not covered, content that covers extraneous subjects, and topics that are not being indexed or annotated properly.

You can also use the same technique for project management purposes. Early in the content development phase, the list of annotations on phrases that don't match the annotations on any topic will inevitably grow, as writers annotate subjects that have not been documented yet. Over time, however, new topics will start to fill in those gaps, and those new topics will contain fewer references to subjects that have not yet been documented. The trend line of the growth of new subjects being annotated versus annotated topics being created will rise and then fall, allowing you to track how close a content set is to completion, even in cases where defining the boundaries in advance is difficult.

Content is one of the hardest assets to audit and inventory. Structured writing, particularly subject-domain structured writing, can greatly aid in establishing an effective audit function for your content. An effective content audit process, in turn, can help avoid the gradual decline of order and reliability that affects so many managed content sets and leads to the slow death by strangulation of so many content management systems.

CHAPTER 40

Change Management

Keeping content in sync with changing subject matter and changing requirements is a major process challenge for all content organizations. Structured writing can do a lot to partition and redirect the management of change and the consequences of change.

Content changes for many reasons. For our purposes it is useful to consider the following sources of change:

- **Subject matter changes:** Content has to change when the subject matter – the real world stuff that it describes – changes.
- **Rhetorical changes:** Content may also change because you decide to express ideas differently. For instance, you may discover a way to present a certain class of content that works better for readers and decide to change all existing instances of that class to fit the new structure. And, as we have seen, outputting to new media requires new rhetoric.
- **Formatting changes:** Content may change because you want to format it differently, either to support new media or to re-brand.
- **Externally-driven changes:** Content may change because of changes external to itself, its style, and its subject matter. For instance, if an item links to another item and that item is removed and replaced with something different, you have to change every item that links to the changed item, even if the subject matter and style of those items has not changed at all.

One important motivation for adopting structured writing is what is often called future proofing. Future proofing means building a system or product with a view to making it able to survive future changes in environments or requirements. Future proofing is difficult because you cannot know with certainty what changes will occur, how likely they are, or what they will cost.

Building a future-proof platform can increase up-front costs, delaying time to market, and possibly causing you to miss a window of opportunity. And you cannot be sure that your investment will ever pay off, since the future you prepared for may not be the future you get.

But if you do not build a future-proof platform, you may not be able to keep up with market developments, causing you to lose an early lead. And you may be forced to make massive and expensive changes when future events render your current system obsolete. Both problems happened frequently when traditional publication systems were confronted by the rapid rise of the web.

The safest approach to future proofing is not to try to anticipate the particular way in which the future will develop. Instead, create content structures that will be of value no matter what happens in the future. Creating content in the subject domain is the best way to practice this kind of future proofing, because writing in the subject domain creates metadata that contains only true statements about the subject matter itself. Those statements will remain true as long as the subject matter remains unchanged. That is as future proof as you can make your content.

For example, suppose you write your ingredient list in reStructuredText as a table:

```
======   ========
Item     Quantity
======   ========
eggs     12
water    2qt
======   ========
```

Later, if you decide to present ingredients as a list, you will have to go back to your content and change the markup. Doing this across a whole collection of recipes will be expensive.

Suppose, instead, that you use subject-domain markup:

```
ingredients:: ingredient, quantity, unit
    eggs, 12, each
    water, 2, qt
```

Now, you don't have to change the content to change the presentation. You just change the publishing algorithm to turn an ingredients record set into a list instead of a table. Thus the subject-domain markup has future proofed your content against this change of presentation. The document-domain reStructuredText markup specified a table, which is not a truth about the subject matter; it's a decision about presentation. That decision can change independent of the subject matter. The subject-domain markup simply specifies that "eggs" is an ingredient and "12" is a quantity. These are truths about the subject matter that will not change. Thus, they are invulnerable to future changes outside of the subject matter itself.

Moving your content from the media domain to the document domain provides a degree of future proofing. By factoring out the formatting details, it protects your content against changes in formatting rules such as what font to use for headings. Moving your content from the document domain to the subject domain provides additional future proofing. By factoring out the content and organization of documents, it allows you to target different publications and create different document designs for different media. For example, it lets you output different beverage suggestions to *Teetotaler's Trumpet* and *Wine Weenie* without making any changes to your source file.

Throughout this book I have described complexity in terms of decisions. Future proofing content is about making it possible to change decisions with the least cost. The best way to reduce the cost of changing a decision is to factor it out of the content and assign it to an algorithm. You can then change the decision by changing the algorithm. By contrast, changing a decision that has been encoded in the content requires you to find and change every piece of content that is affected.

Structured writing imposes specific structures on content for specific purposes. It does not make content magically immune to change nor does it guarantee that you will not have to rewrite your content or change its structure to accommodate changes in subject matter or business requirements. You can, however, design content structures that help you manage specific, foreseeable changes. If you are lucky, the structures you create may also allow you to adapt content for unforeseen circumstances, particularly if your content is stored in the subject domain. But this is a bonus. You cannot guarantee any content or structure will work for things you have not foreseen. However, different domains, by their very nature, provide different degrees of future proofing:

- Content in the media domain encodes all, or nearly all, of your content decisions in the content itself. Few if any decisions are factored out into algorithms, and, therefore, the cost of changing these decisions is high.

- The document domain factors out most formatting decisions from the content, reducing the cost of changing formatting decisions. But it still encodes all, or nearly all, presentation and rhetorical decisions, meaning changes in those areas are still expensive.

- The management domain is a mixed bag. In some cases, it can be used to factor content or data into separate files, which can make that content easier to locate and change. On the other hand, management-domain structures often encode management decisions in the content, meaning you may have to find and change many instances for certain management decision changes. However, management-domain structures at least make the decisions explicit, making them easy to locate.

- The subject domain factors out most presentation decisions – and many rhetorical and management decisions – entirely, while isolating and labeling other rhetorical decisions, making it much easier to locate the instances that need to be updated when a change occurs.

Changes in content happen all the time. Many of them are entirely predictable, and you can use structured writing to help manage those changes. For instance, companies re-brand from time to time, which often requires formatting changes for all publications. If your content is in media-domain structures, the effort to change to a new appearance could be significant. However, if your content is in the document domain, changing to a new appearance is simply a matter of changing the publishing algorithm.

It should be noted, though, that while changing the publishing algorithm is less work than changing the formatting of a large body of content, it is also more complex work. It requires a skill set that is not as widely available as the ability to change fonts in a word processor. It also cannot be done incrementally. Once the new algorithm is complete, you can convert all of your content to the new look almost instantly. But until it is complete, no content can be converted. A structured writing system is not something you can set up once and walk away from. If designed properly, it transfers complexity (and therefore decisions) from writers to information architects and content engineers, and, therefore, you need to maintain the availability of those skills to your team. Hiring someone to write a bunch of algorithms and expecting that those algorithms will never need to change ignores where complexity is being directed in your new system.

Moving complexity to algorithms is the heart of the productivity and quality gains that you get from structured writing, but transferring complexity to an algorithm means transferring it to the people who write and maintain your algorithms – your information architects} and content engineers. In many cases, your change management strategy will depend on your ability to handle certain kinds of change by updating an algorithm rather than updating thousands of pieces of content. If you have a fixed and invariant tool set, you essentially cut yourself off from the possibility of partitioning complexity, particularly the complexity of change, by updating your tools. This will greatly limit the extent to which you can bring your currently unhandled complexity under management, which in turn limits your ability to improve your rhetoric.

Moving to the document domain or the subject domain (or even to a disciplined use of styles in a word processor) will allow you to handle font and layout changes. But what if the re-branding goes further? Suppose it involves changing the names of products or even the company. Should your structured writing approach explicitly support that change? Some organization like to mandate that writers insert a variable rather than the actual name for the company name and all product names. That way, when a product name or the company names changes all you have to do is redefine the variables.

```
We here at >($company-name) do not recommend using
our product to catch roadrunners.
```

I am skeptical of the value of this practice. Writers must remember to use the variable every time, which interrupts their chain of thought, slows down their writing, and uses up some of their precious attention, thus affecting content quality. And it is virtually impossible to ensure compliance. Writers can forget and write the names out normally, which means that when a change occurs, you have to search for those instances anyway. Then there are issues with historical usage of the names (where you don't want the change to happen), inflections (if the new or old names end in 's'), and articles (if the initial letter changes from consonant to vowel or vice versa).[1]

Company and product names are distinct strings that are easy to search for when you need to make a change. The overhead of creating and maintaining variables is greater than the overhead of doing a search and replace through the content. Using variables, in other words, creates more complexity than it partitions and transfers. And doing a search and replace allows you to make intelligent choices about historical usage, inflections, and articles. If your content is in text form in a repository that allows you to do a search and replace across multiple files, search and replace is probably easier and more reliable than using variables. And you will need to do search and replace, anyway, if there is a name change that you did not anticipate and, therefore, did not encode in a variable.

However, you may still need some markup for company and product names. You may want to format them differently or link from them to more information about the product or company. Rather than use a variable, I prefer to use an annotation like this:

```
We here at {Acme Corporation}(company) do not recommend
using our product to catch roadrunners.
```

This second approach identifies the words "Acme Corporation" as a company name. Creating this markup requires no extra thought from writers. They do not have to remember what the appropriate variable name is. (They do have to remember company as an annotation type, but that is a type, not an individual name, and if your markup is well designed, your annotation types should be few and memorable.) And the same markup can be used to format the company name and to generate links to information about the company.

---

[1] These problems occur in English; other languages may have different problems.

This does not guarantee that writers will always remember to add the annotation, or that they will always spell the company name correctly. There is no way to guarantee that writers will always remember a free-floating annotation; the best you can do is make annotations easy to create. However, as discussed in Chapter 39, you can use the `company` annotation to find all the phrases marked as company name, sort them, and look for variants. This allows you to go back and fix incorrect spellings. And it allows you to identify the ways in which writers misspell the company name and search your content for those misspellings. This improves your success rate, catching both misspellings and failures to annotate. You should perform this kind of content hygiene operation regularly. Subject-domain annotation removes a distraction for writers and makes this operation easier to do. (For more on this, see Chapter 26.)

At another level, re-branding may require a change in tone or voice. You may decide to go from professional and reserved to friendly and jocular. There is no way, of course, for any structured writing process to recast content from formal to funny. You can't future proof your content against every kind of change.

Out with the old, in with the new

One easily overlooked source of change is the ongoing creation of new content and the editing of old content. This is a particular concern with web-based content, because you can add, edit, or delete content at any time. You don't have to wait for a major publication release to roll out changes. Each can roll out when it is ready.

However, each time you roll out a change, you affect the information architecture of the entire content set. Adding, editing, or deleting just one topic can have a widespread impact:

- There may be topics that link to the deleted topic.
- There may be topics that should link to a newly added or changed topic.
- There may be topics that should no longer link to a changed topic.
- Topics in a category may now have a new neighbor or may have lost one.
- Any top-down navigation tools need to be updated for the topic changes.
- Deleted topics may leave holes in the information set that need to be filled.
- New topics or edited topics may mention subjects that are not adequately covered by existing topics, revealing the need for yet more topics.

- Deleted or edited topics may leave other topics orphaned, needing to be removed or edited to serve a current purpose.
- Events in the world can change the status of a whole set of topics; for instance, an upcoming event becomes a previous event the moment it has taken place.

When adding, editing, or deleting a topic can have ripple effects through the whole content set, and when these additions, deletions, and edits happen on a daily basis, it is vital to have algorithmic support for change management. Managing all the effects by hand is doomed to failure.

Content management systems often have change management features that can be helpful. For instance, many of them will inform you if changing or deleting an existing topic will break existing links. They will also help you find topics on related subjects or manage the membership of categories and the navigation aids that are based on them.

What they won't do is tell you things such as which pieces of existing content should be linking to the new content you just added. The only way to discover that is if the existing content contains subject-domain annotations that relate to the subject for the new topic. With these in place, a linking algorithm (as described in Chapter 18) can discover these relationships automatically.

As should be coming clear by this point, change management is an aspect of all the structured writing algorithms. For any algorithm to keep working over time, and as content changes, the structures that support those algorithms have to stay conformant. Therefore, as you design structures, you should be thinking seriously about what it will take to maintain them in a conformant state when changes happen.

Change management, therefore, relies heavily on conformance to constraints and to the structures that express those constraints. This is a reciprocal relationship. To ensure that your content remains conformant, you must manage change successfully. But to manage change successfully, you need content to be conformant, so you can reliably identify required changes and, as far as possible, execute them algorithmically.

For example, consider a movie review site that contains movie reviews as well as biographies of actors and directors. Suppose you have a review of *Rio Bravo* marked up like Figure 40.1.

```
movie-review:
    movie-title: Rio Bravo
    review-text:
        In {Rio Bravo}(movie), {the Duke}(actor "John Wayne")
        plays an ex-Union colonel.
```

Figure 40.1 – Movie review for *Rio Bravo*

At first, you do not have a biography for John Wayne on your site, so none of the reviews that mention John Wayne have a link to a biography. Then, you add a John Wayne biography to your collection (see Figure 40.2):

```
topic:
    title: Biography of John Wayne
    subjects:
        type: actor
        value: John Wayne
    body:

        John Wayne was an American actor known for westerns.
```

Figure 40.2 – Markup for a biography of John Wayne

Now, every movie review that mentions John Wayne in the text or lists him as a star should link to this biography. That might be fifty pages of your site that now should link to the topic you just added. That is a lot of change to process just because of one added page. However, if you used the linking technique described in Chapter 18, which is based on subject annotations, then all you have to do is rebuild your content set, and those links will be created automatically. There is no separate change management step required to create those links. That step has been entirely factored out.

As I noted in Chapter 29, the heart of conformance is designing structures that are easy to conform to. Change management is not something you can tack on as an afterthought, but nor is it something that necessarily requires a separate set of structures. Content that is highly conformant and highly auditable is easy to change consistently, which in turn helps maintain conformance. The heart of the problem lies in designing content for conformance.

CHAPTER 41
Repeatability

However you design your content process, the key to its success is repeatability. Making it work once is great, but real success is making it work reliably day after day.

One of the biggest challenges to repeatability in content processes is determining whether you are repeating the right things and repeating them successfully. Content is one of the hardest products to test. One of the biggest barriers to consistently producing quality content is being able to test whether your content persuades, informs, entertains, or enables the reader to act as intended.

The problem has two parts:

- It is hard to observe the effect of content. You aren't there to watch people read it.
- Even when you can observe the effect of content on the reader, how do you know what aspects of your content's rhetoric achieved the effect you observed?

Making these observations on individual pieces of content is of limited value. Yes, you can do A/B testing – putting up multiple versions of content, observing which versions work best, and taking down content that does not perform well. And you can test other changes until you finally create a piece of content that performs well. But then, how do you reproduce that success with the next piece of content? How do you discover the rhetorical pattern that contributed to its success and should be emulated in future content?

Measuring content is not like measuring minivans or cans of peas, where every example is supposed to be exactly alike. Every piece of content is supposed to be different. It is great to find that you have a successful ad or blog post or manual topic, but to repeat that success, you have to look beyond the individual item and see what it has in common with other successful items. You need to find, and reproduce, the common rhetorical patterns shared by many different pieces of successful content.

Some part of the success of individual items doubtless lies in characteristics that they alone possess – your review of a Harry Potter movie is going to get more hits than your review of an art-house flop, for reasons entirely unrelated to the quality of your reviews – but much of it also comes from meeting specific user needs in accessible ways, and, in that case, there is often a pattern that can be repeated. Did your review of Harry Potter get more or fewer views than the next site's review of Harry Potter? Do your reviews of art-house flops get more views than theirs? What is it about your reviews that makes them more popular, regardless of the popularity of the movie?

Maybe it's your acerbic wit, of course, which is not easy for another writer to reproduce, but maybe it is the rhetorical pattern of your reviews – how you present the information that moviegoers really care about in a format that is clear and easy to read. If so, that pattern should be repeated across all your reviews.

Many content types, such as recipes, have generic rhetorical patterns that everyone knows and uses, but you can create more specific versions of these rhetorical patterns that will make your content work better than your rivals' content. With recipes, maybe there is a pattern that makes it particularly easy for your readers to choose which dishes they can prepare successfully. If so, you should repeat that pattern across all of your recipes. While literary charm doubtless counts for something – and for more in some forms of content than in others – saying the right things in the most accessible way is still the bread and butter of content quality, and structured writing is the best way to deliver quality repeatably.

Trying to derive lessons from a single piece of successful content is little more than an educated guess. You are abstracting from a single data point. To determine which aspects of your content contribute the most to its success, you need to observe multiple samples that exhibit the same features. When you see successful results from multiple pieces of content that exhibit the same features, you have some assurance that those features are what make the content successful. Once you know this, you can reliably produce new content that will be similarly successful. You have achieved repeatability.

It is beyond the scope of this book to examine specific content testing methodologies. But meaningful testing requires repeatability. The focus here is on how to achieve repeatability so you can test effectively using whatever testing methodology you choose.

To achieve repeatability, you need to partition those rhetorical elements that you want repeated and constrain your writers to follow them. But it isn't enough to constrain rhetorical patterns after the fact, when you already know what works. If you don't have multiple examples of the same pattern to test, you can't draw useful conclusions about which elements of the pattern work and which don't. You not only have to constrain in order to repeat, you have to constrain in order to measure. Without constraints, you don't know what you are measuring.

In order to generalize your measurements, so you can draw conclusions that apply to more than one piece of content, you need to make sure that each piece of content you are measuring has the same structures and features. That is the only way to be sure that it is the structures and features, rather than the individual texts, that are driving the results you measure. You need consistent conformant content structures that express those features of a text that affect its quality. Until you have that, you have no reliable basis on which to extrapolate findings from your measurements. In order to achieve repeatability in content creation, therefore, you need structured writing.

The most fundamental part of content quality is to give readers the information they need in a form they can use. A confounding aspect of content quality is that every reader is different. They are doing different things, they have different experiences, and they have different vocabularies.

For writers in corporate environments, lack of knowledge about the readers, their tasks, and their backgrounds, is often the biggest problem they face in determining what content is required and what form it should take. While direct contact with customers is undoubtedly the best way to address this problem, you can learn what content to create without ever meeting your readers, as long as you can measure the performance of your content, generalize your results, and repeat the structures that perform best.

If the bread and butter of content quality is to provide the right pieces of information in the most accessible way, and if you have some means of measuring the impact of your content, then you can use strictly constrained patterns and testing to establish what information and which presentation of that information works best for your audience. While it is best to seed this process based on direct knowledge of your readers, ultimately, tests and measurements are the most reliable way to learn what content works for your readers, even if you don't know exactly why it works.

Without known-good patterns and reliable tests, only personal knowledge of your readers will help you craft content that meets their needs. But this means that every writer must do reader research and information design work every time, with a limited body of information and few, if any, opportunities to test the design. The time required for each writer to do this research and design work is a huge overhead for the organization and a huge amount of repeated effort. If writers don't spend the time to do it right, the content probably won't meet the reader's needs. As much as we talk about the potential cost savings of information reuse, the potential cost savings and quality improvements from reusing known, good rhetorical patterns verified by testing is enormous.

Organizations often look to content reuse as the principle source of cost savings in the content system. But content reuse can be expensive to implement. Cost savings are not guaranteed, and quality problems can result from an overzealous approach to reuse. Repeatability, which you can think of as the reuse of patterns rather than individual pieces of content, can be a huge time-saver and one that, if applied correctly, can bring big quality gains. And the kinds of constrained patterns that provide repeatability also direct the complexity of information design away from contributors, opening the way to greater collaboration by bringing in writers who are experts in the subject matter but not information design and who do not have the time or inclination to use complex content management or reuse systems. You should consider carefully whether repeatability, rather than reuse, should be the first place you look to reduce costs in your content system.

Every structured writing domain provides support for repeatability. Word processor style sheets ensure repeatability in the formatting of headings and lists. Document-domain languages such as DITA and DocBook provide repeatability in document structures (but only to the extent that they constrain the use of such structures). However, to get repeatability in the rhetorical structure of content, you need to turn to the subject domain.

Finally, while I have focused on rhetorical repeatability in this chapter, partitioning complexity to make sure it is handled by a person or process that has the knowledge, skills, and resources to handle it contributes to repeatability at every point in the content process. And here again, we see the importance of conformance to known good patterns, because the effectiveness and reliability of algorithms depends entirely on the quality of the content and metadata that is passed to them.

CHAPTER 42
Timeliness

Information changes quickly these days, and readers no longer have any patience with outdated content. However, it is difficult to ensure that content is always timely.

- How do you detect when content is out of date?
- How do you push updated content to readers quickly?
- How do you make sure updates in one place don't break content in another place?

These are all sources of complexity both in the content process and in many of the algorithms we have already looked at. If the partitioning and redirection of complexity is poor in any of these algorithms, or if the algorithms leak complexity by not fully and appropriately handling the responsibilities assigned to their partition, this will reduce your ability to deliver in a timely fashion.

Traditional paper publication could endure inefficient processes because publication was a rare event. Indeed, publication was treated like a wedding day. All the effort and coordination of the preceding months went into making that day work, and normal life ceased several days before the event. The complexity of preparing for and then executing the events of the day was enormous and was tolerable only because publication was rare.

For modern web publishing, a wedding-day model of publication is untenable. Publication must be simple and quick, because every day is publication day. You cannot treat every day like a wedding day, with weeks of elaborate preparation. Instead, you need to maintain your systems and your content in a constant state of readiness to publish. If you attempt frequent publication without correctly partitioning and distributing complexity in your content, much of the complexity will get dumped on the reader in the form of outdated, inaccurate, or inconsistent content. Structured writing helps you address all of these issues.

Algorithms and timeliness

Timeliness rests on executing the entire content process – from ideas in a writer's head to dots on a page or screen in front of a reader – in the shortest time possible, while maintaining quality. Thus, all of the structured writing algorithms contribute to timeliness.

- **Separating content from formatting:** Formatting content by hand takes time and is subject to unintended variation through human error. If you want to publish quickly and with consistent formatting, you need to factor out the formatting from the content and hand the formatting to an algorithm. This will allow you to shorten the time it takes to publish by making formatting and output virtually instantaneous. It also helps speed up the writing process and improve content quality because writers don't have to divide their attention between writing and formatting. However, you need to ensure that the structures you create to factor out formatting are not more complicated than the structures they replace.

- **Single sourcing:** If you deliver content to multiple media, you need to avoid having to prepare content separately for each output. The less you have to manipulate content for each output, the better. A differential single sourcing approach helps ensure high quality in each medium without slowing down the publishing process.

- **Content reuse:** Content reuse can improve timeliness. If you can create a new piece of content by pulling in pieces of existing content, you may be able to publish faster. However, be careful; do not assume that this is an automatic win. The reuse process takes time. A lot of its payoff comes from avoiding re-translation of content, not from reducing end-to-end authoring time. Complex reuse systems also require writers to be conversant with the tools and the content set (so they can find reusable content efficiently). The availability of writers with these skills may affect your ability to move quickly. Reuse can improve timeliness, under the right circumstances, but don't forget that simplicity is a virtue when you need to act quickly.

- **Eliminating duplication:** Maintaining a single source of truth can be a huge win for timeliness – if you can achieve it. If you can ensure that there is only one source for a particular truth (however you define truth for this purpose), then you can simply deliver content from that source when and where it's required.

- **Linking:** One of the most challenging aspects of adding and removing content is managing the links. New content should link to any relevant content in the current content set, and current content should link to any relevant new content. When you remove content, all links to that content should be updated, either to remove the link or to link to something else. Any structured writing approach that manages links will help, but the most efficient way to deal with this challenge is to use the subject-domain linking approach detailed in Chapter 18.

- **Publishing:** Automating publication can allow you to release content more quickly. This is particularly important if you need to release content to multiple media or if you have made changes that affect multiple pieces of content. Continuous publication requires your publishing processes to be mostly hands off and highly reliable. Merely automating the publishing build does little for you if your output requires extensive manual quality assurance. You need to

build reliability into your content and processes from the beginning so that you can confidently press the publication button without needing to look at the output.

- **Generating content:** Changes to subject matter don't just affect individual pieces of content. They affect the overall information architecture. If you know that your content is going to change on a regular basis, it makes sense to create an information architecture that is highly adaptable to change. Many of the artifacts that implement your architecture – the links, lists, categories, and menus that shape the structure and navigation of your content – can be generated with algorithms based on existing metadata (particularly in the subject domain). The more you use algorithms to generate these artifacts, the easier it will be to adapt your architecture to changes in subject matter or business requirements.

  However, this is not just about structured writing. It is also about an approach to architecture that allows for algorithmic generation of the architectural pieces of your content set. A bottom-up information architecture is particularly effective at handling rapid and constant change. Implementing a bottom-up architecture, which relies heavily on well-structured topics connected by rich linking, benefits enormously from structured writing techniques. By changing the way content is organized and linked, these techniques can allow you to add and remove individual pages from a content set without fear of breaking things.

- **Extract:** The extract algorithm is not just a great shortcut for generating content, it is also a great way to keep up with changes in the real world. If the data source you draw content from is updated, repeating the extraction and republishing brings your content up to date. You can also run the extract algorithm dynamically, pulling content from the source when a reader requests it. If you have a fully automated publishing process that does not require any user intervention, you can effectively publish live data right out of the source file.

- **Merge:** In many cases the extract algorithm works with the merge algorithm to combine content your team has written with content extracted from an external source. In this case, you can track changes in the external source to flag when your local content needs to change.

Process and timeliness

As with algorithms, your processes must be working well to maintain quality and provide timely delivery.

- **Content management:** Every change in your content set is a content management action. Unless you design your system carefully, the cost of each content management action will increase as your content set grows, because each action must account for its impact on more and more resources. Using structured writing techniques to automate aspects of content management can help keep the cost of content management from escalating and make each content management action easier and more reliable.

- **Collaboration:** Collaboration can improve the timeliness of content by making more writers available to contribute content and by moving content creation closer to the source of information changes. Structured writing can be a double-edged sword when it comes to collaboration. Automating publishing and implementation of information architecture, makes it easier to coordinate the efforts of multiple contributors. By supporting composability, structured writing can help ensure that the contributions of different collaborators work well together. And by enforcing content constraints, it can help ensure that all contributors create content that does the job it is supposed to do. However, many structured writing systems are difficult to learn and use, making it difficult for collaborators to contribute. They can also be expensive, making it uneconomical to include occasional contributors as writers. Both of these problems can be avoided by focusing on the functional lucidity of your structures and markup, which not only avoids the heavy learning curve of structured writing but also removes the need for complex editing tools.

- **Auditing:** A huge part of timeliness is knowing when you need new content or edits to old content. The world changes all the time, but unless you have an efficient way to determine the impact of changes on your content set, you can't respond to them in a timely manner. Taking a structured writing approach to auditing, as outlined in Chapter 39, can help ensure that you always know when changes need to be made and that you can always find the content that needs to be changed.

- **Change management:** Although the audit algorithm will alert you to the need for changes, you still need to manage the execution of those changes in such a way that they happen quickly and without compromising quality. Using structured writing techniques to facilitate change management can help ensure that you can make the required changes quickly and reliably.

- **Repeatability:** Repeatability is key to timely delivery. Repeatability is the ability to make sure that you are doing the same thing every time. If your processes don't support repeatability, you can't deliver quickly without the risk of introducing variability into your content and compromising its quality.

The way you select and integrate the other structured writing algorithms and processes has a big impact on your ability to deliver in a timely fashion. Optimizing pieces individually may not give you the time savings you are looking for or may not maintain your quality standards. You need to carefully consider where bottlenecks can occur in your overall process and identify any points in the chain where complexity may go unhandled and end up being dumped on readers.

Although your content structure design activities depend mostly on the specific algorithms you want to support, you also need to consider how those structures and algorithms affect the timeliness and quality of the entire process. For example, complex document-domain/management-domain reuse structures may save time by avoiding duplicate work, but you need to consider the time it takes writers to use these features and any quality traps that lurk in their use.

You also need to consider whether such features leave you overly dependent on writers with specialized training. The most technically efficient system is of little value if the only person who can execute it is on vacation. All structured writing systems require special skills in certain key roles, but it pays to avoid setting up your system in a way that creates skill bottlenecks, especially for the skills required to execute quick content changes.

One of the key features of the subject domain is that subject-domain structures support multiple algorithms with the same markup. This can be an enormous benefit in ensuring timeliness. Not only does it simplify the writer's job when changes need to be made quickly, it hands more of the management and production phases over to algorithms, which are always faster than people. It also means that content updates require no knowledge of the management systems, which helps avoid a skills bottleneck for content changes.

CHAPTER 43
Translation

Translation is a major source of complexity for organizations that deliver content in multiple languages. Complexity comes not only from the translation process but also from integrating the translated content back into your publishing process.

Translation is a huge subject in its own right, and I have neither the space nor the expertise to do it full justice. Therefore, I will make just a couple of points about how structured writing can help partition and transfer certain aspects of the complexity of translation.

Avoiding trivial differences

Trivial differences in how the same thought is worded make little difference to a reader, but they can run up translation costs. Structuring content to avoid these trivial differences can reduce translation costs.

There are two principle ways you can do this with structured writing:

- Reuse the same piece of content every time the same thought is expressed. However, this approach comes with complexities of its own. The smaller you make the pieces of content, the more reuse you can get, but the more pieces you have to manage and the more pieces writers have to look through to find reusable content. And it is hard to constrain and maintain rhetorical quality when writers write in tiny fragments.
- Factor out the content that repeats the thought. I have shown several examples in this book. When we moved the recipe example into the subject domain, we factored out the titles of the recipe sections, eliminating any trivial differences in the titles that writers might introduce. In Chapter 13 we factored out a repeated safety warning by adding an `is-it-dangerous` field to the `procedure` structure. This removed the need for writers to think about reusing the warning, and it also factored out any issues regarding the file name for the different language versions. (Something similar could be achieved using the keys approach described in the same chapter.) The modeling example in Chapter 17 factored out large amounts of text, enabling you to generate procedure text from a model. Translating the model and the common text supplied by the algorithm would allow you to do the same thing for other languages.

Isolating content that has changed

When you revise text, such as when you bring out a new version of a product, some of the text changes, but much of it remains the same. You can save time and money on translation if you only translate the content that has changed. However, to do this you need a way to isolate the content that has changed from that which has not, and you need a way to integrate the changed material back into your content after it has been translated. Structured writing allows you to clearly define structures in your content set and deliver just those structures that have changed.

Continuous translation

If you write a book and then send it for translation, the translated version will be ready to release long after your first language version is ready. This delay can result in missed opportunities. To bring out first language and translated versions simultaneously, or close to it, you need translation to occur simultaneously with the development of the first language content.

Continuous translation is not just about getting all versions of a book released at the same time. On the web, you may be releasing new pieces of content every day. If you maintain translated versions of that content, you want them to go out at the same time. You may also want to provide links between the different language versions of the same content because searches and links do not always send readers to content in their preferred language. A structured approach to publishing and information architecture can accommodate these requirements.

Design

System Design

Unless you are writing directly on paper, chiseling into stone, or drawing letters by hand in a paint program, you are using structured writing techniques to create content. You are using tools that have been designed to partition and distribute part of the complexity of content creation in some way, and you are observing a discipline – a set of constraints – imposed by that partitioning. It is not a matter of whether your system is structured or not, but how and for what purpose it is structured and whether that purpose is being achieved.

Therefore, the question you need to address is whether your current partitioning ensures that all of the complexity in your process is directed to a person or process with the skills, resources, and time required to handle that complexity. Or is complexity going unhandled in your system and compromising efficiency and rhetorical quality?

If this complexity is not appropriately handled, your rhetoric will be compromised and your customer satisfaction and bottom line will suffer. At that point, the question becomes, how do you change your approach to structured writing to handle complexity better?

Do not approach this problem piecemeal. Complexity cannot be destroyed – it can only be redirected – and every tool you adopt introduces new complexity, which must also be partitioned and redirected. Attacking one piece of complexity in isolation directs that complexity away from the area you attacked, but that complexity goes somewhere, and if you don't think about where it goes and how it will be handled, you can easily end up with more unhandled complexity in your system than you started with.

Because structured writing has the ability to move the complexity around the system, imposing it on one role at the expense of another can lead to piecemeal rather than comprehensive solutions as different groups or different interests try to attack their part of the problem. For instance, there is a long history of IT departments choosing content management systems because they were easy for the IT department to install and administer, only to have those systems become deeply unpopular because they pushed complexity out to writers and other users. On the other hand, some groups of writers want to write in uber-simple formats, such as Markdown, despite the difficulties that the limited structure and capabilities of those formats create for the overall publishing process. Each group tries to simplify its own work, heedless of the expense to others.

Here is the inescapable fact: complexity is irreducible. It can be moved but it cannot be destroyed. The only way to remove complexity from a process is to stop doing something. Therefore, the first question you should ask is: what you are doing that you should simply stop doing? What is not creating any value?

But let's assume that you have done that. Once you've stopped doing things that don't need to be done, simplifying a function in one part of your process will always move complexity somewhere else in the system. Moving complexity somewhere else in the system is fine as long as you hand that complexity to an appropriate person or algorithm. The point is not to eliminate complexity, but to make sure it is handled correctly. The enemy is not complexity, it is unhandled complexity.

How do you design a content system that ensures that as much complexity as possible is handled by a person or algorithm that has the time, skills, and resources to handle it? You start by understanding the sources of complexity in your content system.

Identifying complexity

Where is the complexity in your content system? It is different for every organization. Although the basic functions of content development are the same for everyone, the amount of complexity these functions generate differs greatly, depending on your needs and circumstances. For example, if you translate content into 35 languages, translation will generate more complexity in your content system than it will in the system of an organization that doesn't translate or translates into just one language. If you have a complex product line in an industry with a complex vocabulary, terminology control will create far more complexity in your system than for an organization with a simple consumer product. To partition and redirect complexity effectively, you need to identify the biggest sources of complexity in your content system. Here are some places to look:

Current pain points

Inconsistency, duplication, delay, error, and failure are escape valves for unhandled complexity. The complexity underlying these problems can be difficult to see, even when it is causing pain. Familiarity with your current processes may make it difficult to see that things could be different. Seeing that there are alternatives often provides the insight into the hidden complexity of the current process. Hopefully the chapters in this book on the structured writing algorithms will help you see your current processes in a different light and, therefore, help you to identify the complexity in them and assess whether it is being adequately handled.

However, to find the sources of complexity, you may have to work your way upstream from where the problem is manifest. Ultimately, all unhandled complexity finds its way to your readers, so quality problems are the most common symptom of unhandled complexity. You may also see the impact of unhandled complexity in processes that seem too expensive or staff that seem overburdened or burnt out.

Let's say that you observe a rhetorical problem: readers are getting lost in the content. They find an initial page, but when they search for more information on the concepts it mentions, they end up back at the same page. What is the root cause of this problem? It could be one of (at least) three things:

- **Incomplete information:** The information should be on the page, but it isn't. In this case, the unhandled complexity is in determining the readers' needs and making sure that needed information is present. You have a repeatability problem.
- **Missing information:** The information is not available anywhere in the content set. In this case, the unhandled complexity is in content planning. You may need to improve your audit capability.
- **Inaccessible information:** The information is in the content set, but there is no link to it. In this case, the unhandled complexity is in managing and expressing relationships between subjects. If writers are unaware of these relationships, you may have a rhetorical-structure problem. If they are aware, then the cost of creating and maintaining links may be too high, leading them to skimp on linking. You may need a different approach to linking.

When you do this analysis, be careful to avoid these two common pitfalls:

- Don't stop your analysis too soon, and don't be blinded by the limits of your current process. Trace the problem back to the actual source of complexity, independent of any tool or process considerations.
- Don't stop with the first big chunk of complexity you find and rush off to buy a tool to address it. You need to identify all the sources of complexity in your system and come up with a comprehensive plan to manage them. Otherwise, you just move complexity around, while adding tool complexity to the mix. You could easily end up worse off. Even your biggest hot-button issues may demand a different solution after you have assessed all of the complexity in your system and created an overall plan to partition and direct that complexity.

Unrealized possibilities

Chances are that there are things you don't do today because they are not possible with your current tools, or any of the tools you are aware of. It is easy to overlook these unrealized possibilities because everyone else is dealing with the same limitations, meaning you may not see those possibilities implemented elsewhere. But don't allow your designs to be limited by what your current tools can execute.

A comprehensive bottom-up linking strategy is a good example. If you are working in the document domain – using document-domain or management-domain linking techniques – creating, managing, and updating links is expensive and complex. That complexity leads writers to minimize the links they create or create them in sub-optimal places. That complexity manifests itself as unrealized potential.

However, if you switch to subject-domain linking based on subject annotations, you partition and distribute the complexity of linking away from writers and towards an algorithm. This radically reduces the barriers to a comprehensive linking strategy, making such a strategy possible. And you don't need a full subject-domain system to use this approach. It works just as well if you add subject annotations to otherwise document-domain information types.

What's working now

Some parts of your current system are working well, successfully partitioning and directing complexity to the right people or processes. However, you still need to inventory this complexity to make sure it will still be handled well in your new process. Also, even if you are satisfied with parts of your current process, don't assume that those parts work optimally – you may just be settling for what you are used to.

For instance, if you currently handle complex tasks with an algorithm – for example, you generate reference information from source code using Doxygen – make sure you add those tasks and solutions to your inventory so that when you design your new system, you still handle that complexity properly.

At the same time, look around the edges of your current success stories for any unrealized potential. For example, if your generated references are published through a separate tool chain that makes them look different from, and not link to, the rest of your content, the unhandled integration complexity will compromise your rhetoric.

Even if some of your current processes handle complexity well, you may need to change the way they distribute complexity in order to better manage complexity in your overall environment. Don't lose sight of complexity just because you are currently handling it well.

Complexity that affects others

More difficult to identify are points where the impact of complexity falls on someone else. You don't feel that pain, because someone else is suffering the consequences. Often this is your readers, but it could be others in your organization – support, sales, field engineering, etc.

For instance, if you ask developers to contribute content to your documentation set, but you are working in a complex document-domain or media-domain tool that they don't have the bandwidth to learn, you could be dumping a lot of authoring complexity on them.

You may get push back from the developers, saying that they are only willing to contribute content in a simplified markup language, such as Markdown. This simply deflects complexity onto someone else, since such languages may not give writers the structure and metadata they need to integrate the content with your publishing system.

Here you have a functional lucidity problem. The complexity of integrating content into your content set is not sufficiently partitioned and directed away from the occasional contributor. Chances are there is also a rhetorical problem here: developers may not know what information is needed, and they may not have received any rhetorical guidance. The solution may be to give them a subject-domain format that precisely specifies the content required. Developers may be willing to use this format (if its functional lucidity is good) precisely because it makes their lives easier by redirecting rhetorical complexity away from them.

Communication overhead

Any partitioning of a task creates communication overheads, because information has to be transferred between partitions. Communication overhead can be a huge source of complexity. Desktop publishing (DTP) was revolutionary in its day because it combined three jobs into one, eliminating communication overhead between writers, designers, and typesetters. This eliminated the communication overhead, but it made the writer's job much more complex.

When designing a system it is important to think about how much communication overhead your partitioning will create. In particular, consider how much out-of-band communication will be required to make your system work. Out-of-band communication means all the conversations and exchanges that are not captured in the content itself.

For instance, in a document-domain content reuse scenario, the markup provides a way to include content, but writers still have to find the content. In addition, they may need to communicate with the person who creates and maintains the content to determine whether it is suitable for reuse and how it might change in the future. This is all out-of-band communication that is required to make reuse work. Content management systems and workflow systems are often used to manage out-of-band communication about content.

Organizations often place a lot of emphasis on improving collaboration by opening communication channels. However, opening communication channels is not the same thing as encouraging or facilitating out-of-band communication. Out-of-band communication, particularly around routine tasks is neither fruitful nor efficient. Instead, the real secret to effective collaboration is to improve your in-band communication, which improves productivity by reducing the need for constant meetings and emails.

You can reduce the overhead created by out-of-band communication by placing more information in-band in your content and by designing your partitioning to avoid the need for a lot of out-of-band communication between partitions. For instance, with the subject-domain approach to linking, subject annotations contain in-band information about subjects mentioned in a document, making it possible for algorithms to find link targets at build time with a minimum of communication overhead among writers.

Your style guide

Traditionally, style guides have been a dumping ground for complexity. Every new rule and requirement created in response to a content problem gets written into an ever-growing style guide. Adding to the style guide does not handle complexity, it merely dumps complexity on writers. As the style guide grows, writers lose the capacity to remember or follow all of the rules and rhetoric suffers.

This does not mean you don't need a style guide. Some style issues cannot be effectively factored out, and writers need guidance to deal with those issues consistently. However, a writer's ability to conform to stylistic constraints is directly proportional to the number of such constraints. Using structured writing techniques to redirect as many style issues as possible away from the writer (and thus from the style guide) increases the conformance to those issues that cannot be partitioned or directed away by giving the writers less to remember.

A great way to inventory the complexities in your system is to go through your style guide looking for rules you could factor out by moving content to a different domain.

Change management issues

The growth in unhandled complexity is often interpreted (especially by the advocates of a system you have just installed) as a change management problem or a training problem. However, even though change management and training are both necessary any time you introduce a new system, design issues are the most likely cause of problems. In particular, problems often come from a system design that redirects complexity without considering where that complexity will land or how it will be handled.

Change management is a scapegoat for process complexity that no one will own. It is common to blame the failure of structured writing systems on writers being unwilling to change. The cure, the backers of the system assure us, is more training and more vigorous promotion of the system and its virtues. But the problem is more likely to be that the system design has dumped new and unmanageable complexity on whatever group is rebelling – almost always the writers, either full-time writers or occasional contributors.

What's coming down the road

The world is in the middle of a transition from paper to hypertext.[1] This involves three types of complexity. First, the complexity of architecting information and publishing to paper. Second, the complexity of architecting information and publishing it to hypertext. Third, the complexity of transitioning your content processes from exclusively focusing on paper and paper equivalents (such as PDFs) to either publishing both or only hypertext.

Most organizations currently produce both paper and hypertext, but most are not doing both well. As noted in Chapter 12 and Chapter 8, there are major differences in how you write, organize, present, and format content for paper and for hypertext, and differential single sourcing requires different tools and techniques than are found in most single sourcing tools today. Most tools allow you to design for one medium and then do a more-or-less best effort attempt to transfer that design to other media.

---

[1] I say hypertext here rather than web because the web is both a delivery vehicle and an information architecture. You can publish to the web by sticking a PDF on a server. This is not hypertext. On the other hand, there are hypertext media that are not on the web, such as hypertext CD-ROMs or help systems. In terms of creating complexity, it is hypertext, not the web itself, that makes a difference. Hypertext requires a different approach to rhetoric and information architecture and requires major differences in the linking and publishing algorithms.

This means that in practice, whether you acknowledge it or not, you have a primary medium and a secondary medium. And while there are differential single sourcing techniques that can improve your ability to address each separately, there are limits to what you can do to address both without rewriting content to fit two different information architectures. In short, unless you plan to go web only (and some organizations are web only for a large part of their content), you have to choose a primary medium.

Shifting to a new primary or exclusive medium can remove some complexity from your system. Reuse, for instance, is a much bigger issue for print than for hypertext, where you can link to common material rather than including it inline. Single sourcing, and differential single sourcing in particular, are larger issues in the transition period than they will be if you stop producing paper formats. Linking and the techniques of a bottom-up information architecture, on the other hand, belong to the hypertext world and are likely to grow in importance as you turn more of your focus to the web and online media.

If you choose your tools and your structures primarily on the basis of the needs of the paper world or the transition period, you are likely to find yourself going through a similar upheaval in just a few years when the transition to hypertext catches up with you.

Partitioning and directing complexity

Once you have identified your major sources of complexity and have a good understanding of what that complexity costs you in terms of process and quality, it is time to decide how to partition and divide that complexity and to choose the structured writing techniques you will use to achieve that partitioning.

In some cases, you can reduce the complexity that falls on each individual in a group by taking a complex operation away from the group and distributing it to one uniquely qualified person. For example, when you separate content from formatting, you take formatting responsibility away from all writers and distribute it to a single designer who writes the formatting algorithm.

In other cases, you can take a complex operation currently being performed by a single person and distribute it out to many contributors. For example, linking is typically the sole responsibility of the person writing the piece that contains the links. But the subject-domain approach to linking, which uses subject annotation rather than link markup, partitions and distributes the linking task in three ways. Writers identify significant subjects in their text, other writers annotate or structure their content so that its subject matter can be identified clearly by algorithms, and an information architect or content engineer writes the algorithm that handles either linking or some other method of handling subject affinities. In another example, the bottom-up approach to terminology management partitions and distributes the task of terminology discovery to writers while distributing terminology management to a terminologist.

Partitioning the writing task

As is no doubt clear by now, I regard the heart of this process to be partitioning and directing complexity away from writers. Writers are the principle source of value in your content process. Writing requires full attention, so any complexity that writers have to deal with diminishes the attention they have available for writing, which always compromises rhetoric and content quality. Achieving functional lucidity for writers should be the first priority of your content system design.

Because attention is a limited resource, there is value in partitioning and distributing tasks even when you perform all of the tasks yourself. Partitioning tasks allows you to give your full attention to each individual task and, thus, perform better than you would if you tried to divide your attention between multiple tasks at the same time.

But functional lucidity is neither fixed nor universal. It depends on each writer's background and skill set. It also depends, to some extent, on the nature of the writing task. Functional lucidity for creating entries in a reference is different from functional lucidity for creating a slide deck or an essay on the architecture of a product. For any activity, familiarity and frequent practice reduce the amount of attention required to do it well. Thus, a full-time professional technical writer should be able to handle more content management responsibility and need less rhetorical guidance than an occasional contributor. On the other hand, a full-time writer might need more rhetorical guidance than a subject matter expert to create highly technical material, because the writer may not understand the technology or the reader's task as well. However, no matter how much complexity they can handle, writers will always benefit from a system that maximizes functional lucidity and partitions complexity appropriately.

That said, there are times when you should direct complexity toward writers. As we saw when we looked at terminology management, writers understand how terminology is being used and what subjects your terminology needs to cover. A terminologist, however skilled, has less of this day-to-day knowledge and may miss subtleties that need to be expressed or be unaware of local terms that would communicate those subtleties most effectively to a particular audience. Subject annotation partitions terminology discovery towards writers without adding any complexity, as long as they are already creating subject annotations for other reasons. And it partitions conformance checking and decision making towards the terminologist.

Partition towards expertise

A good principle is to always partition and direct complexity towards expertise. If expertise is distributed, direct complexity outwards. If expertise is centralized, direct complexity inwards. For example, bottom-up terminology management partitions language choices towards writers – who understand the context for the terms they use – while allowing discovery and conformance to flow back to terminologists – who have the skills to look at the broader issues of conformance across the content set.

In many cases, you want to direct complexity to algorithms. To do this, you need to clarify the rules that an algorithm must follow, partitioning complexity between processes that follow consistent rules and, therefore, can be handled by an algorithm and processes that require human attention.

Using subject-domain annotations to drive linking is a good example of this. Finding a resource to link to for every significant subject mentioned in a content set is complex and tedious. The task becomes even harder when content changes all the time and has to be continuously published. Even so, the basic rules for link discovery are consistent. Where subject X is mentioned, create a link to the best resource or resources on subject X. The problem is identification. How do you identify subjects that are worth linking to and how do you identify the best resources for those subjects?

Forming links by hand is complex and time consuming. However, if you annotate phrases and topics by subject, you can provide algorithms with the information they need to create links accurately. But there is a wrinkle to this. If you don't control the terminology you use to talk about significant subjects, simply naming the subjects might not be enough for an algorithm. Inconsistent or ambiguous names can cause an algorithm to misidentify subjects and topics. Adding type information to subject annotations (such as identifying *Rio Bravo* as a reference to a movie) removes the ambiguity and lays the groundwork for appropriate terminology management. This is a complicated piece of complexity management and partitioning, but it allows you to remove a lot of complexity from writers while greatly improving linking, change management, and terminology management.

Focus on complexity, not effort

Complexity is by no means the same thing as effort. Well-designed structured writing systems that support appropriate algorithms can reduce overall effort and significantly improve quality. However where you place complexity matters. Even if a task requires less effort, adding complexity changes how those assigned to the task work and how they need to be qualified and trained. It is important to appreciate how the distribution of complexity and effort affects the dynamics and composition of your team.

For example, managing links using keys, as described in Chapter 18, makes it easier to change link targets in reuse scenarios, compared to editing links by hand. However, creating links in the first place is much more complex, requiring more knowledge and skill. By contrast, managing linking using the subject-domain method of subject annotations (described in the same chapter) reduces both complexity and effort for writers, but it distributes more complexity to the processing system and those who maintain it. Each method for managing links – by hand, by keys, or by subject annotations – represents a different distribution of effort and complexity.

In the end, the question of where you distribute complexity in your system is at least as important as the question of how much effort you avoid. The wrong distribution of complexity compromises quality by undermining both the productivity of those saddled with misplaced complexity and the reliability of your algorithms. Not only does badly distributed complexity compromise quality, it also undermines cost saving efforts.

Distributing complexity away from writers is key because when a structured writing system distributes complexity towards writers, it doesn't merely add a new and complex task, it imposes that complexity directly on the activity of writing itself. Giving writers more complexity than they can handle not only affects the quality of their content, it also affects the correctness of the structures they create. Algorithms that reduce effort rely on accurate structures. If overloaded writers create poor content structures, algorithms cannot work properly, which will result in even more effort spent fixing problems.

You will probably achieve a greater reduction in effort by focusing on partitioning complexity rather than attempting to reduce effort directly. Indeed, it is sometimes worth investing in activities that seem like additional effort – for example, creating structures and algorithms to improve repeatability and conformance or creating specific subject-domain structures for a new subject – instead of knocking out content in a generic document-domain format. Avoiding up-front effort may seem like a win, but in many cases, the down-stream effort created by less reliable content and algorithms will be far greater.

Indivisible complexity

There are complex tasks which by their very nature have to be done by one mind. It is important to respect this indivisible complexity when planning your content system. Splitting functions that belong together, such as writing semantic blocks and combining them into sound readable rhetorical blocks, can cause severe quality problems. Remember that partitioning complexity always requires you to create a structure that transmits all of the information required for the next partition to do its job. If there is too much required information to be passed, or if the required information can't be easily expressed in a standard way, then you haven't reduced the complexity of the first partition sufficiently, and complexity will fall through the cracks. That which cannot be cleanly partitioned should be kept together.

Subtract current tool complexity

All tools introduce complexity into the content system. We choose them because they allow us to partition and distribute complexity better, including their own complexity. But when you are inventorying the complexity of your system, the complexity introduced by your current tools should not be part of the inventory. Those tools may be going away as a result of your system redesign, so their complexity should not be part of the inventory of complexity in your content system.

Avoid tool filters

When you design a process, it is important not to see things through the filter of your current tools. Every tool reflects its designer's view of how the complexity of the content system should be partitioned and directed. Thus, because the way you partition and direct complexity defines your process, every tool encapsulates a process. When you buy a tool, you buy the process it encapsulates. Long practice with a tool shapes how you view process, and, after a while, it is hard to imagine a process in any other terms.

Thus, all too often when we write the specifications for a new tool, we essentially end up asking for our old tool with some particular improvement we think will make our lives better. But often, the improvement we are seeking is not compatible with the current tool. (If it were, the vendor would probably have included it in their ongoing attempts to drive upgrade sales through new features.) If your old tool won't cut it, chances are you need a different process from the one incorporated in that tool, and you need to break through your tool filter to envision a new process. That is why this book has focused on algorithms and structures rather than tools and systems – to help you overcome the tool filter in your process design decisions.

Over the years I have seen many requirements documents for proposed structured writing systems that essentially said that the proposed system must work exactly like Microsoft Word. This is not surprising when the people writing the requirements have used nothing but Word to create content for years. The tools you know shape how you work and what you think of as possible. As Henry Ford is supposed to have said about the Model T, "If I asked customers what they wanted, they would have said faster horses." Even when we are dissatisfied with our current tools, we tend to want the same basic tool only more so. That is why so many structured writing tool vendors advertise that their editor looks and feels "just like Microsoft Word" and why some vendors create tools that modify Word itself.

But Microsoft Word sits on the boundary between the media and document domains. Using Word, or something that looks like Word, for structured writing is usually an attempt to move writers a little bit further into the document domain without disrupting their familiar work environment. But the WYSIWYG authoring interface invites writers to slide back into the media domain by hiding structure and showing only formatting, which should be factored out in the document domain.

It is not surprising, then, that the most popular structured writing tools to date predominantly use document-domain structures that are, like DocBook, very loosely constrained. It is much easier to write an XML document in a WYSIWYG editor if the underlying structures are minimally constrained, since you can insert whatever bit of formatting you want anywhere you want, just as you can with Word. Even DITA – which, while fundamentally in the document domain, is more constrained and capable of being constrained further – tends to be used in its generic out-of-the-box form with tools that provide a Word-like WYSIWYG interface.

Thus even when the decision-making process is based on business requirements rather that specific tools, it is often tacitly driven by existing tool sets and ways of doing things, because those existing tools and processes shape our view of what the business requirements actually are. We don't ask for a better way to get from Des Moines to Albuquerque, we ask for a faster horse that eats fewer oats. This is why it is important to forget about your current tools and their associated processes and focus on the sources of complexity in your system and how you can partition and distribute that complexity effectively.

Selecting domains

Once you have a good idea of where the complexity lies in your content system and how you would like to partition and direct that complexity, it is time to decide which domains you want to work in. This is not as simple as picking one of the three and using it for everything. There are some pieces of content for which no meaningful subject-domain markup makes sense – for example, content that has no repeatable pattern of either subject matter or rhetoric. For such content you need generic markup in the document or media domain. Some content needs to be laid out by hand with an artist's eye for design. That can only happen in the media domain. In practice, your content system is likely to include a mix of subject-domain, document-domain, and media-domain content, with some measure of the management domain thrown in where needed. In fact, major public languages like DITA and DocBook contain structures from all four domains.

Also, the places where complexity is hurting your content are not necessarily the same for all content types. The complexities that attend the maintenance of a reference work are likely to be different from those that attend the creation of a full-color print ad in five languages. Different types of content require different partitioning and direction of complexity and, therefore, require different structures from different domains.

In an ideal world, the choice of domains would be simple. Given the complexities you have identified for partitioning and redirection, choose the domain that accomplishes the desired partitioning. But, in practice, it is more complicated because structures and the algorithms that process them are themselves sources of complexity.

For example, DITA may address the complexities of content reuse, but it also introduces complexities into the writing process in the form of complex markup and management-domain intrusions. DITA also introduces significant content management complexity, both because it produces so many small artifacts and because writers need a way to find reusable content. This creates a retrieval problem, which creates terminology management complexity. Handing these new complexities requires new tools, typically a DITA-aware structured editor and a DITA-aware component content management system, along with new roles and extensive training. And even with those things in place, writers must deal with a lot of conceptual and management complexity. And you still have issues with rhetorical conformance because there is no way to constrain rhetorical blocks larger than a DITA topic.

All of that additional complexity, and the cost of the tools, may be worth it if you can realize big enough gains from reuse without falling into its quality traps. For many organizations, the savings from reduced translation costs swing the needle to the positive side rather than any benefits from reuse alone. But some organizations have reported that they simply don't realize the amount of reuse that would justify the expense and complexity of their DITA systems.

DITA provides every document- and management-domain reuse algorithm in the book, but at a high cost in additional introduced complexity. You might find you are better off with either a reuse system that doesn't impose the kind of information typing constraints that are fundamental to DITA or a simple document-domain system that provides a less comprehensive suite of reuse features but injects less complexity into your process. Or you might be better off with subject-domain reuse tactics that are less comprehensive than DITA-aware document- and management-domain tools but which actually remove complexity from the writer rather than adding it, improving both functional lucidity and conformance. It is usually better to optimize the overall management of complexity across your entire system rather than to optimize one function at the expense of others.

Another source of complexity to consider when you choose domains is the amount of in-house development needed to implement them. Media-domain tools such as FrameMaker or public document-domain tools, such as those that support DocBook and off-the-shelf DITA, all come with ready-made structures and algorithms. To implement a subject-domain strategy for your own content, you have to develop some structures and algorithms yourself. This obviously adds a level of complexity to your process.

However, even with media-domain and public document-domain tools, you will almost certainly have to do some structure and algorithm development, probably on an ongoing basis. Developing and maintaining FrameMaker style sheets is structure and algorithm development, and supporting tools such as FrameScript transfer mundane formatting or management tasks from writers to algorithms but require effort to develop those algorithms. If you want your DITA or DocBook output to match your brand, you need to develop formatting algorithms or modify the existing style sheets to create the look you want. And if you do any kind of DITA specialization or DocBook customization, you will need the same development skills that would you would need to develop your own subject-domain structures and algorithms.

As we have seen throughout this book, the subject domain provides the most comprehensive set of structures and algorithms for addressing structured writing requirements. In particular, it does the best job of partitioning and directing complexity away from writers, which potentially opens your system up to a wider range of contributors. It also provides for higher levels of conformance, which in turn means more reliable algorithms. And with the subject domain, most algorithms work on the same set of structures, as opposed to the document and management domains, where each algorithm requires new structures. Finally, while there are more of them, subject-domain structures tend to be small with few permutations of structure, meaning that their algorithms have less complexity to handle. All this helps contain development complexity.

Nonetheless, even if you, for example, identify the subject domain as the best way to create the bottom-up information architecture you want, you might decide that writing and maintaining the necessary structures and algorithms introduces too much complexity. In that case, you might be better off with a solution based on a wiki or a simplified markup language such as Markdown. This would throw more of the responsibility for linking and repeatability onto writers, but that might be the right balance to strike in how complexity is partitioned and directed in your organization. Scale can play a big role in this decision. At a large scale, managing linking and repeatability by hand becomes onerous, while developing and maintaining subject-domain structures is amortized over a larger body of content.

We can usefully map the various content management domains in terms of two properties: complexity and diversity. Complexity measures the size of a language and its level of abstraction. Diversity measures the number of different languages required to represent a set of content (see Figure 44.1).

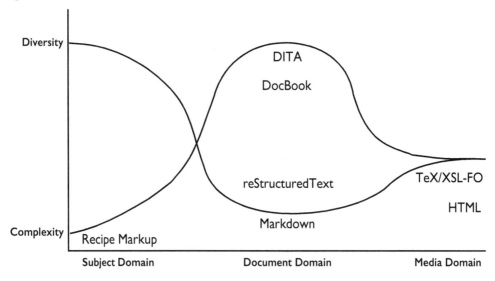

Figure 44.1 – Diversity and complexity of markup languages in the three domains

As you can see in Figure 44.1, the well-known document-domain markup languages, such as DITA and DocBook, are by far the most complex, especially when they are used out of the box with no customization. Complexity is not merely a matter of the number of elements in a schema. It is also a matter of the number of different combinations or permutations of those elements that are allowed. If, as is common, a schema allows writers to insert any one of dozens of different elements into many different places in a document, then they have to understand the possible permutations and their consequences. And information architects and content engineers must anticipate and handle all those permutations in the algorithms they write.

Because of their size and loose structure, there is little you cannot express in these languages, but they are also a huge source of complexity that can be difficult to partition and redirect successfully. Simpler document-domain languages are less complex but also less capable. Some document designs simply cannot be expressed in Markdown or similar languages.

However, the upside of the document domain is that it is very homogeneous. You can express nearly any document design using DocBook or out-of-the-box DITA. This does not mean that these languages support an efficient partitioning and redirecting of content system complexity; it simply means that you can use either one of these languages to produce nearly anything other than layout-specific designs (which require the media domain). This is obviously appealing because you need only a single tool set and a single set of training. No matter how complicated the tools and the training may be or how much those complications limit participation by occasional contributors, at least you have only one thing to worry about.

The media domain, by contrast, is both moderately complex and moderately diverse. At the core of its diversity is the paper/hypertext divide. But there are multiple forms of paper (including virtual ones such as ebooks) and multiple forms of online media, which adds to its complexity. The initial motivation for most structured writing, separating content from formatting, was to tackle the challenge of publishing in diverse media.

The move away from the media domain accelerated as we entered the transition between paper and the hypertext. Prior to the beginning of this shift, most organizations delivered only to paper and, therefore, could work in the media domain without problems. However, they could not take advantage of structure and algorithms in handling the rhetoric and process of the content system.

While there is no sign that paper or paper-like electronic media are going away entirely, we have reached the point where more and more content is created and delivered only on the web. Thus, media-domain tools, and simple document-domain tools like Markdown that are specific to the presentation needs of one medium, are gaining in popularity. Again, the use of the media domain cuts you off from the assistance of algorithms in performing the synthesis. This means you can't use structure to constrain your rhetoric or algorithms to check your conformance and you can't vary your presentation for different audiences or media, since these things happen in the subject and document domains. However, because so much information delivery is once again targeted at a single medium, the media domain is becoming increasingly viable.

The subject domain is far less complex than the media domain or the document domain. Our recipe markup is less complex than even Markdown. But the subject domain is far more diverse. There are hundreds of different subjects, each with its own unique structure. And presenting the same subject to different audiences may require a different subject-domain structure to capture a different rhetoric. Obviously, you don't need all the subject-domain structures in the world. You only need enough to cover your subject matter for your audience, but that is still be more than you will need if you chose to write in the document domain or the media domain.

On the other hand, most subject-domain structures are simple. That does not necessarily mean that they have a small number of elements, though that is almost always the case. But it does mean that there are few permutations of those elements, which makes it easier to validate and process subject-domain content. Because subject structures are concrete rather than abstract and because they use the language of the subject matter, they tend to be clear and intuitive to writers. Writers need little or no training to use them. And subject-domain structures can often be expressed in lightweight syntax, removing the need for expensive editing software.

However, the diversity of structures is itself a source of complexity. You can't buy subject-domain structures and algorithms off the shelf. You have to create and maintain them yourself. This does not mean you have to maintain a complete custom publishing tools chain. You only have to manage the processing from subject domain to document domain. You can make any existing document-domain language your document-domain layer and use its tool chain for the rest of your publishing system. By extension, if you already have a document-domain system in place, you can add a subject-domain layer on top of it without changing the rest of your system.

Creating and maintaining subject-domain structures and algorithms is relatively simple, because the structures themselves are simple and well constrained with few permutations. It is also made significantly simpler because most structured writing algorithms are supported by the same subject-domain structures, unlike the document and management domains, where each algorithm requires a different structure.

For instance, subject annotations can be used for linking, presentation, auditing, and terminology management, whereas link markup in the document domain is useful for link management alone. And because it has good functional lucidity, the subject domain tends to produce more reliable structures, so there are fewer error conditions to worry about.

However, you will need to add new structures and algorithms from time to time, sometimes in relatively short time frames. This means that you need to keep information architecture and content engineering skills available to your organization. Although this is definitely a source of complexity in a subject-domain system, it is not as onerous as it sounds, and any large document-domain system will also require these skills.

Ultimately, then, there is no one right domain that everyone should be working in, and certainly there is no one right tool that everyone should be using for content creation. Rather, each of the three (plus one) domains of structured writing offers different capabilities for partitioning and directing the irreducible complexity of the content system. Which you choose should ultimately depend on the nature and type of complexity you are dealing with and the resources available to address that complexity. None of the structured writing domains are destinations, and none are desirable in themselves. They are merely means to an end: a content system in which all of the complexity of content creation is handled by a person or process with the appropriate skills, time, and resources to handle it, resulting in efficient process and effective rhetoric.

Selecting tools

Once you have decided how to distribute complexity in your content system, you need to choose the languages, systems, and tools that will allow you to implement that partitioning most efficiently. This may be a recursive process, since tools introduce complexity of their own that must be partitioned and distributed to make sure each tool does not introduce unhandled complexity. As you begin to select tools for your system, pause to consider how the complexity that each tool introduces will be handled. This may require changing other procedures or introducing other tools. If a tool introduces too much downstream complexity or the complexity it introduces is too hard to handle, you might need to consider a different tool or even a different strategy.

It is a simple fact of life that as you move content from the media domain towards the document and subject domains, the number of available off-the-shelf tools becomes fewer and the need to configure or extend those tools to get the result you need becomes greater. The reason for this is simple. The further you move your content towards the subject domain, the more you must rely on context-dependent algorithms to move it back to the media domain for publishing.

This is, after all, the point of the exercise. You move functions from people to algorithms to ensure that the complexity of the content system is handled appropriately and efficiently. The objective is to produce better content in less time and at less cost, and that is accomplished by handing parts of the work over to algorithms. The further you go along the continuum from media domain to subject domain, the more particular the algorithms become to your organization, your audience, and your subject matter. Thus, you have to become more engaged with the design of algorithms and the structures they require.

The need to take more responsibility for structures and algorithms is not a downside. It is what you are aiming for. You want to transfer effort and complexity from humans to algorithms, which means you need to transfer complexity to those who write algorithms. The need to employ people to write and maintain those algorithms is not a drawback. It is not that structured writing requires you to hire people to do these things. It is that partitioning some of the complexity of content creation away from writers and to content engineers and information architects requires that you adopt structured writing as a means to this end. Partitioning complexity better means dividing responsibilities among different specialists. Structured writing is merely the means used to achieve this distribution while meeting all of your rhetorical and technical requirements for output.

This should not be taken lightly. It is the point of the exercise and, therefore, needs to be taken seriously and approached deliberately. Developing this capability within your organization is far more important than your choice of tools or even domains. Understanding the capabilities you need in your organization comes back to how you partition the complexity of your system. Remember, the goal is to enhance rhetoric and process by partitioning and directing complexity so that every part of that complexity is handled by a person or algorithm with the skills, time, and resources to handle it.

Once you decide what complexity you want to partition and direct away from your writers, you can identify the tasks your information architects and content engineers will need to handle. Quite simply, then, you are looking for people with the skills to handle those tasks, and you need to give them the time and resources they need to do the job. The roles and qualifications for these people will differ for each organization, since each organization faces different levels of complexity and will partition and distribute complexity differently.

As we have noted, most off the shelf content management systems and most public content formats tend towards the use of a single format for all content, usually using a combination of the document and management domains. I have explored the disadvantages of that model throughout this book. However, this does not mean that all such systems become ineligible for use in your content process. The use of a rich document domain format packaged with sophisticated publishing capabilities in multiple media can make perfect sense, as long as you are not boxed into using that format for all authoring. Such a system may perform very well as the middle part of your publishing system, with a separate subject-domain authoring layer which can feed content from many sources. (If you go this route, though, do remember to make adequate provision for differential single sourcing to all the media you need to support.)

Counting the costs and savings

Most of what is written about structured writing focuses on cost savings, especially savings from content reuse. I have chosen, instead, to focus this book on managing complexity and enhancing rhetoric. Focusing on cost reduction, while it makes for an easy sell to those who must ultimately fund any structured writing project, often leads organizations to single out one particular cost, to the exclusion of other costs and opportunities. Organizations that fall into this trap sometimes introduce complexity into their systems that eats up all of the anticipated savings while damaging the rhetoric of the content.

A focus on cost reduction also carries with it an implicit admission that you can't think of ways to enhance the value of what you do – increasing value trumps reducing costs, and it has more upside. A focus on complexity, on the other hand, addresses both cost and quality at the same time. Unhandled complexity, or complexity handled by the wrong person or process, not only reduces quality, it also costs money. You may not be able to produce a neat (and misleading) spreadsheet that equates content reused with dollars saved, but a focus on comprehensive management of complexity can yield cost savings while enhancing value.

The economics of this decision are complex. You may decide that the cost of creating and maintaining the most appropriate algorithms and structures is not worth the cost or quality improvements they promise. But you should make that decision with a full appreciation of the benefits those algorithms are capable of delivering. Whatever you decide, make sure you understand how complexity is distributed in your system, make sure that the people you distribute complexity to have the skills, time, and resources to handle that complexity, and be conscious of the effect it will have on their productivity and the quality and reliability of their work.

Structured writing is a tool for managing the complexity of your content system and making sure all of that complexity gets handled by people with the right skills and resources. The ultimate objective is to enhance both process and rhetoric. Getting there may require your team to learn new skills, or it may require you to bring in staff with the needed skills. Don't regard bringing in those skills as a downside of structured writing. Instead, look at it as a method to better handle the complexity of content creation. Structured writing is simply the technique you use to partition and transfer complexity around your system so that everyone on the team can do their jobs better.

Index

X

Colophon

About the Author

Mark Baker is the author of *Every Page is Page One: Topic-based Writing for Technical Communication and the Web* as well as other books on content and content technologies and dozens of articles on technical communication, content strategy, and structured writing. He has worked as a technical writer, tech comm manager, director of communications, trainer, programmer, copywriter, and consultant and has spoken frequently at industry conferences. He has designed, built, and used multiple structured writing tools and systems, including the one used to write this book. He blogs at http://everypageispageone.com and tweets as https://twitter.com/mbakeranalecta. For more information, see http://analecta.com.

About XML Press

XML Press (http://xmlpress.net) was founded in 2008 to publish content that helps technical communicators be more effective. Our publications support managers, social media practitioners, technical communicators, and content strategists and the engineers who support their efforts.

Our publications are available through most retailers, and discounted pricing is available for volume purchases for business, educational, or promotional use. For more information, send email to orders@xmlpress.net or call us at (970) 231-3624.

www.ingramcontent.com/pod-product-compliance
Lightning Source LLC
La Vergne TN
LVHW062300060326
832902LV00013B/1971